Robust Control Optimization with Metaheuristics

Robust Control Optimization with Metaheuristics

Philippe Feyel

iSTE

WILEY

First published 2017 in Great Britain and the United States by ISTE Ltd and John Wiley & Sons, Inc.

ISTE Ltd
27-37 St George's Road
London SW19 4EU
UK

www.iste.co.uk

John Wiley & Sons, Inc.
111 River Street
Hoboken, NJ 07030
USA

www.wiley.com

Library of Congress Control Number: 2016958347

British Library Cataloguing-in-Publication Data
A CIP record for this book is available from the British Library
ISBN 978-1-78630-042-3

MIX
Paper from responsible sources
FSC® C013604

Contents

Preface

In industry, control engineers have to design a unique control law valid on a single prototype with a sufficient degree of robustness to satisfy a complex specification on many systems. To this end, the development methodology employed consists of an experimental iterative process (trial and error phase) that is heavily reliant on engineers' own level of expertise.

In this book, we try to make the methodology for computing controllers that are more efficient and more direct with a less costly development time by calculating a final structured controller using a direct optimization on a high-level specification system.

The complexity of high-level specifications drives us to the use of metaheuristics: these optimization techniques do not require gradient formulation, the only constraint being the possibility of evaluating the specification. Thus, in this work, we propose to reformulate robust control problems for stochastic optimization: we show how to synthesize structured controllers for control problems such as H_∞ synthesis, μ-synthesis or LPV/LFT synthesis, showing that the interest of the formulated approach lies in its flexibility and the consideration of exotic complex constraints.

Since evolutionary algorithms have proved to be so effective and competitive, we have used them as the foundation for a new method for synthesizing robust and structured controllers with respect to any form of optimization criteria. The validation of this work was performed on the industrial example of the line of sight stabilization problem in addition to several academic problems.

Philippe FEYEL
November 2016

Introduction and Motivations

I.1. Developing control engineering in an industrial framework

The problem of inertial stabilization involves creating an image in which orientation and quality do not depend on its carrier. To do this, optronic sensors are carried on a mechatronic servo device inertially stabilized with gyrometers or gyroscopes, which gives the viewfinder particular features for observation, detection and identification. As an example, we list viewfinders for tanks, helicopters, periscopes, missile guidance, etc.

Owing to the fundamental principle of the dynamic for a solid in rotation, the absolute rate Ω_a of the line of sight with inertia J is governed by:

$$J\frac{d\Omega_a}{dt} = C_{mot} - C_{ext} \qquad [I.1]$$

Therefore, inertial stabilization is a disturbance rejection problem: with a servo-loop, a useful torque C_{mot} is produced (generally provided by an electric motor) to compensate external disturbances C_{ext} being applied to the load at each moment. Several architectural solutions can be found in [MAS 08] and [HIL 08].

To be able to use its features, the viewfinder should have an adequate range. This range is directly linked to the viewfinder's stabilization performance, especially in mechanical environments and relatively harsh conditions of use and should be compatible with the optronic sensors' definition; the latter are characterized by an integration time (T_i) (the time during which the image is acquired and electronically generated) and by the size of their pixel detectors (IFOV)[1].

Hence, to have an adequate range, the absolute angular performance of the line of sight[2] $\theta_a = \int \Omega_a$ has to be compatible with the IFOV and the integration time T_i, in response to all disturbances.

The problem of inertial stabilization consists of rejecting mainly two types of external disturbance:

– The first is the friction torque Γ_f created during relative motion between the viewfinder carrier and the line of sight. This disturbance appears when the viewfinder carriers' orientation changes at low frequencies. In order to avoid image shaking during observation (and therefore provide visual comfort), we should make the line of sight angular dynamic movement in response to Γ_f compatible with the IFOV. Similarly, the steady-state error caused should be null. More particularly, the dynamic of the friction rejection should be compatible with the dynamic of the host vehicle's oscillation regime, which is only a few hertz, ten at the very most. The error in response to the friction should therefore be null (or at least very weak) after a time T_f. To validate the performances against Γ_f, an *oscillating table* simulating the host vehicle's angular motion at low frequencies is generally used.

– The second disturbance Γ_v comes from the mechanical distortions of the viewfinder's structure which, through its flexibility, transmits to the line of sight part of the mechanical environment to which the viewfinder is subject and indeed amplifies it. In order to enable the viewfinder to operate its observation features, we should limit the blurring caused by this high-frequency disturbance[3] to a value compatible with the IFOV during the integration time T_i. To validate performances against Γ_v, a shaker simulating the host vehicle's high-frequency vibrations is generally used.

In addition to these performance requirements, other requirements should be taken into consideration when designing the servo-loop:

– because the viewfinder is an embedded device, and/or to preserve the integrity of the motorization, a requirement on the stabilization stage's energy consumption is necessary (on the maximum instantaneous power, on the maximum current or tension, etc.);

1 Instantaneous field of view.

2 Considering null precession rate.

3 High frequency compared with the first disturbance, which is of much lower frequency.

– because the viewfinder is a mass system used in variable environments, the design for the control law should be made with robustness constraints (stability robustness, performance robustness) considering uncertainties.

Thus, the automation engineer should design a single control law that they will validate on a single prototype, with a degree of robustness sufficient to satisfy a complex specification on a large number of systems. In reality, this is the objective sought by any automation engineer working in an industrial framework, wishing to develop the most effective and robust control law possible in the shortest possible length of time.

The general methodology for developing a servo controller on a prototype viewfinder can be summarized in Figure I.1. In reality, it is a good example of the methodology used for solving automation problems in an industrial context. In fact, it has the four usual phases of servo-loop development:

– declining high-level complex specifications using simplifications, linearization, etc. in order to build some linear frequency shapes;

– modeling the system and its uncertainties;

– synthesizing control laws;

– experiments.

However, this methodology is sub-optimal and time-consuming, and therefore expensive, as:

– The final control law, which is obtained at stage 7 through inevitable simplifications (linearization, etc.) and a repetitive (stages 5 and 6) and expensive experimental process, should satisfy an often complex high-level specification (stage 1). These simplifications are needed to use control techniques (stage 4), which in fact require the engineer to have a degree of expertise.

– Based on the standard form for control, modern control techniques such as H_∞ synthesis used at stage 4 enable a (potentially multi-variable) controller to be determined, which is optimal (in the sense of the H_∞ norm) to make linear closed-loop or open-loop shapes being satisfied (stage 3). These shapes are materialized via setting the frequency weights on the signals to be monitored and on the exogenous inputs (the standard approach), or directly on the system's inputs and outputs (the loop-shaping approach). Thus, the most complex task of stage 4 is not computing the controller itself, but declining the simplified specifications into a judicious choice of weights; this adjustment is in itself a repetitive process requiring the engineer to have a high level of expertise, especially as they should ask a number of questions, for example:

- For a SISO[4] problem, how should we choose the structure of the weights to be set, the number of poles/zeros and the order?

- In the MIMO[5] case where we generally add as many weights as available measures and commands, the number of transfers to be shaped becomes very high so that we generally proceed to a sequential and decoupled tuning of the weights of the different channels by assuming that the weights are diagonal structured: what about the optimality of the tuning under these simplifying structural hypotheses? Which is the most adapted structure, diagonal or full?

- What is the best trade-off attainable between performance and robustness?

- We note that some very powerful robust synthesis techniques such as μ-synthesis become sub-optimal through their resolution method, which can naturally restrict their use.

- Finally, modern synthesis methods generate an often high-order controller; in general, we proceed to a post-reduction of the latter before its implementation. This post-reduction can degrade performance and robustness so much that it is necessary to take account of the order constraint beforehand. Moreover, there is also the question of finding the best controller structure for the problem posed: is it a cascaded multi-loop structure, or a multi-variable feedback correction, or a decentralized correction?

Here, we have listed some examples of questions that the engineer is led to ask. This list is not exhaustive. In reality, the right answer to these questions depends on the engineer's level of expertise and on his or her ability, and most of the time, we have to proceed to simplifications in order to move forward in the process of synthesis, and simplifications making the final solution costly and sub-optimal for the primary objective, which is to satisfy the high-level system specification of stage 1.

Thus, the aim of this work is to make servo controller synthesis in the industrial framework better adapted through being more direct and therefore less time-consuming to develop, by computing a final (structured) controller by directly tackling the high-level specification system.

4 Single input, single output.

5 Multiple inputs, multiple outputs.

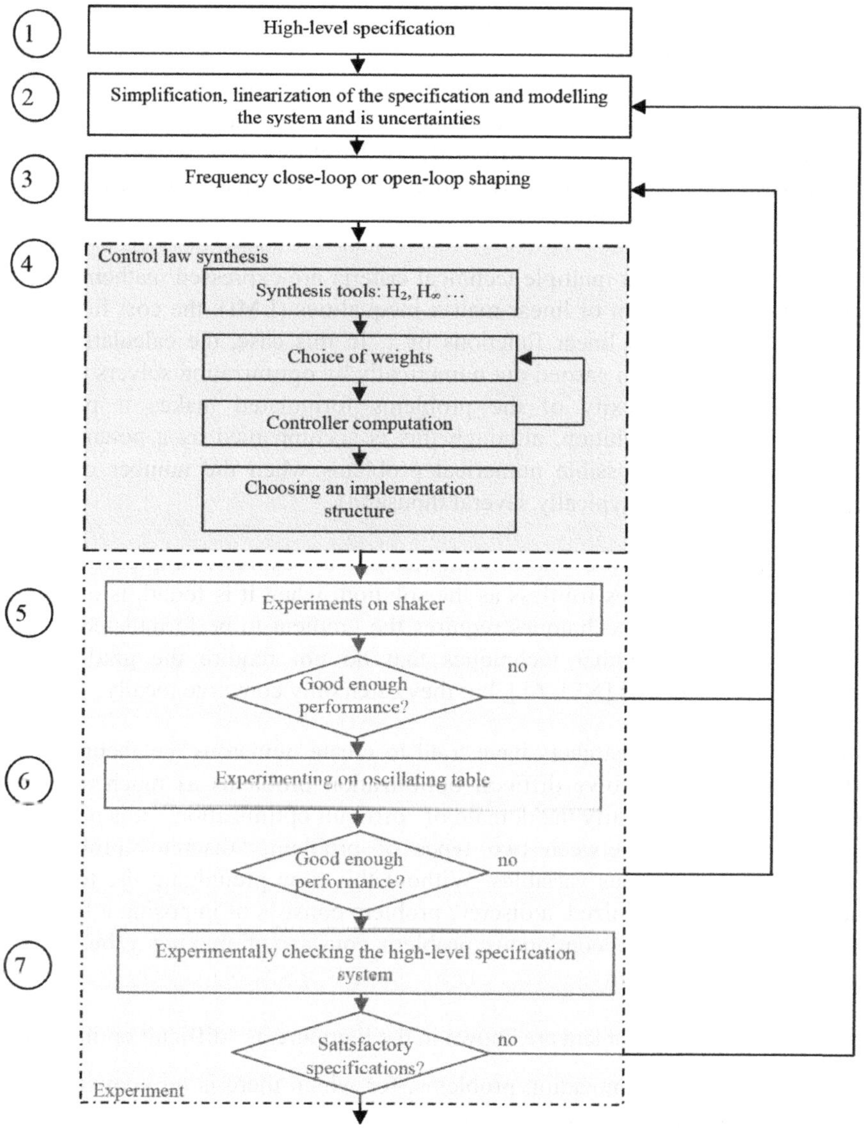

Figure I.1. *Method for developing controllers in viewfinders*

I.2. The place of optimization

Optimization always plays a major role in solving automation problems, which most of the time means calculating a stabilizing controller that meets specifications.

The problem to be solved can often be expressed in the general form of an optimization problem in which an objective or cost function (or indeed several) $f(x)$ that needs to be minimized in relation to the parameters involved x is defined. The definition of such a problem is often completed by constraints on x. The decision variables (the parameters to be determined) are therefore the variables that define the searched controller (for example, the matrices of its state-space representation or the coefficients of its transfer function).

[SCH 97] shows that multiple technical criteria are expressed mathematically in a simple way in the form of linear matrix inequalities (LMI); the cost function and the constraints are then linear functions of x. In this case, the calculation for the optimal controller is then carried out numerically by optimization solvers under LMI constraints. The convexity of the problems formulated makes it possible to determine the global solution, although this is accompanied by a potentially high calculation time and possible numerical problems when the number of decision variables is significant (typically several thousand).

When the criteria are more complicated to express, implementing such techniques usually proves fruitless as the solution, when it is found, is only locally optimal. Using descent techniques requires the gradient to be formulated, which is not always possible. Other techniques that do not require the gradient to be formulated exist, such as [NEL 65], but they often only converge locally.

For some years, researchers have tried to create numerous metaheuristics with a shared ambition: to solve difficult optimization problems as much as possible [DRÉ 06]. To try to identify the domain of "difficult optimization", it is necessary to make the distinction between two types of problem: "discrete" problems and problems with continuous variables. Without this time prejudging the form of the function $f(x)$ to be minimized, a discrete problem consists of imposing a finite set of possible values on x; a continuous problem consists of making x belong to an interval of values.

Thus, two sorts of problem are known in the literature as "difficult optimization":

– Some discrete optimization problems, for which there is no knowledge of an exact polynomial algorithm (in other words, whose calculation time is proportional to a power of n, where n is the number of unknowns in the problem). This is particularly the case with problems called "NP hard", for which it is conjectured or known that there is no algorithm whose resolution time is bounded by an n-degree polynomial.

– Some optimization problems with continuous variables, for which no algorithm enabling us to definitely locate a global optimum (in other words, the best possible solution) in a finite number of calculations is known.

Metaheuristics appear to be a valuable industrial tool for solving the optimization problems posed here, as these methods do not prejudge the shape of the cost function $f(x)$ to be minimized, which finally makes it possible to take account of some requirements in any form, especially the constraints on the system's dispersion, on any system, in fact simplifying the phase of declining or reformulating the specifications in an adapted mathematical framework and therefore reducing the impact of the designer's experience on the robustness on the final result: it is not necessary to formulate the gradient, the only constraint being the capability to evaluate the cost function $f(x)$, which can take any form. Unlike *simplex* methods, these are stochastic, which avoids the combinatorial explosion of possibilities for finding the global minimum of $f(x)$. [SIA 02] clearly shows the advantage of optimizing using metaheuristics, which makes it possible to remedy the problem of trapping in a "classic" iterative algorithm in a local minimum.

The principle of a classic iterative improvement algorithm is as follows: we start from an initial configuration x_0, chosen randomly. We then attempt a basic modification, often called a movement, and we compare the values of the objective function $f(x)$ before and after modification. If the change leads to a decrease in the objective function $f(x_1) < f(x_0)$, then it is accepted and the configuration x_1 serves as a starting point for the next test. In the opposite case, we return to the previous condition before making another attempt. The process is repeated until any attempt gives a bad result. Figure I.2 shows that the classical methods, the descent method or the method of the steepest descent often leads only to a local minimum x_n, which is the best possible solution, taking the initial condition x_0 into account. Even by applying the procedure several times with different initial conditions, we cannot be certain of finding the optimal configuration x_{opt}.

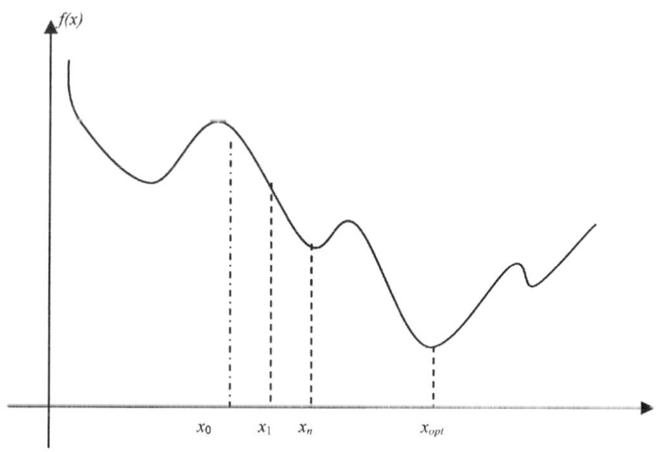

Figure I.2. *The multimodal function to be minimized*

To overcome the obstacle of local minimums, another idea has been shown to be so fruitful that it forms the basis of numerous metaheuristics: this is to allow ascents from time to time, in other words to accept a temporary degradation in the situation when the standard configuration is changed. This occurs when we pass from x_n to x_n', as shown in Figure I.3. A mechanism for controlling the degradations, specific to each metaheuristic, enables divergence in the process to be avoided. This makes it possible to extract the trap that a local minimum represents, and start to explore another, more promising, "valley".

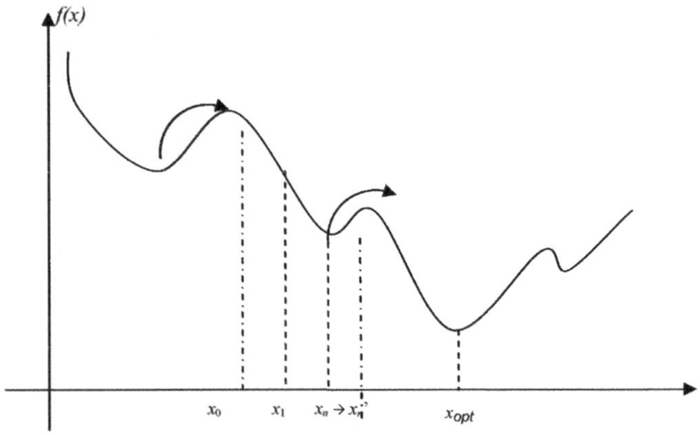

Figure I.3. *Extracting local minimums*

This is the principle that we wish to use to establish a controller for a complex high-level specification system.

As an example, [SAN 08] shows an interesting and promising application of the particle swarm method for automatically tuning weighting functions for a 4-block H_∞ synthesis scheme. Studies show especially how to simultaneously take time and frequency criteria for setting weights into account: the metaheuristic considered is therefore seen as an overlay of the synthesis method used.

The work is organized as follows: in the first instance, we describe the main metaheuristics used. Then, in the second section, we reformulate modern control problems for stochastic optimization: by envisaging structured problems in H_∞ synthesis, μ-synthesis and LPV-synthesis, we show the great flexibility offered by the use of stochastic optimization to take account of complex constraints (multiple plant synthesis, mixed H_2/ H_∞ synthesis, pole placement constraints, etc.). The third section, based on the results obtained, introduces a generic method for synthesizing robust and structured controllers for the criteria of any form. After translating the high-level criteria into the form of a generic cost function, we suggest a solution to two dual problems: on the one hand, optimizing synthesis weights in the framework of H_∞ synthesis by loop-shaping, and, on the other hand, directly optimizing the structured controller by introducing static scalings in place of the frequency weights. Finally, the fourth part describes the numerical toolbox *HinfStoch* developed by the author from theoretical results obtained throughout the book and is downloadable on the editor's website as well as on the author's personal page: https://philippe.feyel.free.fr.

The entire presentation is illustrated not only with industrial examples such as that of the line of sight inertial stabilization problem, but also with academic examples.

I.3. Notations and definitions

Below, we recap on some basic notions and expressions used frequently in the different chapters of this work.

I.3.1. *Linear time-invariant systems (LTISs)*

I.3.1.1. *Representation of LTISs*

A n-order linear time-invariant system with m inputs and p outputs is described by a state-space representation defined by the following system of differential equations:

$$\begin{aligned} \frac{dx}{dt} &= Ax(t) + Bu(t), \;\; x(t_0) = x_0 \\ y(t) &= Cx(t) + Du(t) \end{aligned} \quad [I.2]$$

where:

- $x(t) \in \mathbf{R}^n$ is the state of the system;
- $x(t_0)$ is the initial condition;
- $u(t) \in \mathbf{R}^m$ is the system input;
- $y(t) \in \mathbf{R}^p$ is the system output;
- $A \in \mathbf{R}^{n \times n}$ is the state-space matrix;
- $B \in \mathbf{R}^{n \times m}$ is the control matrix;

– $C \in \mathbf{R}^{p \times n}$ is the observation matrix;

– $D \in \mathbf{R}^{p \times m}$ is the direct transfer matrix.

For a given initial condition $x(t_0)$, the evolution of the system's state and its output is given by:

$$x(t) = e^{A(t-t_0)}x(t_0) + \int_{t_0}^{t} e^{A(t-\tau)}Bu(\tau)d\tau$$
$$y(t) = Cx(t) + Du(t) \quad [I.3]$$

The system is stable (in the sense that it has bounded input/bounded output) if the eigenvalues of A have a strictly negative real part, in other words if:

$$\max_{i \in [1,\ldots n]} \mathrm{Re}(\lambda_i(A)) < 0 \quad [I.4]$$

where $\lambda_i(A)$ is the ith eigenvalue of A.

For a zero initial condition, the system's input/output transfer matrix is defined in Laplace form by:

$$H(s) = C(sI - A)^{-1} B + D \quad [I.5]$$

which for convenience is written as:

$$\left[\begin{array}{c|c} A & B \\ \hline C & D \end{array}\right] := C(sI - A)^{-1} B + D \quad [I.6]$$

or indeed:

$$[A, B, C, D] := C(sI - A)^{-1} B + D \quad [I.7]$$

When $H(\infty)$ is bounded, H is called proper[6]. When $H(\infty)=0$, then the system is called strictly proper and $D = 0$.

6 In the case of a SISO transfer, this amounts to saying that the degree of the numerator is less than or equal to the degree of the denominator.

Finally, an infinite number of state-space representations are possible for a single transfer matrix. In fact, we consider the linear transformation $T \in \mathbf{R}^{n \times n}$, with T to be invertible, such that:

$$\bar{x} = Tx \qquad [I.8]$$

Then, the initial state-space representation becomes:

$$\begin{aligned} \frac{d\bar{x}}{dt} &= TAT^{-1}\bar{x}(t) + TBu(t) \\ y(t) &= CT^{-1}\bar{x}(t) + Du(t) \end{aligned} \qquad [I.9]$$

The corresponding transfer function is:

$$CT^{-1}\left(sI - TAT^{-1}\right)^{-1} TB + D = C\left(sI - A\right)^{-1} B + D = H(s) \qquad [I.10]$$

I.3.1.2. *Controllability and observability of LTISs*

The system H or the pair (A, B) is said to be controllable if, for any initial condition $x(t_0) = x_0$, for any $t_1 > 0$ and for any final state x_1, there is a continuous command $u(.)$, which can change the state of the system to $x(t_1) = x_1$.

We determine controllability by checking that, for any $t > t_0$, the controllability Grammian $W_c(t)$ is positive definite:

$$W_c(t) := \int_{t_0}^{t} e^{A\tau} BB^T e^{A^T\tau} d\tau \qquad [I.11]$$

An equivalent condition is that the matrix $\begin{pmatrix} B & AB & A^2B & \dots & A^{n-1}B \end{pmatrix}$ must be full line rank, i.e. n.

The system H or the pair (C, A) is observable if, for any $t_1 > 0$, the initial state $x(t_0) = x_0$ can be determined by the past values, the control signal $u(t)$ and the output $y(t)$ in the interval $[t_0, t_1]$.

We determine observability by checking that, for any $t > t_0$, the observability Grammian $W_o(t)$ is positive definite:

$$W_o(t) := \int_{t_0}^{t} e^{A^T\tau} C^T C e^{A\tau} d\tau \qquad [I.12]$$

An equivalent condition is that the matrix:

$$\begin{pmatrix} C \\ CA \\ CA^2 \\ \vdots \\ CA^{n-1} \end{pmatrix} \qquad [I.13]$$

must be full column rank, i.e. n.

I.3.1.3. *Elementary operations on the LTISs*

Consider H, the transfer system:

$$\left[\begin{array}{c|c} A & B \\ \hline C & D \end{array}\right] := C(sI - A)^{-1} B + D \qquad [I.14]$$

The transpose of H is defined by the system:

$$H^T(s) = B^T\left(sI - A^T\right)^{-1} C^T + D^T := \left[\begin{array}{c|c} A^T & C^T \\ \hline B^T & D^T \end{array}\right] \qquad [I.15]$$

The conjugate of H is defined by the system:

$$H^*(s) = H^T(-s) = B^T\left(-sI - A^T\right)^{-1} C^T + D^T := \left[\begin{array}{c|c} -A^T & -C^T \\ \hline B^T & D^T \end{array}\right] \qquad [I.16]$$

If D is invertible, the inverse of H is defined by the system:

$$H^{-1} = \left[\begin{array}{c|c} A - BD^{-1}C & -BD^{-1} \\ \hline D^{-1}C & D^{-1} \end{array}\right] \qquad [I.17]$$

Now, consider two systems, H_1 and H_2, whose respective state-space representations are:

$$H_1 = \left[\begin{array}{c|c} A_1 & B_1 \\ \hline C_1 & D_1 \end{array}\right] \quad H_2 = \left[\begin{array}{c|c} A_2 & B_2 \\ \hline C_2 & D_2 \end{array}\right] \qquad [I.18]$$

The serial connection of H_1 with H_2 (or product of H_1 by H_2) leads to the system:

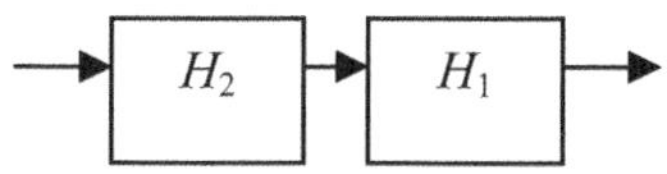

$$
\begin{aligned}
H_1H_2 &= \left[\begin{array}{c|c} A_1 & B_1 \\ \hline C_1 & D_1 \end{array}\right]\left[\begin{array}{c|c} A_2 & B_2 \\ \hline C_2 & D_2 \end{array}\right] \\
&= \left[\begin{array}{cc|c} A_1 & B_1C_2 & B_1D_2 \\ 0 & A_2 & B_2 \\ \hline C_1 & D_1C_2 & D_1D_2 \end{array}\right] = \left[\begin{array}{cc|c} A_2 & 0 & B_2 \\ B_1C_2 & A_1 & B_1D_2 \\ \hline D_1C_2 & C_1 & D_1D_2 \end{array}\right]
\end{aligned} \qquad [I.19]
$$

The parallel connection (or addition) of H_1 and H_2 leads to the system:

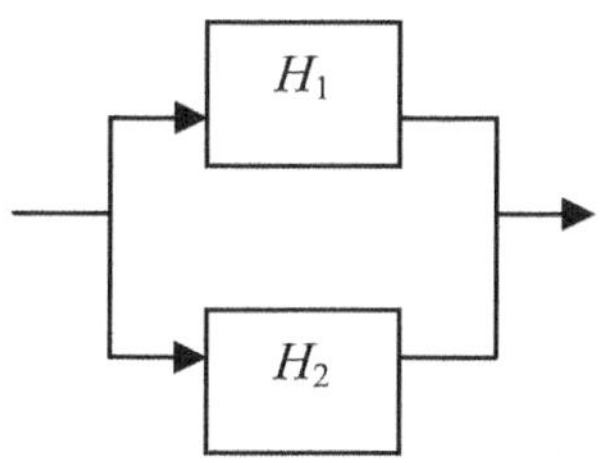

$$
\begin{aligned}
H_1 + H_2 &= \left[\begin{array}{c|c} A_1 & B_1 \\ \hline C_1 & D_1 \end{array}\right] + \left[\begin{array}{c|c} A_2 & B_2 \\ \hline C_2 & D_2 \end{array}\right] \\
&= \left[\begin{array}{cc|c} A_1 & 0 & B_1 \\ 0 & A_2 & B_2 \\ \hline C_1 & C_2 & D_1 + D_2 \end{array}\right]
\end{aligned} \qquad [I.20]
$$

The looping of H_2 in retroaction to H_1 leads to the system:

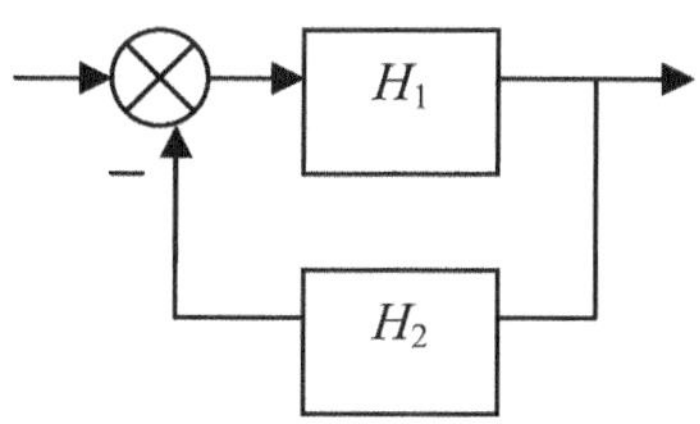

$$(I+H_1H_2)^{-1}H_1 = \left[\begin{array}{cc|c} A_1 - B_1D_2R_{12}{}^{-1}C_1 & -B_1R_{21}{}^{-1}C_2 & B_1R_{21}{}^{-1} \\ B_2R_{12}{}^{-1}C_1 & A_2 - B_2D_1R_{21}{}^{-1}C_2 & B_2D_1R_{21}{}^{-1} \\ \hline R_{12}{}^{-1}C_1 & -R_{12}{}^{-1}D_1C_2 & D_1R_{21}{}^{-1} \end{array}\right] \qquad [I.21]$$

with $R_{12} = I + D_1D_2$ and $R_{21} = I + D_2D_1$.

Many notions about linear time invariant systems are explained in [ZHO 96].

I.3.2. *Singular values*

I.3.2.1. *Definition*

The singular values of a transfer matrix $H(s)$ of dimension $p \times m$ are defined as the square roots of the eigenvalues of the product of its frequency response $H(j\omega)$ by its conjugate:

$$\sigma_i(H(j\omega)) = \sqrt{\lambda_i\left(H(j\omega)H(-j\omega)^T\right)} = \sqrt{\lambda_i\left(H(-j\omega)^T H(j\omega)\right)} \qquad [I.22]$$
$$i = [1,\ldots,\min(m,p)]$$

Singular values are positive or null real numbers and can be classified. The largest singular value, also called the maximum singular value, is written as $\overline{\sigma}(H)$ and the smallest, also called the minimum singular value, is written as $\underline{\sigma}(H)$.

$$\overline{\sigma}(H(j\omega)) = \sigma_1(H(j\omega)) \geq \sigma_2(H(j\omega)) \geq \ldots \geq \underline{\sigma}(H(j\omega)) \qquad [I.23]$$

In the case of a monovariable system (i.e. $m = p = 1$), the unique singular value is equal to the gain of the frequency response:

$$\overline{\sigma}(H(j\omega)) = \underline{\sigma}(H(j\omega)) = |H(j\omega)| \qquad [I.24]$$

The singular values thus extend the notion of gain established with monovariable systems to multivariable systems. Therefore, H will be called high gain if $\underline{\sigma}(H)$ is large and low gain if $\overline{\sigma}(H)$ is small.

I.3.2.2. *Properties*

In this work, we often use the following properties:

$$\begin{aligned}
&\overline{\sigma}(H)=0 \Leftrightarrow H=0\\
&\overline{\sigma}(H^*)=\overline{\sigma}(H)\\
&\overline{\sigma}(H^T)=\overline{\sigma}(H)\\
&\sigma_i(\alpha H)=|\alpha|\sigma_i(H)\\
&\text{if } H^{-1} \text{ exists, } \underline{\sigma}(H)\overline{\sigma}(H^{-1})=\underline{\sigma}(H^{-1})\overline{\sigma}(H)=1\\
&\overline{\sigma}(H)=\max_{\substack{x\in C^m\\ x\neq 0}}\frac{\|Hx\|_2}{\|x\|_2}\\
&\underline{\sigma}(H)=\min_{\substack{x\in C^m\\ x\neq 0}}\frac{\|Hx\|_2}{\|x\|_2}
\end{aligned} \qquad [I.25]$$

In the case of two parallel systems, we will retain the following properties:

$$\begin{aligned}
&\overline{\sigma}\begin{pmatrix}H_1\\ H_2\end{pmatrix}\leq\overline{\sigma}(H_1)+\overline{\sigma}(H_2)\\
&\max\left(\overline{\sigma}(H_1),\overline{\sigma}(H_2)\right)\leq\overline{\sigma}\begin{pmatrix}H_1\\ H_2\end{pmatrix}\leq\sqrt{2}\max\left(\overline{\sigma}(H_1),\overline{\sigma}(H_2)\right)\\
&\overline{\sigma}\begin{pmatrix}H_1 & 0\\ 0 & H_2\end{pmatrix}=\max\left(\overline{\sigma}(H_1),\overline{\sigma}(H_2)\right)
\end{aligned} \qquad [I.26]$$

In the case of two serial systems, an important property is:

$$\begin{aligned}
&\sigma_i(H_1)\underline{\sigma}(H_2)\leq\sigma_i(H_1H_2)\leq\sigma_i(H_1)\overline{\sigma}(H_2)\\
&\qquad\qquad\text{or}\\
&\underline{\sigma}(H_1)\sigma_i(H_2)\leq\sigma_i(H_1H_2)\leq\overline{\sigma}(H_1)\sigma_i(H_2)
\end{aligned} \qquad [I.27]$$

In particular, we shall use the following specific cases that will retain the following particular cases:

$$\begin{aligned}
&\underline{\sigma}(H_1)\overline{\sigma}(H_2)\leq\overline{\sigma}(H_1H_2)\leq\overline{\sigma}(H_1)\overline{\sigma}(H_2)\\
&\overline{\sigma}(H_1)\underline{\sigma}(H_2)\leq\overline{\sigma}(H_1H_2)\leq\overline{\sigma}(H_1)\overline{\sigma}(H_2)\\
&\underline{\sigma}(H_1)\underline{\sigma}(H_2)\leq\underline{\sigma}(H_1H_2)\leq\overline{\sigma}(H_1)\underline{\sigma}(H_2)\\
&\underline{\sigma}(H_1)\underline{\sigma}(H_2)\leq\underline{\sigma}(H_1H_2)\leq\underline{\sigma}(H_1)\overline{\sigma}(H_2)
\end{aligned} \quad [I.28]$$

In the case of the sum of two systems, an important property is:

$$\left|\sigma_i(H_1)-\overline{\sigma}(H_2)\right|\leq\sigma_i(H_1+H_2)\leq\sigma_i(H_1)+\overline{\sigma}(H_2) \quad [I.29]$$

In particular, we will retain the following two particular cases:

$$\begin{aligned}
&\left|\overline{\sigma}(H_1)-\overline{\sigma}(H_2)\right|\leq\overline{\sigma}(H_1+H_2)\leq\overline{\sigma}(H_1)+\overline{\sigma}(H_2)\\
&\underline{\sigma}(H_1)-\overline{\sigma}(H_2)\leq\underline{\sigma}(H_1+H_2)\leq\underline{\sigma}(H_1)+\overline{\sigma}(H_2)
\end{aligned} \quad [I.30]$$

which leads to:

$$\underline{\sigma}(H)-1\leq\underline{\sigma}(H+I)\leq\underline{\sigma}(H)+1 \quad [I.31]$$

or again:

$$\underline{\sigma}(H)-1\leq\frac{1}{\overline{\sigma}\left((H+I)^{-1}\right)}\leq\underline{\sigma}(H)+1 \quad [I.32]$$

Finally, we will retain the following property:

$$\overline{\sigma}(H_1)<\underline{\sigma}(H_2)\Rightarrow\underline{\sigma}(H_1+H_2)>0 \quad [I.33]$$

The interested reader can find further discussion about inequalities on singular values in [MER 04].

I.3.3. *Subspace* R H_∞ *and norm* H_∞

I.3.3.1. *Definition*

We use the notation L_∞^n to represent the set of vectorial functions $f(s)$, $s \in C$ of dimension n and bounded on the imaginary axis, i.e. which satisfies:

$$\|f\|_\infty = \sup_{\omega} \|f(j\omega)\|_2 < +\infty \qquad [I.34]$$

where $\| \|_2$ is the Euclidian norm.

H_∞^n is the subspace of the analytical functions of L_∞^n bounded in $\mathbf{C}^+$.

$\mathbf{R}L_\infty^{p\times m}$ is the subspace of rational proper transfer matrices of dimensions $p \times m$ with real coefficients, and without pole on the imaginary axis.

$\mathbf{R}H_\infty^{p\times m}$ is the subspace of rational stable proper transfer matrices of dimensions $p\times m$ with real coefficients (in other words without a pole in $\mathbf{C}^+$).

For any $H \in \mathbf{R}H_\infty^{p\times m}$ system, the H∞-norm of H is defined by:

$$\|H\|_\infty = \sup_{\omega \in R} \bar{\sigma}(H(j\omega)) \qquad [I.35]$$

This is therefore the highest value of the system's gain on the set of pulsations.

I.3.3.2. *Properties*

The set of properties valid for the maximum singular value is also valid for the norm H_∞.

In particular in this work, we will frequently use the following properties:

$$\begin{aligned} &\|H_1(s)H_2(s)\|_\infty \le \|H_1(s)\|_\infty \|H_2(s)\|_\infty \\ &\sup\left(\|H_1(s)\|_\infty, \|H_2(s)\|_\infty\right) \le \left\| \begin{pmatrix} H_1(s) \\ H_2(s) \end{pmatrix} \right\|_\infty \\ &\sup\left(\|H_1(s)\|_\infty, \|H_2(s)\|_\infty\right) \le \left\| \begin{pmatrix} H_1(s) & H_2(s) \end{pmatrix} \right\|_\infty \end{aligned} \qquad [I.36]$$

Notably, this implies that:

$$\left\| \begin{matrix} H_1(s) & H_3(s) \\ H_2(s) & H_4(s) \end{matrix} \right\|_\infty \leq \gamma \Rightarrow \begin{cases} \|H_1(s)\|_\infty \leq \gamma \\ \|H_2(s)\|_\infty \leq \gamma \\ \|H_3(s)\|_\infty \leq \gamma \\ \|H_4(s)\|_\infty \leq \gamma \end{cases} \qquad [I.37]$$

Finally:

$$\left\| \begin{pmatrix} H_1(s) \\ H_2(s) \end{pmatrix} \right\|_\infty \leq \sqrt{\|H_1(s)\|_\infty^2 + \|H_2(s)\|_\infty^2} \qquad [I.38]$$

I.3.4. *Linear fractional transformation (LFT)*

I.3.4.1. *Definition*

Suppose that a complex matrix P is partitioned as follows:

$$P = \begin{pmatrix} P_{11} & P_{12} \\ P_{21} & P_{22} \end{pmatrix} \in \mathbb{C}^{(p_1+p_2)\times(q_1+q_2)} \qquad [I.39]$$

and suppose that two other complex matrices are $\Delta_l \in \mathbb{C}^{q_2\times p_2}$ and $\Delta_u \in \mathbb{C}^{q_1\times p_1}$.

Assuming that $(I - P_{22}\Delta_l)^{-1}$ exists, the lower linear fractional transformation (or *LFT*) is defined by:

$$F_l(P, \Delta_l) = P_{11} + P_{12}\Delta_l (I - P_{22}\Delta_l)^{-1} P_{21} \qquad [I.40]$$

which corresponds to the following block diagram where the matrix Δ_l loops P "from below":

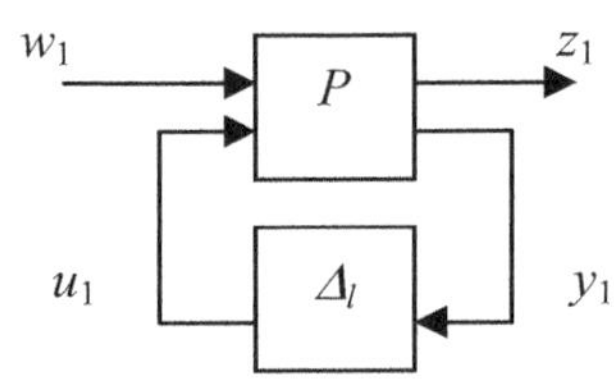

$$\begin{pmatrix} z_1 \\ y_1 \end{pmatrix} = P \begin{pmatrix} w_1 \\ u_1 \end{pmatrix} = \begin{pmatrix} P_{11} & P_{12} \\ P_{21} & P_{22} \end{pmatrix} \begin{pmatrix} w_1 \\ u_1 \end{pmatrix} \qquad [I.41]$$
$$u_1 = \Delta_l y_1$$

Assuming that $(I - P_{11}\Delta_u)^{-1}$ exists, the upper linear fractional transformation (or *LFT*) is defined by:

$$F_u(P, \Delta_u) = P_{22} + P_{21}\Delta_u (I - P_{11}\Delta_u)^{-1} P_{12} \qquad [I.42]$$

which corresponds to the following block diagram where the matrix Δ_l loops P "from above":

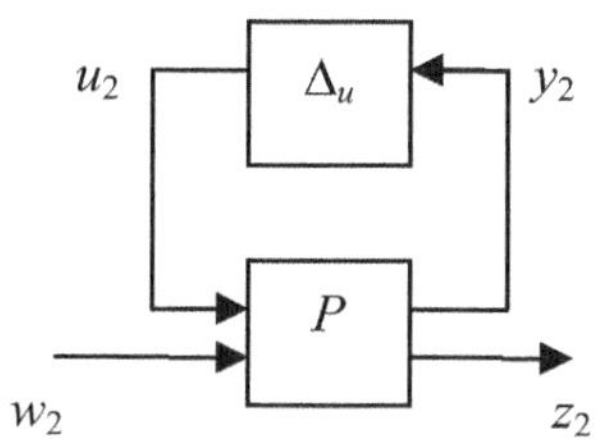

$$\begin{pmatrix} y_2 \\ z_2 \end{pmatrix} = P \begin{pmatrix} u_2 \\ w_2 \end{pmatrix} = \begin{pmatrix} P_{11} & P_{12} \\ P_{21} & P_{22} \end{pmatrix} \begin{pmatrix} u_2 \\ w_2 \end{pmatrix} \qquad [I.43]$$
$$u_2 = \Delta_u y_2$$

It is also worth remembering that if H_3 is invertible, then by definition:

$$\begin{aligned} (H_1 + H_2 Q)(H_3 + H_4 Q)^{-1} &= F_l(M, Q) \\ (H_3 + H_4 Q)^{-1}(H_1 + H_2 Q) &= F_l(N, Q) \end{aligned} \qquad [I.44]$$

with:

$$\begin{aligned} M &= \begin{pmatrix} H_1 H_3^{-1} & H_2 - H_1 H_3^{-1} H_4 \\ H_3^{-1} & -H_3^{-1} H_4 \end{pmatrix} \\ N &= \begin{pmatrix} H_3^{-1} H_1 & H_3^{-1} \\ H_2 - H_4 H_3^{-1} H_1 & -H_4 H_3^{-1} \end{pmatrix} \end{aligned} \qquad [I.45]$$

I.3.4.2. *Properties*

A basic property of LFT is that the association of several LFTs remains an LFT.

Suppose that M and Q are partitioned as follows:

$$M=\begin{pmatrix} M_{11} & M_{12} \\ M_{21} & M_{22} \end{pmatrix},\ Q=\begin{pmatrix} Q_{11} & Q_{12} \\ Q_{21} & Q_{22} \end{pmatrix} \quad \text{[I.46]}$$

The upper and lower LFTs are linked by the following equality:

$$F_u(M,\Delta)=F_u(N,\Delta) \quad \text{[I.47]}$$

with:

$$N=\begin{pmatrix} 0 & I \\ I & 0 \end{pmatrix} M \begin{pmatrix} 0 & I \\ I & 0 \end{pmatrix}=\begin{pmatrix} M_{22} & M_{21} \\ M_{12} & M_{11} \end{pmatrix} \quad \text{[I.48]}$$

The inversion of an LFT is an LFT:

$$\left(F_u(M,\Delta)\right)^{-1}=F_u(N,\Delta) \quad \text{[I.49]}$$

with:

$$N=\begin{pmatrix} M_{11}-M_{12}M_{22}{}^{-1}M_{21} & -M_{12}M_{22}{}^{-1} \\ M_{22}{}^{-1}M_{21} & M_{22}{}^{-1} \end{pmatrix} \quad \text{[I.50]}$$

The sum of two LFTs is an LFT:

$$F_u(M,\Delta_1)+F_u(Q,\Delta_2)=F_u(N,\Delta) \quad \text{[I.51]}$$

with:

$$N=\left(\begin{array}{cc|c} M_{11} & 0 & M_{11} \\ 0 & Q_{11} & Q_{12} \\ \hline M_{21} & Q_{21} & M_{22}+Q_{22} \end{array}\right),\ \Delta=\begin{pmatrix} \Delta_1 & 0 \\ 0 & \Delta_2 \end{pmatrix} \quad \text{[I.52]}$$

The product of two LFTs is an LFT:

$$F_u(M,\Delta_1)F_u(Q,\Delta_2)=F_u(N,\Delta) \quad [I.53]$$

with:

$$N=\left(\begin{array}{cc|c} M_{11} & M_{12}Q_{21} & M_{12}Q_{22} \\ 0 & M_{11} & Q_{12} \\ \hline M_{21} & M_{22}Q_{21} & M_{22}Q_{22} \end{array}\right),\ \Delta=\begin{pmatrix}\Delta_1 & 0 \\ 0 & \Delta_2\end{pmatrix} \quad [I.54]$$

Suppose that G is partitioned as follows:

$$G=\left[\begin{array}{c|cc} A & B_1 & B_2 \\ \hline C_1 & D_{11} & D_{12} \\ C_2 & D_{21} & D_{22} \end{array}\right] \quad [I.55]$$

The looping of two LFTs together is itself an LFT:

$$F_l\left(F_u(G,\Delta_1),F_u(Q,\Delta_2)\right)=F_u\left(F_l\left(G,F_u(Q,\Delta_2)\right),\Delta_1\right)=F_u(N,\Delta) \quad [I.56]$$

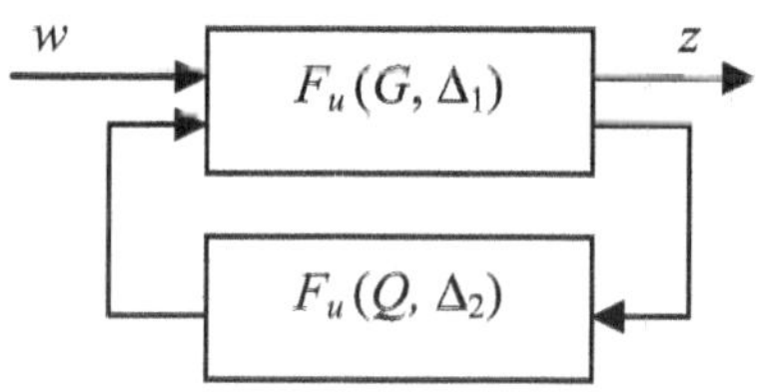

with:

$$N=\left(\begin{array}{cc|c} A+B_2Q_{22}L_1C_2 & B_2L_2Q_{21} & B1+B_2Q_{22}L_1D_{21} \\ Q_{12}L_1C_2 & Q_{11}+Q_{12}L_1D_{22}Q_{21} & Q_{12}L_1D_{21} \\ \hline C_1+D_{12}L_2Q_{22}C_2 & D_{12}L_2Q_{21} & D_{11}+D_{12}Q_{22}L_1D_{21} \end{array}\right) \quad [I.57]$$

$$\Delta=\begin{pmatrix}\Delta_1 & 0 \\ 0 & \Delta_2\end{pmatrix}$$

Finally, we will finish with the Redheffer's theorem: if $\|M(s)\|_\infty<\gamma$ and $\|\Delta(s)\|_\infty<\gamma^{-1}$, then $\|F_l(M(s),\Delta(s))\|_\infty<\gamma$.

1

Metaheuristics for Controller Optimization

1.1. Introduction

Stochastic optimization using metaheuristics is well adapted for solving problems for which it is difficult to find a global optimum or good local optima using classical methods. This type of optimization has three characteristics that are often decisive in global optimization:

– optimization using metaheuristics does not require us to know the gradient of the function to be minimized, the only constraint being that we must be able to evaluate the latter, which can therefore have any form;

– it is not necessary to use a "good" initial point, the initialization being carried out at random in the search space;

– finally, this type of optimization is stochastic, which makes it possible to overcome the combinatorial explosion of possibilities and limits trapping in the local optima.

In [DRÉ 06], [YAN 10a] and [FIS 13], you will find descriptions of metaheuristic principles inspired by analogies with physics, ethology, biology, etc.

Despite the main drawback, which is the pertinence of the solution found and which can be evaluated statistically using several successive runs (which creates a generally high calculation time), this type of optimization proves useful for difficult optimization problems and is shown here, adapted for controller synthesis.

As for the calculation time, this can prove to be substantial as most of the algorithms proceed to evaluate the cost function for each individual in a population at each iteration. One way of reducing the calculation times is to parallelize them at each iteration with specialized numerical tools [MAT 10, LUS 09]; it is therefore possible to distribute the calculation charge over several processors or cores in parallel by allocating a group of individuals from the population to each processor.

In this chapter, we introduce the metaheuristics used in the rest of the book as they have given "good results" for all the automation problems we discuss. By "good result", we mean that the optimization has a good degree of reproducibility and a good rate of convergence, and provides a good optimum.

The set of these algorithms is classed into three large families:

– evolutionary approaches;

– swarm approaches;

– hybrid evolutionary/swarm approaches.

To introduce these algorithms, we consider the problem of finding X_{opt} so that $X_{opt} = \underset{X \in \Psi}{\operatorname{argmin}}\left(f\left(X\right)\right)$, where $\psi \subset \mathrm{R}^n$ is the search space for X and n is the problem's dimension. In the following, we set $\Psi = \left[X_{\min}; X_{\max}\right]$. The function $rnd(x, y)$ designs a random number uniformly generated in the interval $[x, y]$.

1.2. Evolutionary approaches using differential evolution

1.2.1. *Standard version*

Differential evolution (DE) [STO 95] is a recent metaheuristic that belongs to the class of evolutionary algorithms (Figure 1.1), which are inspired by the theory of the evolution of species (Figure 1.2).

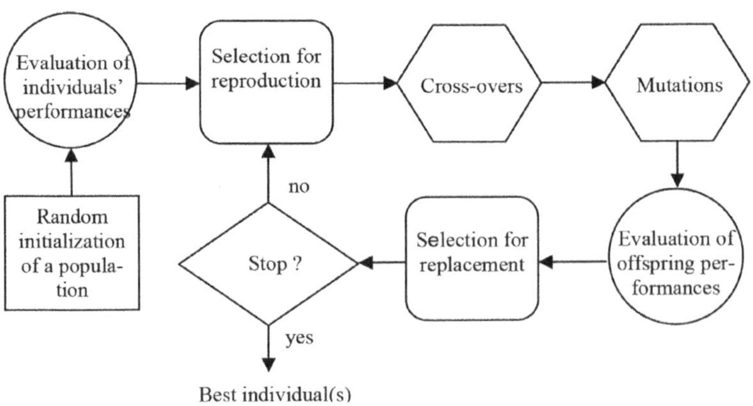

Figure 1.1. *Principle of evolutionary algorithms*

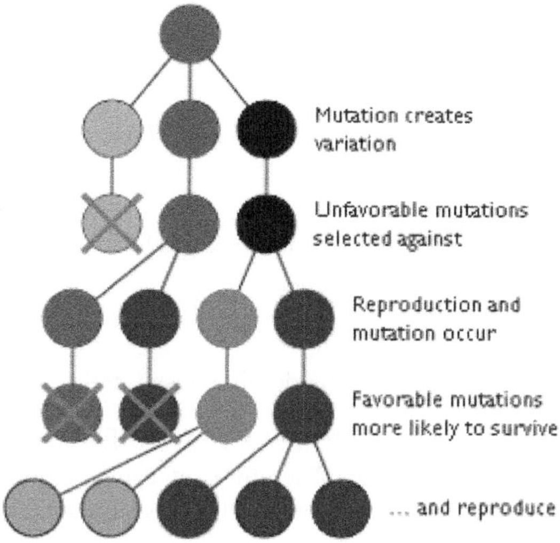

Figure 1.2. *Darwin's theory. For a color version of this figure, see www.iste.co.uk/feyel/optimization.zip*

Below, we describe the classical version of the algorithm.

Initializing the algorithm

We build an initial population P from N individuals.

$$P = \{X_1, \ldots, X_i, \ldots, X_N\} \quad [1.1]$$

Each individual is defined by its n genes:

$$X_i = (x_{i1}, \ldots, x_{ij}, \ldots, x_{in}) \quad [1.2]$$

n is therefore the problem's dimension.

Assuming that $x_j \in \left[X_{\min j}, X_{\max j}\right]$, the jth gene of the ith individual is randomly chosen in its definition interval:

$$x_{ij}^{(0)} = X_{\min j} + rnd(0,1)\left(X_{\max j} - X_{\min i}\right),\ j = 1, \ldots, n,\ i = 1, \ldots, N \quad [1.3]$$

The k^{ith} iteration consists of successively carrying out the three stages described below.

Describing the algorithm

– Step 1: mutation

Usually, two mutation schemes are used: the *rand* one and the *best* one. They are described below.

$$\begin{aligned} &\text{DE/Rand/1} \rightarrow U_i^{(k+1)} = X_{a_i}^{(k)} + F\left(X_{b_i}^{(k)} - X_{c_i}^{(k)}\right) \\ &\text{DE/Best/1} \rightarrow U_i^{(k+1)} = X_{best}^{(k)} + F\left(X_{a_i}^{(k)} - X_{b_i}^{(k)}\right) \end{aligned} \quad [1.4]$$

where $X_{a_i}, X_{b_i}, X_{c_i}$ are different and randomly chosen in the population and X_{best} is the best individual obtained since the start of the optimization, and $F \in [0,2]$ is a number called the mutation factor. Therefore, at each iteration, a number N of mutants U_i are generated according to one of the previous schemes (chosen once for all at the start of the optimization).

The *rand* scheme promotes diversity, whereas the *best* scheme encourages convergence. As the problems that we face are high-dimensional, we will here prioritize exploration and therefore the *rand* scheme.

– Step 2: cross-over

By crossing $U_i^{(k+1)}$ with $X_i^{(k)}$, a new individual $V_i^{(k+1)}$ is generated with genes defined in the following rule:

$$v_{ij}^{(k+1)} = \begin{cases} u_{ij}^{(k+1)} & if \ \ rnd(0,1)_{ij} \leq C_r \ \ or \ if \ \ j = j_i \\ & x_{ij}^{(k)} \ \ else \end{cases} \quad [1.5]$$

where $C_r \in [0,1;0,9]$ is the cross-over rate and n is an integer number randomly chosen in $\{1,2,...,N\}$.

– Step 3: selection

The population is finally modified with the individuals defined by:

$$X_i^{(k+1)} = \begin{cases} V_i^{(k+1)} si \ \ f\left(V_i^{(k+1)}\right) \leq f\left(X_i^{(k)}\right) \\ \qquad X_i^{(k)} \ \ ifnot \end{cases} \quad [1.6]$$

If the stop criterion is satisfied, then return the best solution found so far, if not, return to stage 1.

The algorithm's pseudo-code

The algorithm's pseudo-code is given below:

Generate an initial population.

While the stop criterion is not satisfied, repeat:

For $i = 1$ to N individuals, make:

Choose randomly 3 individuals $X_{a_i}, X_{b_i}, X_{c_i}$

Mutation:

Create a mutant U_i from individual X_i and/or $X_{a_i}, X_{b_i}, X_{c_i}$ (depending on the mutation scheme chosen at the beginning)

Cross-over: [1.7]

Create the trial V_i from the mutant U_i and X_i

Selection:

If $f(U_i) \leq f(X_i)$ then we replace $X_i \leftarrow U_i$

End For

Calculate X_{best}

End While

Algorithm stop criterion

Several stop criteria can be used. We will use:

– the maximum number of iterations, knowing that, at each iteration, the fitness function is evaluated P times;

– a maximum number of fitness evaluations;

– premature convergence of the algorithm, detected when all the individuals tend to be identical, that is to say, when the following relation is satisfied:

$$\exists X_{best} : \frac{\max_{i=1,\dots,P} \left\| X_i^{(k)} - X_{best} \right\|}{\left\| X_{max} - X_{min} \right\|} < \varepsilon \qquad [1.8]$$

In most cases we will choose $\varepsilon = 10^{-5}$.

Parameters settings

One of the algorithm's advantages is the low number of setting parameters. Here, we will use the classical settings of [STO 96]: $F = 0.75$ and $C_r = 0.8$.

Although a population size N between $5n$ and $10n$ is generally advised, the algorithm has provided the best results using the same idea as in (Q)PSO, i.e. a number of individuals equal to:

$$N = floor\left(10 + \sqrt{n}\right) \qquad [1.9]$$

Variants of the algorithm

In the suite, $X_{a_i}, X_{b_i}, X_{c_i}, X_{d_i}, X_{e_i}$ are different individuals and are randomly chosen in the population and X_{best} is the best individual in the population obtained since the beginning of the optimization.

There are many variants of DE [MEZ 08], which most of the time consist of defining particular mutation schemes. As examples we cite:

– DE/Best/2:

$$\text{DE/Best/2} \rightarrow U_i^{(k+1)} = X_{best}^{(k)} + F\left(X_{a_i}^{(k)} - X_{b_i}^{(k)} + X_{c_i}^{(k)} - X_{d_i}^{(k)}\right) \qquad [1.10]$$

– DE/Rand/2:

$$\text{DE/Rand/2} \rightarrow U_i^{(k+1)} = X_{a_i}^{(k)} + F\left(X_{b_i}^{(k)} - X_{c_i}^{(k)} + X_{d_i}^{(k)} - X_{e_i}^{(k)}\right) \qquad [1.11]$$

– DE/Current-to-Rand/1:

$$\text{DE/Current-to-rand/1} \rightarrow U_i^{(k+1)} = X_i^{(k)} + F\left(X_{a_i}^{(k)} - X_{b_i}^{(k)}\right) + F'\left(X_{c_i}^{(k)} - X_i^{(k)}\right) \quad [1.12]$$

– DE/Current-to-Best/1:

$$\text{DE/Current-to-best/1} \rightarrow U_i^{(k+1)} = X_i^{(k)} + F\left(X_{a_i}^{(k)} - X_{b_i}^{(k)}\right) + F'\left(X_{best}^{(k)} - X_i^{(k)}\right) \quad [1.13]$$

The *rand* scheme promotes diversity (and therefore exploration of the search space), whereas the *best* scheme encourages convergence but often leads to sub-optimal solutions. This is why, with the aim of benefitting from the advantages of these two basic schemes, we suggest using the mutation scheme [1.14] which we call DE/Rand-to-best/1:

$$\text{DE/Rand/2} \rightarrow U_i^{(k+1)} = X_{a_i}^{(k)} + F\left(X_{b_i}^{(k)} - X_{c_i}^{(k)} + X_{d_i}^{(k)} - X_{e_i}^{(k)}\right) \quad [1.14]$$

F' is another mutation factor that we choose, equal to $1 - F$ with $F' = 0.25$ here. This represents a good balance between diversity and convergence.

As we will see further on, this mutation scheme proves to be very effective for solving control problems.

The versions retained

We will keep the two versions of DE which have proven to be the most effective in our study: *Rnd_DE* and *Rnd2Bst_DE*, and which correspond, respectively, to the mutation schemes DE/Rand/1 and DE/Rand-to-best/1.

1.2.2. *Perturbed version*

DE has proven very effective for solving numerous complex problems [PRI 05]. Compared with many other metaheuristics, DE is very simple to implement. Moreover, its low number of setting parameters makes the algorithm relatively simple to set to achieve good results. However, like other metaheuristics, DE can converge prematurely, particularly when the population is small. One simple idea therefore consists of increasing the size of the population but at the cost of greater calculation times. Thus, some studies have been designed to maintain the algorithm's diversity. For example, the adaptive variants of DE [BRE 06] are useful, but they often introduce several additional setting parameters. Other ideas involve hybridizing the DE with other metaheuristics such as cuckoo search (CS) [WAN 12], which we will describe further on: generally, algorithms that are fairly effective but much more complex to implement are obtained. More recently in [FAJ 12], a random perturbation mechanism applied to V_i, after the cross-over step has been considered, is envisaged. Linked to different selection strategies, this mechanism seems to be effective, but at the cost of a much more complex selection phase than in the original version of DE.

To limit the premature convergence of DE, we suggest completing the mutation phase of DE with a simple mechanism consisting of perturbing the mutants U_i according to the following rule:

$$U_i^{(k+1)} = U_i^{(k+1)} + \text{rnd}(-1,1) X_{f_i} \quad if \ \text{rnd}(0,1) < p_m \qquad [1.15]$$

where p_m is the probability of perturbing the mutants' and X_{f_i} is randomly chosen in the population, different to X_i and distinct from all the elements used in the mutation scheme used to generate U_i. A typically suitable value for p_m is 0.025.

We will call this extension *Perturbed Differential Evolution* (PDE), which can be linked to all mutation schemas. The only difference between DE and PDE lies in a simple additional rule. In Appendix F, you will find a comparison of performances between DE and PDE.

In this study, we will, in particular, keep variant ***Rnd2Bst_PDE***, which consists of the combination of mutation scheme DE/Rand-to-best/1 with PDE and which has given especially interesting results.

1.3. Swarm approaches

1.3.1. *Particle swarm optimization algorithm*

Principle

The particle swarm optimization algorithm (Particle Swarm Optimization – PSO) [CLE 06, BRA 07, KEN 95] is a metaheuristic inspired by the social behavior of flocks of birds (Figure 1.3) or schools of fish (Figure 1.4).

Figure 1.3. *A flock of birds*

Figure 1.4. *A school of fish*

N_P particles move through the search space with their own speed. They are capable of recalling where they had their best performance. In the *LocalBest* version of PSO presented here, for a given particle, we define its neighbors (or friend particles) as a subset of particles with which it can communicate. Therefore, at each instant, each particle knows the characteristics of the best of its neighbors. The following expressions are used:

– $X_p^{(k)}$ (resp. $V_p^{(k)}$): position (resp. velocity) of the particle p at iteration k;

– $B_p^{(k)} = \arg\min\left(f\left(X_p^{(k-1)}\right), f\left(X_p^{(k)}\right)\right)$: best position found by the particle p until iteration k;

– $\Omega\left(X_p^{(k)}\right) \subset \{1,2,...,N_P\}$: set of particles defining the neighborhood of particle p;

– $G_p^{(k)} = \underset{X \in \left\{B_i^{(k)}, i \in \Omega\left(X_p^{(k)}\right)\right\}}{\arg\min} f(X)$: best position found by the neighborhood of particle p until iteration k.

Each particle moves in the search space according to the following transition rule:

$$\begin{aligned} V_p^{(k+1)} &= w.V_p^{(k)} + c_1 \otimes \left(B_p^{(k)} - X_p^{(k)}\right) + c_2 \otimes \left(G_p^{(k)} - X_p^{(k)}\right) \\ X_p^{(k+1)} &= X_p^{(k)} + V_p^{(k+1)} \end{aligned} \qquad [1.16]$$

where

– $\otimes$ is the element-wise product;

– w is the inertia factor;

– $c_1 = rnd(0, \bar{c}_1)$, $c_2 = rnd(0, \bar{c}_2)$ are the acceleration factors.

Therefore, each particle of the swarm moves depending on its own velocity and its best position yet obtained, as well as the position of the best of its neighbors (Figure 1.5).

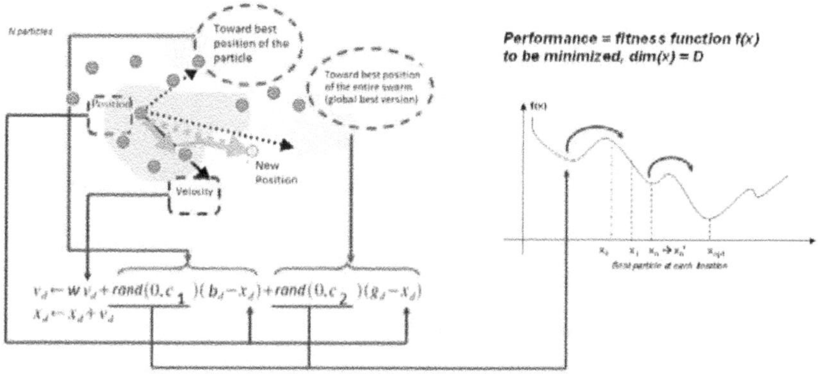

Figure 1.5. *PSO Principle. For a color version of this figure, see www.iste.co.uk/feyel/optimization.zip*

In the case of the *GlobalBest* version of the algorithm, $G_p^{(k)}$ is replaced by $G^{(k)} = \underset{X \in \left\{B_i^{(k)}, i=\{1,2,\ldots,N_P\}\right\}}{\arg\min} f(X)$ which is the best position that the swarm has found at iteration k.

While the *GlobalBest* version favors convergence, the *LocalBest* version favors exploration and proves to be better adapted to high-dimensional problems: we will therefore work on this version here.

Particles confinement

To ensure the algorithm's convergence, the particles' position and speed should be confined:

$$\begin{aligned} V_p^{(k+1)} &= \min\left(\max\left(V_p^{(k+1)}, V_{\min}\right), V_{\max}\right) \\ X_p^{(k+1)} &= \min\left(\max\left(X_p^{(k+1)}, X_{\min}\right), X_{\max}\right) \end{aligned} \quad [1.17]$$

where

– $[X_{\min}, X_{\max}]$ is the search space interval;

– $[V_{\min}, V_{\max}]$ limits a particles' maximum speed of movement and:

$$\begin{aligned} V_{\max} &= \delta\left(X_{\max} - X_{\min}\right),\ \delta \in \left]0,1\right] \\ V_{\min} &= -V_{\max} \end{aligned} \quad [1.18]$$

Initialization of the algorithm

Initializing the algorithm means randomly choosing particles in the search space interval. It is the same for their velocity:

$$\begin{aligned} X_p^{(0)} &= X_{\min} + rnd(0,1)\left(X_{\max} - X_{\min}\right) \\ V_p^{(0)} &= V_{\min} + rnd(0,1)\left(V_{\max} - V_{\min}\right) \end{aligned} \quad [1.19]$$

The algorithm's pseudo-code

The algorithm's pseudo-code is given below:

Randomly generating particle positions and velocities in the search space using [1.19].

While the stop criterion is not satisfied, repeat:

For $p = 1$ to N_P particles:

Compute the velocity V_p of the particle,

$V_p = \min(\max(V_p, V_{min}), V_{max})$,

Compute the position X_p of the particle,

$X_p = \min(\max(X_p, X_{min}), X_{max})$ [1.20]

If $f(X_p) < f(B_p)$ then $B_p = X_p$, end if

End For

Compute $G_p = \underset{X \in \{B_i, i \in \Omega(X_p)\}}{\arg\min} f(X)$, $p = 1$ à N_P

Compute $G = \underset{X \in \{B_i, i = \{1,2,\ldots,N_P\}\}}{\arg\min} f(X)$

End while

Canonical setting of the algorithm

The standard version of PSO (*Std_PSO*) is defined hereafter with the following settings (called canonical) for which, generally, we obtain good results:

– Swarm size $N_P = floor(10+\sqrt{n})$, where n is the number of optimization variables, in other words the problem's dimension;

– $w = 0.729$; $\bar{c}_1 = \bar{c}_2 = 1.494 = 2.05 \times 0.729$;

– $\dim\left(\Omega\left(X_p^{(k)}\right)\right) = 3$;

– $\delta = 0.9$.

Several topologies exist for the choice of the neighborhood $\Omega\left(X_p^{(k)}\right)$, but, for the sake of simplicity, we often use a cyclic neighborhood (*social ring*) defined once at initialization. Therefore, the neighborhood of particle i is defined by the following set of particles:

$$\Omega(X_i) = \left(X_{(i-1)\ \mathrm{modulo}(N_P)}\ ;\ X_i\ ;\ X_{(i+1)\ \mathrm{modulo}(N_P)}\right) \quad [1.21]$$

Explanations for these settings can be found in [CLE 02], [CLE 12] and [CLE 04].

The algorithm's stop criterion

Several stop criteria can be used. We will retain:

– the maximum number of iterations, knowing that, at each iteration, the fitness function is evaluated N_P times;

– a maximum number of fitness evaluations;

– premature convergence of the algorithm, which is detected when all the particles tend to be identical, that is to say, when the following relation is satisfied:

$$\exists G : \frac{\max\limits_{i=1,\ldots,N_P} \left\| X_i^{(k)} - G \right\|}{\left\| X_{max} - X_{min} \right\|} < \varepsilon \quad [1.22]$$

In most cases, $\varepsilon = 10^{-5}$ will be chosen.

Variants of the algorithm

The previous metaheuristic forms a basis from which many versions have been developed [SED 09]. As the idea is not to substitute a controller setting problem with another parameter setting problem, it is a good policy to not make the algorithm more complicated.

In order to limit the algorithm's premature convergence, we will keep the *TVRndIW_PSO*[1] version in which the inertia factor has been varied at each iteration according to the law [EBE 01]:

$$w_k = rnd\left(\frac{1}{2},1\right) \qquad [1.23]$$

The inertia factor's mean value therefore remains around 0.75. By slightly modifying the inertia factor at each iteration, we thus create diversity, which makes the algorithm more effective for high-dimensional problems. The other setting parameters remain unchanged from the canonical version.

The versions retained

We will keep the two versions of PSO (*LocalBest* version) that have proven to be most effective in our study: *Std_PSO* and *TVRndIW_PSO*.

1.3.2. *Quantum particle swarm algorithm*

Principle

An important property of PSO [CLE 02] is that the algorithm's convergence is reached if the *i*th particle converges on its local attractor, whose coordinates are defined in an *n*-dimensional search space (*n* being the problem's dimension) by $P_i^{(k)} = \left(p_{i1}^{(k)}, p_{i2}^{(k)}, \ldots, p_{ij}^{(k)}, \ldots, p_{in}^{(k)}\right)$ with:

$$p_{ij}^{(k)} = \frac{c_1 b_{ij}^{(k)} + c_2 g_j^{(k)}}{c_1 + c_2} = \phi b_{ij}^{(k)} + (1-\phi) g_j^{(k)} \qquad [1.24]$$

φ is a random number distributed uniformly in [0.1] and:

$$\begin{aligned} B_i^{(k)} &= \left(b_{i1}^{(k)}, b_{i2}^{(k)}, \ldots, b_{ij}^{(k)}, \ldots, b_{in}^{(k)}\right) \\ G^{(k)} &= \left(g_1^{(k)}, g_2^{(k)}, \ldots, g_j^{(k)}, \ldots, g_n^{(k)}\right) \end{aligned} \qquad [1.25]$$

1 TVRndIW: time varying randomly inertia weight.

In quantum particle swarm optimization (QPSO) [SUN 12a], the swarm is a quantum system, and each particle has a quantum behavior with its quantum state modeled by a wave function $\psi\left(\vec{X},t\right)$ that should meet the Schrödinger equation:

$$i\hbar\frac{\partial\psi\left(\vec{X},t\right)}{\partial t}=\hat{H}\left(\vec{X}\right)\psi\left(\vec{X},t\right) \qquad [1.26]$$

where $\hbar$ is the Planck constant and $i=\sqrt{-1}$. The Hamiltonian operator is defined for a particle with a mass m in a potential field $V\left(\vec{X}\right)$ by:

$$\hat{H}\left(\vec{X}\right)=-\frac{\hbar^2}{2m}\nabla^2+V\left(\vec{X}\right) \qquad [1.27]$$

Moreover, $Q=\left|\psi\right|^2$ represents the particle's probability density function to be in $\vec{X}$ and is such that:

$$\int Q(\vec{X})d\vec{X}=1 \qquad [1.28]$$

In the case of movement in a one-dimensional space and in a case where the abscissa point x_0 is the center of the potential well:

$$\begin{aligned}&V\left(X\right)=-\gamma\delta\left(X-x_0\right)=-\gamma\delta\left(Y\right)\\&\left.\begin{aligned}&\delta\left(Y\right)-1 \text{ si } Y=0\\&\delta\left(Y\right)=0 \text{ si } Y\neq 0\end{aligned}\right\}\end{aligned} \qquad [1.29]$$

where ε is the potential well's intensity and the function $\delta(x)$ is equal to 1 when x is equal to 0, and 0 if not. By denoting E the particle's energy ($E < 0$), the Hamiltonian is written as:

$$\hat{H}\left(Y\right)=-\frac{\hbar^2}{2m}\frac{d^2}{dY^2}-\left(\gamma\delta(Y)+E\right) \qquad [1.30]$$

Therefore, at equilibrium, the particle's state satisfies the following stationary Schrödinger equation (we retain the time dependency of [1.26]):

$$\hat{H}\left(Y\right)\psi\left(Y\right)=0 \qquad [1.31]$$

In other words:

$$\frac{d^2\psi}{dY^2}+\frac{2m}{\hbar^2}\left(\gamma\delta(Y)+E\right)\psi\left(Y\right)=0 \qquad [1.32]$$

A solution such as $\psi \to 0$ when $|Y| \to \infty$ is:

$$\psi(Y)=\frac{1}{\sqrt{L}}e^{-|Y|/L},\ L=\frac{2E}{\gamma},\ E=\frac{\hbar^2}{2m} \qquad [1.33]$$

and:

$$Q(Y)=\frac{1}{L}e^{-2|Y|/L} \qquad [1.34]$$

It is verified that:

$$\begin{aligned}\int_{-\infty}^{+\infty}\frac{1}{L}e^{-2|Y|/L}dY &= \int_{-\infty}^{0}\frac{1}{L}e^{2Y/L}dY+\int_{0}^{+\infty}\frac{1}{L}e^{-2Y/L}dY \\ &=\left[\frac{1}{2}e^{2Y/L}\right]_{-\infty}^{0}+\left[-\frac{1}{2}e^{-2Y/L}\right]_{0}^{+\infty} \\ &=1\end{aligned} \qquad [1.35]$$

The corresponding probability distribution is written as:

$$F(Y)=1-e^{-2|Y|/L} \qquad [1.36]$$

The ith particle has the coordinates $X_i=(x_{i1},x_{i2},\ldots,x_{ij},\ldots,x_{in})$.

Inspired by analysis of the traditional PSO's convergence, the ith particle moves at iteration k around a potential well centered on $p_{ij}^{(k)}$ for the jth dimension (Figure 1.6).

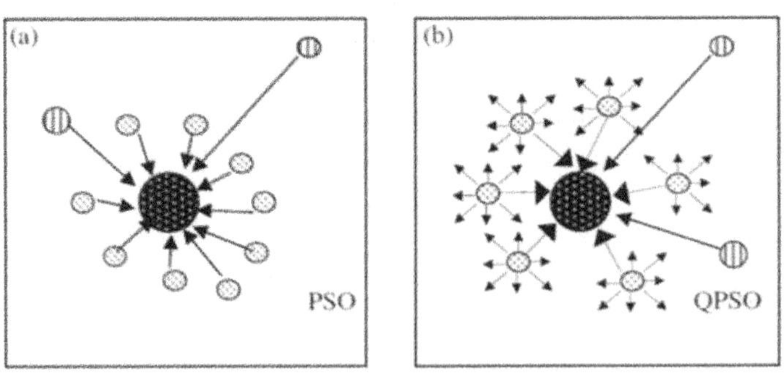

Figure 1.6. *The particles' movement in PSO (a) and QPSO (b)*

The corresponding wave function at iteration $k + 1$ is written as:

$$\psi\left(x_{ij}^{(k+1)}\right)=\frac{e^{\frac{-\left|x_{ij}^{(k)}-p_{ij}^{(k)}\right|}{L_{ij}^{(k)}}}}{\sqrt{L_{ij}^{(k)}}} \quad [1.37]$$

The corresponding probability distribution is written as:

$$F\left(x_{ij}^{(k+1)}\right)=1-e^{-2\frac{\left|x_{ij}^{(k)}-p_{ij}^{(k)}\right|}{L_{ij}^{k}}} \quad [1.38]$$

from which we infer that:

$$x_{ij}^{(k)} = p_{ij}^{(k)} \pm \frac{L_{ij}^{(k)}}{2}\ln\left(\frac{1}{1-F\left(x_{ij}^{(k+1)}\right)}\right) \quad [1.39]$$

Let us imagine that uniformly distributed draws are made for F, thus $F \approx U(0,1)$ and so $1-F \approx U(0,1)$. By denoting $u=1-F$, it becomes:

$$x_{ij}^{(k)} = p_{ij}^{(k)} \pm \frac{L_{ij}^{(k)}}{2}\ln\left(\frac{1}{u}\right), \quad u=U(0,1) \quad [1.40]$$

Therefore, in [SUN 12a] and [SUN 12b], we suggest replacing the traditional PSO's evolution law [1.16] with:

$$x_{ij}^{(k+1)} = p_{ij}^{(k)} \pm \beta\left|C_j^{(k)} - x_{ij}^{(k+1)}\right|\ln\left(\frac{1}{u_{ij}^{(k+1)}}\right) \quad [1.41]$$

where $C^{(k)}=\left(C_1^{(k)},C_2^{(k)},\ldots,C_n^{(k)}\right)$ is the mean of the best positions found by the particles at iteration k, that is to say:

$$C^{(k)}=\left(\frac{1}{P}\sum_{i=1}^{P}b_{i1}^{(k)},\frac{1}{P}\sum_{i=1}^{P}b_{i2}^{(k)}\ldots,\frac{1}{P}\sum_{i=1}^{P}b_{in}^{(k)}\right) \quad [1.42]$$

$u_{ij}^{(k+1)}$ is a random number distributed uniformly in [0.1] and β is the contraction–expansion coefficient (CEC) and is used to control the algorithm's rate of convergence.

Particles confinement

To ensure the algorithm's convergence, the particles' position should be confined in the search space:

$$X_p^{(k+1)} = \min\left(\max\left(X_p^{(k+1)}, X_{\min}\right), X_{\max}\right) \quad [1.43]$$

Initializing the algorithm

Initializing the algorithm consists of randomly choosing particles in the search space interval:

$$X_p^{(0)} = X_{\min} + rnd(0,1)\left(X_{\max} - X_{\min}\right) \quad [1.44]$$

The algorithm's pseudo-code

The algorithm's pseudo-code is given below:

Randomly generate particle positions in the search space

While the stop criterion is not satisfied, repeat:

  Compute the best mean position C with [1.42],

  For $i = 1$ to N_P particles, do:

    if $f(X_i) < f(B_i)$ then $B_i = X_i$ end if

    For $j = 1$ to n dimensions, do:

      $\varphi = rnd(0.1)$; $u = rnd(0.1)$;

      $p_{ij} = \varphi \, .b_{ij} + (1-\varphi).g_j$

      if rnd(0,1) > 0,5 [1.45]

        $x_{ij} = p_{ij} + \beta.\text{abs}(C_j - x_{ij}).\log(1/u)$

      else

        $x_{ij} = p_{ij} - \beta.\text{abs}(C_j - x_{ij}).\log(1/u)$

      end if

      $x_{ij} = \min(\max(x_{ij}, X_{min\_j}), X_{max\_j})$

    End for

  End for

  Find $G = \underset{X \in \{B_i, i=\{1,2,\ldots,N_P\}\}}{\arg\min} f(X)$

End while

Setting the algorithm

One advantage of this algorithm is its low number of setting parameters. The only parameter to set is β. The drawback is that the algorithm's behavior is very sensitive to β. However, in general, a constant value of β around 0.75 leads to good results [SUN 12a]. This is the *StaticCEC_QPSO* version.

Following the same idea as the traditional PSO, the size of the swarm is fixed at $N_P = floor(10 + \sqrt{n}\,)$, where n is the number of optimization variables, that is to say the problem's dimension.

The algorithm's stop criterion

Several stop criteria can be used. We will retain:

– the maximum number of iterations, knowing that, at each iteration, the cost function is evaluated N_P times;

– a maximum number of fitness evaluations;

– the algorithm's premature convergence, which is detected when all the particles tend to be identical, that is to say, when the following relationship is satisfied:

$$\exists G : \frac{\max\limits_{i=1,\dots,N_P} \left\| X_i^{(k)} - G \right\|}{\left\| X_{max} - X_{min} \right\|} < \varepsilon \qquad [1.46]$$

In most cases, we will choose $\varepsilon = 10^{-5}$.

Variants of the algorithm

Many versions of QPSO exist [SUN 12a]. Most consist of varying β according to particular laws. For example, we can create diversity at each iteration by randomly varying β as follows:

$$\beta(k) = rnd\left(\beta_{\min}, \beta_{\max}\right); \ \beta_{\max} = 1{,}0 \ ; \ \beta_{\min} = 0{,}5 \qquad [1.47]$$

We will retain version *TVDecCEC_QPSO* [SUN 12a, SUN 12b], in which β decreases during an N-iterations optimization process according to the following law:

$$\beta(k) = \beta_{\max} - \frac{\beta_{\max} - \beta_{\min}}{N} \times k \ ; \ \beta_{\max} = 1{,}0; \beta_{\min} = 0{,}5 \qquad [1.48]$$

The versions retained

Finally, we will retain the two versions of QPSO that have proven to be most effective in our study: *StaticCEC_QPSO* and *TVDecCEC_QPSO*.

1.3.3. *Artificial bee colony optimization algorithm*

Principle

Artificial bee colony (ABC) optimization [KAR 07, KAR 08] is a recent metaheuristic inspired by the natural model of the behavior of honey bees when they search for food.

The bees' process of searching for food is based on a very effective movement mechanism. It enables them to draw the attention of the other bees in the colony to food sources found, with the aim of collecting diverse resources. In fact, the bees use a set of wriggling dances as a means of communication with each other. These dances enable the bees to share information on the direction, distance and quantity of nectar with their peers. The collective collaboration and knowledge of bees in the same colony is based on the exchange of information on the quantity of nectar in the food source found by the various members.

In a bee colony optimization algorithm, a source of nectar corresponds to a possible solution of the problem to be solved. The colony of artificial bees consists of three types of bees: the "employed" bees, the onlooker bees and the scouts:

– the employed bee exploits the food source found. It relies on its memory and tries to make changes to its current position (solution) to find a new position (i.e. food source);

– the onlooker waits for the employed bees to return to the dance field to observe their dances and collect information on the sources of nectar that they have found;

– the scout bee exploits the search space by launching a random search for a new food source.

Initializing the algorithm

The initial population is formed of a number of N_{FS} food sources randomly generated in the search space. Assuming that the population's *i*th food source is represented by $X_i = (x_{i1},\ldots,x_{ij},\ldots,x_{in})$, where n is the problem's dimension, each food source is generated by:

$$x_{ij} = X_{\min j} + rnd(0,1)\left(X_{\max j} - X_{\min j}\right),\ j = 1,\ldots,n,\ i = 1,\ldots,N_{FS} \qquad [1.49]$$

These food sources are randomly assigned to N_O employed bees in the hive and the values of the corresponding cost functions are evaluated. Here, we will assume that $N_O = N_{FS}$.

Similarly, we will suppose that the number of onlooker bees N_S is equal to N_O.

Finally, each food source has an abandonment counter initialized at 0.

The k^{ith} iteration consists of successively carrying out the three phases described below.

Algorithm description

– Employed bee phase:

In this phase, each employed bee generates a new food source $x_{new,}$ chosen in the neighborhood of the current position found:

$$x_{newij} = x_{ij} + rnd(-1,1)\left(x_{ij} - x_{kj}\right), \quad i = 1,\ldots,N_O \qquad [1.50]$$

where $k \in \{1,2,3,\ldots,N_{FS}\}$, so that $k \neq i$ and $j \in \{1,2,3,\ldots,n\}$ are chosen at random.

We then proceed to a selection after evaluating the cost function for this new solution. X_{new} replaces X_i in the population if $f(X_{new}) \leq f(X_i)$ (and its abandonment counter is re-initialized), otherwise X_i is retained and its abandonment counter is incremented.

– Onlooker phase:

In this stage, the onlookers recover information on the quantity of nectar in the source x_i from the employed bees. The probability p_i that the onlookers choose the source x_i is determined as follows:

$$p_i = \frac{fit_i}{\sum_{k=1}^{N_{FS}} fit_k} \qquad [1.51]$$

where fit_i is the quantity of nectar in the ith food source X_i.

In the literature, we also find the following expression [OZT 12]:

$$p_i = 0{,}1 + \frac{0{,}9\, fit_i}{\max\limits_k \left(fit_k\right)} \qquad [1.52]$$

The quantity of nectar linked to the source X_i is determined by:

$$fit_i = \begin{cases} \dfrac{1}{1+f(X_i)}, \; if \; f(X_i) \geq 0 \\ \\ 1+\left|f(X_i)\right|, \; if \; f(X_i) < 0 \end{cases} \quad [1.53]$$

According to [1.51] and [1.52], it is clear that the higher the fit_i is, the greater the probability of selecting the corresponding source X_i is.

Once the ith source is selected by the onlookers, they modify it using [1.50].

If the source thus modified is better than X_i, then the modified source replaces X_i in the population (its abandonment counter is then reinitialized), otherwise X_i is retained and its abandonment counter is incremented.

– Scout phase:

The ith food source is abandoned if it cannot be improved after a set number of tests, T_{limit} decided beforehand, (in other words, if its abandonment counter exceeds T_{limit}), and the corresponding employed bee then becomes a scout bee. The scout bee then generates its food source randomly:

$$x_{ij} = X_{\min j} + rnd(0,1)\left(X_{\max j} - X_{\min j}\right), \; j = 1,\ldots,n \quad [1.54]$$

Pseudo-code of the algorithm

The pseudo-code of the algorithm is given below:

For each employed bee

Generate an initial solution X_i for each worker bee with [1.49],

Evaluate the cost function $f(X_i)$

Evaluate the associated quantity of nectar (fit_i) using [1.53],

Initialize an abandonment counter $C_i = 0$.

The stop criterion is not satisfied, repeat:

Employed Phase

For each employed bee

Generate a new position X_{new} with [1.50], [1.55]

Evaluate the corresponding cost function $f(X_{new})$,

Evaluate its quantity of nectar (fit_{new}) using [1.53],

If $f(X_{new}) \leq f(X_i)$, then

$$X_i \leftarrow X_{new}$$

$C_i = 0$

Else

$C_i = C_i + 1$

End If

End For

Evaluate the probability of transmitting information associated with each source p_i using [1.51],

Onlooker phase

For each onlooker

if rnd(0,1) $< p_i$

Generate a new position X_{new} with [1.50],

Evaluate the corresponding cost function $f(X_{new})$,

Evaluate its quantity of nectar (fit_{new}) using [1.53]

If $f(X_{new}) \leq f(X_i)$, then

$X_i \leftarrow X_{new}$

$C_i = 0$

Else

$C_i = C_i + 1$

End If

Scout phase

For each scout

If $C_i \geq T_{limit}$

Generate a new position with [1.54],

Evaluate the corresponding cost

Evaluate its quantity of nectar (fit_i) using [1.53],

$C_i = 0$

Retain the best solution X_{best}

End while

Stop criterion

Several stop criteria can be used. We will retain:

– the maximum number of iterations, knowing that, at each iteration, the fitness function is evaluated P times;

– a maximum number of evaluations of the cost function;

– the algorithm's premature convergence, which is detected when all the bees tend to be identical, in other words when the following relation is met:

$$\exists X_{best} : \frac{\max\limits_{i=1,\ldots,P} \left\| X_i^{(k)} - X_{best} \right\|}{\left\| X_{max} - X_{min} \right\|} < \varepsilon \qquad [1.56]$$

In most cases, we will choose $\varepsilon = 10^{-5}$.

Variants of the algorithm

The algorithm's parameters are the number of food sources N_{FS}, the number of employed bees N_O, the number of onlookers N_S and, finally, the limit T_{limit} at which the food source should be abandoned.

The standard version of the algorithm consists of taking:

$$N_{FS} = N_O = N_S \qquad [1.57]$$

Therefore, there remain only two parameters to set.

Although a population size N_{FS} about $10n$ is generally advised, the algorithm has given its best results using the same idea as in (Q)PSO, in other words:

$$N_{FS} = floor\left(10 + \sqrt{n}\right) \qquad [1.58]$$

Finally, the literature recommends choosing [KAR 08]:

$$T_{limit} = n \times N_{FS} \qquad [1.59]$$

1.3.4. *Cuckoo search algorithm*

Principle

The CS is a recent metaheuristic [YAN 09a] inspired by the reproduction method of some species of cuckoo. In fact, their reproduction strategy has the peculiarity that the females lay their eggs in other species' nests (where the eggs have a similar appearance (Figure 1.7)). The eggs can therefore be hatched by substitute parents. Moreover, when the cuckoo eggs succeed in hatching in the host nest (they hatch more rapidly), the cuckoo chicks have a reflex that causes them to throw the eggs of the host species out of the nest and even imitate the cry of the host chicks with the aim of being fed by the host species.

Figure 1.7. *A cuckoo egg (gray) in a dunnock nest. For a color version of this figure, see www.iste.co.uk/feyel/optimization.zip*

However, it is possible for the cuckoo eggs to be discovered; in this case, the substitute parents remove them from the nest, or abandon the nest and start their brood elsewhere. This metaheuristic is therefore based on the cuckoo species' parasitic behavior, linked to a "Levy flight" displacement specific to certain birds and certain species of flies.

The CS algorithm is based on the following rules:

– each cuckoo lays only one egg at a time and places it in a nest chosen randomly;

– the best nests with high-quality eggs (solutions) are kept for subsequent generations;

– the number of host nests is fixed and the egg laid by the cuckoo can be discovered by the host species with a probability $p_a \in [0,1]$. In this case, the host bird either removes the egg from the nest, or leaves the nest and builds a new one. To simplify, this last hypothesis can be approached by replacing a fraction p_a of n nests with new ones.

In CS, each egg in a nest represents a solution and each cuckoo can lay a single egg (which represents a solution), the aim is to use the new and potentially best solution to replace a poorer solution in a nest. Although the algorithm can be extended to a more complex case where each nest contains several eggs representing a set of solutions, here we use the simplest version where each nest only contains a single egg. In this case, there is no longer a distinction between egg, nest and cuckoo, and each nest corresponds to an egg that also represents a cuckoo.

Levy flight

In the framework of CS, the cuckoos' displacement step is determined by Levy flight (Figure 1.8). The Levy flight is a random walk in which the steps have a length with a particular probability distribution (Levy distribution), the steps' direction being isotropic and random. The Levy flight is a class of random walk in which the jumps are distributed according to the Levy distribution, which consists of a power law with infinite variance and infinite mean of type:

$$Levy(\beta) \sim y = x^{-\beta}, \; 1 < \beta \leq 3 \qquad [1.60]$$

In the case of CS, the use of the Levy flight improves and optimizes the search: new solutions are generated by a Levy random walk around the best solution yet obtained, which accelerates the global search.

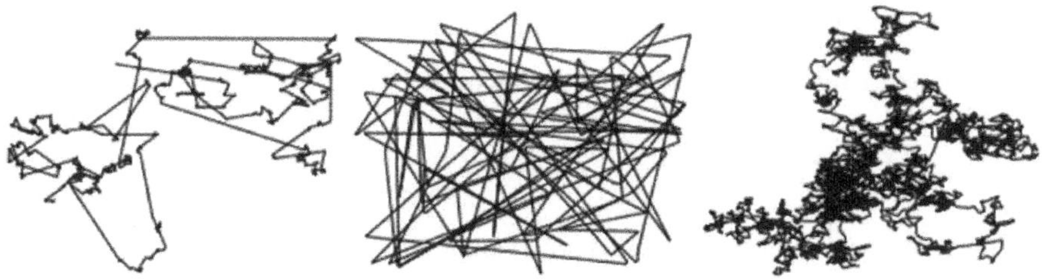

a) b) c)

Figure 1.8. *Levy flight a), simple random walk b) and Brownian movement c)*

From the point of view of implementation, generating a random number with the Levy flight follows two stages: the choice of a random direction and the generation of a step that ought to obey the Levy distribution. A direction can be generated from a uniform distribution, whereas generating steps is more delicate. There are several methods for arriving at this, but one of the simplest and effective methods consists of using the Mantegna formulae to determine step s:

$$s = \frac{u}{|v|^{1/\beta}} \qquad [1.61]$$

where u and v are normal distributions centered such that:

$$u = N\left(0, \sigma_u^2\right), \quad v = N\left(0, \sigma_v^2\right) \qquad [1.62]$$

with:

$$\sigma_u^2 = \frac{\Gamma(1+\beta)\sin(\pi\beta/2)}{\Gamma((1+\beta)/2)\beta 2^{(\beta-1)/2}}, \quad \sigma_v^2 = 1 \qquad [1.63]$$

where $\Gamma(z)$ is the gamma function:

$$\Gamma(z) = \int_0^{+\infty} t^{z-1} e^{-t} dt \qquad [1.64]$$

Initializing the algorithm

The initial population consists of a number N_H of host nests randomly generated in the search space. We recall that the numbers of host nests, cuckoos and eggs are equal.

Assuming that the *i*th host nest is represented by $X_i = (x_{i1},...,x_{ij},...,x_{in})$, where *n* is the problem's dimension, each nest is generated by:

$$x_{ij} = X_{\min j} + rnd(0,1)\left(X_{\max j} - X_{\min j}\right),\ j = 1,\ldots,n,\ i = 1,\ldots,N_H \qquad [1.65]$$

The corresponding cost functions are evaluated.

Describing the algorithm

– Step 1: global search

The first step consists of carrying out a global search. To do this, new cuckoo (solution) tests are generated from existing cuckoos (solution) by performing the Levy flight. This consists of using the following evolution law for each cuckoo:

$$\left.\begin{aligned} X_{new\,i} &= X_i + \alpha \otimes Levy(\beta) \\ F_{new\,i} &= f\left(X_{new\,i}\right) \end{aligned}\right\},\ i = 1,\ldots,N_H \qquad [1.66]$$

where $\alpha > 0$ is the size of the displacement step, which depends on the problem considered and such that $\alpha \approx O(L/10)$, where L is the characteristic scale of the problem considered. The product $\otimes$ is an element-wise product.

In order to know where/if the new *i*th cuckoo will deposit an egg, it randomly chooses a host nest X_j $(j \neq i)$ with the cost function F_j.

The new cuckoo lays an egg if:

$$F_{new\,i} < F_j \qquad [1.67]$$

In this case, solution $X_{new\,i}$ replaces solution X_j in the population.

– Step 2: local search

The new cuckoo eggs have a probability p_a of being discovered.

In this case ($rnd(0,1) < p_a$), new eggs are generated. Several methods can be envisaged for this. For CS, we will retain the following, where we proceed to a hybridization between elements randomly chosen in the population:

$$\begin{aligned} x'_{new\,ij} &= x_{ij} + rnd(0,1)\left(x_{pj} - x_{mj}\right),\ j = 1,\ldots,n,\ i = 1,\ldots,N_H \\ F'_{new\,i} &= f\left(X'_{new\,i}\right) \end{aligned} \qquad [1.68]$$

where $m \in \{1,2,3,\ldots,N_H\}$ and $p \in \{1,2,3,\ldots,N_H\}$, such that $p \neq m$ are randomly chosen.

A selection identical to that of the previous step is then made.

Pseudo-code of the algorithm

The algorithm's simplified pseudo-code is given below:

Generate an initial population of N_H host nests X_i

While the stop criterion is not satisfied, repeat:

Generate a cuckoo using Levy Flight [1.66] and evaluate its cost function F_i,

Choose a nest, X_j, at random from among the host nests N_H,

If $F_i < F_j$ [1.69]

Replace F_j with this new solution,

End If

A fraction p_a of the worst nests is abandoned and others are generated instead [1.68], and make a selection.

Identify the best solution X_{best}

End while

Setting the algorithm

We will retain the following standard settings for CS [YAN 09a]:

– The eggs' probability of discovery: $p_a = 0.25$

– Size of the displacement step: $\alpha = \dfrac{\|X_{\min} - X_{\max}\|}{10}$

– Sizing of the Levy flight: $\beta = 3/2$

There are no empirical rules for choosing the number of nests. However, following the same principles as PSO, the algorithm has given its best results with a number of nests equal to:

$$N_H = floor\left(10 + \sqrt{n}\right) \quad [1.70]$$

The stop criterion of the algorithm

Several stop criteria can be used. We will retain:

– the maximum number of iterations, knowing that, at each iteration, the fitness function is evaluated P times;

– a maximum number of fitness evaluations;

– the algorithm's premature convergence, which is detected when all the cuckoos tend to be identical, in other words when the following relation is satisfied:

$$\exists X_{best} : \frac{\max\limits_{i=1,\ldots,P} \left\| X_i^{(k)} - X_{best} \right\|}{\left\| X_{max} - X_{min} \right\|} < \varepsilon \qquad [1.71]$$

In most cases, we will choose $\varepsilon = 10^{-5}$.

Variants of the algorithm

Various existing versions consist of modifying the local and/or the global exploration phase. We will retain the CS/DE version [WAN 12], which consists of hybridizing the algorithm with DE. More precisely, the principle consists of mutation and cross-over operations of DE/Rand/1 in the CS global search phase to respectively generate new solutions instead of the Levy flight and potentially to select them. This generation/selection mechanism is given by [1.72].

For i = 1 is N_H, do

Randomly select $r_1 \neq r_2 \neq r_3 \neq i$,

$X_v = X_{r1} + F \times (X_{r2} - X_{r3})$,

$r_4 = floor(1 + rnd(0,1) \times N_H)$

For j = 1 to n, do

If $rnd(0,1) \leq C_r$ or j == r_4 then

$X_u(j) = X_v(j)$

Else $X_u(j) = X_i(j)$

End If [1.72]

End for

If $F(X_u) < F(X_{r4})$ then

$X_{r4} = X_u$

$F(X_{r4}) = F(X_u)$

End If

End For

We find again the DE mutation and cross-over operators. We will retain the following standard settings: $F = 0.75$ and $C_r = 0.8$.

The CS/DE pseudo-code is given below:

Generating an initial population of N_H host nests X_i

While the stop criterion is not satisfied, repeat:

Generate a new cuckoo population by [1.72] [1.73]

Proceed with the local search (stage 2)

Identify the best solution X_{best}

End while

The versions retained

We will retain the two versions of CS that have proven to be most effective in our study: *CS* and *CS/DE*.

1.3.5. *Firefly algorithm*

Principle

The firefly algorithm (firefly – FF) [YAN 09b] is inspired by fireflies' brilliant phosphorescence and their behavior of attraction. In particular, bright fireflies attract one another (the brightest attracts the others), and the further apart they are, the less they attract. Some hypotheses are realized to simplify the algorithm and improve its efficiency:

– each firefly is assumed to be unisex, which means that it can be attracted by any other firefly;

– the fireflies are only attracted by other fireflies that are brighter and this attraction decreases with the distance that separates them;

– luminosity is determined by the cost function to be minimized.

At iteration k, the position of the ith firefly attracted by the jth firefly is revealed by the following evolution law:

$$X_i^{(k+1)} = X_i^{(k)} + \beta_{ij}\left(X_j^{(k)} - X_i^{(k)}\right) + \alpha_k \varepsilon_i^{(k)} \quad [1.74]$$

where β_{ij} is the attraction coefficient of the *i*th firefly to the *j*th firefly and α_k is a parameter. In most cases:

$$\alpha_k = \alpha_0^{\ k} \qquad [1.75]$$

ε_i is usually a normal distribution but, as in *CS*, the *Levy flight* principle can also be used.

The attraction is assumed to follow the following law:

$$\beta_{ij} = \begin{cases} \beta_0 e^{-\gamma r_{ij}^2} & \text{si } f(X_i) \geq f(X_j) \\ 0 & \text{si } f(X_i) < f(X_j) \end{cases} \qquad [1.76]$$

where ε is a parameter (absorption coefficient) and r_{ij} is the distance between the fireflies X_i and X_j, defined as the Euclidean norm:

$$r_{ij} = \left\| X_i - X_j \right\| = \sqrt{\sum_k \left(x_{ik} - x_{jk} \right)^2} \qquad [1.77]$$

Pseudo-code of the algorithm

The algorithm's pseudo-code is given below:

Generate randomly an initial population of N_L fireflies X_i (i=1,…,N_L) in the search space,

The intensity of light I_i at X_i is determined by $f(X_i)$,

define the light absorption coefficient ε,

While the stop criterion is not satisfied, repeat:

    For i = 1 to N_L fireflies

        For j = 1 to N_L fireflies

            if ($I_i > I_j$) then

                Move firefly i towards firefly j with [1.74] [1.78]

            End If

            The attraction varies with the distance r via exp($-r^2$),

            Evaluate the new solutions and reveal the intensities.

        End For j

    End For i

    Identify the best solution X_{best}

End while

Setting the algorithm

We will retain the following standard settings:

– Sizing of the Levy flight: $\beta = 3/2$

– Absorption coefficient: $\gamma = \frac{1}{d_{\max}^2}$, $d_{\max} = \|X_{\max} - X_{\min}\|$

– Initialization: $\beta_0 = 0,01L$, $L = \frac{1}{D}\sum_k (X_{\max k} - X_{\min k})$

– Initialization: $\alpha_0 = 0,9$

It is shown that, in some conditions, *FF* tends towards particular variants of *PSO*. Thus, we can use the same rules for setting the number of fireflies:

$$N_L = floor\left(10 + \sqrt{D}\right) \quad [1.79]$$

Stop criterion of the algorithm

Several stop criteria can be used. We will retain:

– the maximum number of iterations, knowing that, at each iteration, the fitness function is evaluated P times;

– a maximum number of fitness evaluations;

– the algorithm's premature convergence, which is detected when all the fireflies tend to be identical, in other words when the following relationship is met:

$$\exists X_{best} : \frac{\max\limits_{i=1,\ldots,P} \left\|X_i^{(k)} - X_{best}\right\|}{\left\|X_{max} - X_{min}\right\|} < \varepsilon \quad [1.80]$$

In most cases, $\varepsilon = 10^{-5}$ will be chosen.

1.4. Summary

The algorithms and the variants retained can be summarized in the following table:

Family	Name of the algorithm	Description
Swarm	Std_PSO	Particle swarm algorithm, *LocalBest* version, canonical settings
	TVRndIW_PSO	Particle swarm algorithm, *LocalBest* version, canonical settings with stochastic inertia factor around the canonical value
	StaticCEC_QPSO	Classical quantum particle swarm algorithm with a constant contraction–expansion coefficient
	TVDecCEC_QPSO	Classical quantum particle swarm algorithm with a contraction–expansion coefficient that decreases over time
	FF	Firefly algorithm
	ABC	Standard bee colony algorithm
	CS	Classical cuckoo search algorithm
Evolutionary	Rnd_DE	Differential evolution algorithm in DE/Rand/1 version
	Rnd2Bst_DE	Differential evolution algorithm in DE/Rand-2-best/1 version
	Rand2Bst_PDE	Differential evolution algorithm in PDE/Rand-2-best/1 version
Swarm/evolutionary hybrid	CS/DE	Cuckoo search algorithm CS hybridized with DE/Rand/1

Table 1.1. *Tested algorithms*

Other algorithms have been tested but without real success (either because they have not converged, or because they are too complex to set). For example, we cite:

– Bat algorithm (BA) [YAN 10b];

– Electromagnetic-like algorithm (ElA) [BIR 03];

– Harmony search (HS) [LEE 05].

2

Reformulation of Robust Control Problems for Stochastic Optimization

2.1. Introduction

In this section, we focus on one of the major drawbacks inherent in advanced, modern robust synthesis methods: the order of synthesized controllers is high and even more so with H_∞ loop-shaping synthesis, as the order of the final controller is at least twice the order of weights. This aspect of synthesis can quickly become problematic; in fact, the required performances of servo-loops increasingly joined with an ever greater need for robustness push the engineer to implement more and more sensors. The need to develop systems at ever lower cost means using calculators with limited capacity, so that the engineer is inevitably forced to limit the order of implemented control. Post-reduction of the controllers is possible, but poses the question of the robustness of the reduced controller: it is therefore necessary to constrain the order of the controller *a priori* during its synthesis.

2.2. H_∞ synthesis

2.2.1. *Full H_∞ synthesis*

2.2.1.1. *Introduction*

We consider the standard form for control depicted in Figure 1.1, where:

– the input e includes m_1 exogenous inputs;

– the output z includes p_1 signals to survey;

– y includes p_2 measures used by the controller;

– u includes m_2 control signals computed by the controller:

$$e(t) \in \mathbf{R}^{m_1},\ u(t) \in \mathbf{R}^{m_2},\ z(t) \in \mathbf{R}^{p_1}, y(t) \in \mathbf{R}^{p_2}$$

$$\begin{pmatrix} z(s) \\ y(s) \end{pmatrix} = P(s)\begin{pmatrix} e(s) \\ u(s) \end{pmatrix} = \begin{pmatrix} P_{11}(s) & P_{12}(s) \\ P_{21}(s) & P_{22}(s) \end{pmatrix}\begin{pmatrix} e(s) \\ u(s) \end{pmatrix} \quad [2.1]$$

The standard H_∞ problem consists of computing a controller $K(s)$, which:

– internally stabilizes the closed loop $F_l(P, K)$ depicted in Figure 1.1;

– minimizes γ such that:

$$\left\| F_l(P,K) \right\|_\infty < \gamma \quad [2.2]$$

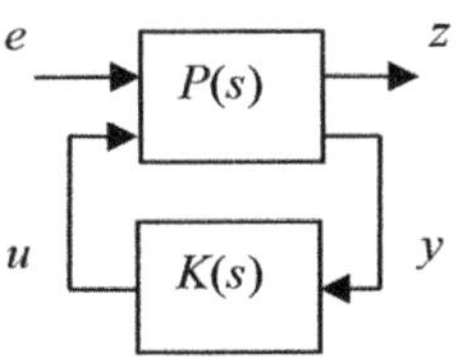

Figure 2.1. *Standard form for H∞ synthesis*

In the full-order case (i.e. when the order n_k of the controller $K(s)$ is equal to n_p), the state-space solution of this problem is perfectly established using mainly the following two methods:

– the Riccati equations approach [DOY 89];

– the Linear Matrix Inequality approach [GAH 94].

2.2.1.2. *Standard Riccati approach*

2.2.1.2.1. Principle

We use the following state-space representation for $P(s)$:

$$\begin{pmatrix} \dot{x}(t) \\ z(t) \\ y(t) \end{pmatrix} = \begin{pmatrix} A & B_1 & B_2 \\ C_1 & D_{11} & D_{12} \\ C_2 & D_{21} & D_{22} \end{pmatrix}\begin{pmatrix} x(t) \\ e(t) \\ u(t) \end{pmatrix} \quad [2.3]$$

In addition, we note:

$$B_P = \begin{pmatrix} B_1 & B_2 \end{pmatrix},\ D_P = \begin{pmatrix} D_{11} & D_{12} \\ D_{21} & D_{22} \end{pmatrix},\ C_P = \begin{pmatrix} C_1 \\ C_2 \end{pmatrix} \qquad [2.4]$$

Below, we summarize the results of standard H_∞ synthesis by solving the Riccati equations.

We note $n_p = \dim(A)$. First, the following hypotheses A1–A4 should be verified as follows:

– A1) (A, B_2) is stabilizable and (C_2, A) is detectable. This hypothesis is common to all the control problems;

– A2) $D_{12} = \begin{pmatrix} 0 \\ I \end{pmatrix}$, $D_{21} = \begin{pmatrix} 0 & I \end{pmatrix}$ and $D_{22} = 0$. Later, we will see that this hypothesis is in no way restrictive.

More precisely, it is necessary that $\operatorname{rank}(D_{12}) = m_2$ and $\operatorname{rank}(D_{21}) = p_2$, where m_2 is the number of control signals and p_2 is the number of measurements;

– A3) $\operatorname{rank}\begin{pmatrix} A - j\omega I & B_2 \\ C_1 & D_{12} \end{pmatrix} = \operatorname{rank}\begin{pmatrix} A - j\omega I & B \\ C & 0 \\ 0 & I \end{pmatrix} = n_p + m_2,\quad \forall \omega \in R$;

– A4) $\operatorname{rank}\begin{pmatrix} A - j\omega I & B_1 \\ C_2 & D_{21} \end{pmatrix} = \operatorname{rank}\begin{pmatrix} A - j\omega I & 0 & B \\ C & I & 0 \end{pmatrix} = n_p + p_2,\quad \forall \omega \in R$

We then note:

$$D_{11} = \begin{pmatrix} D_{1111} & D_{1112} \\ D_{1121} & D_{1122} \end{pmatrix}$$

$$\begin{aligned}
&\dim(D_{11}) = (p_1) \times (m_1) \\
&\dim(D_{1111}) = (p_1 - m_2) \times (m_1 - p_2) \\
&\dim(D_{1112}) = (p_1 - m_2) \times (p_2) \\
&\dim(D_{1121}) = (m_2) \times (m_1 - p_2) \\
&\dim(D_{1122}) = (m_2) \times (p_2)
\end{aligned} \qquad [2.5]$$

With hypotheses A1–A4 satisfied, there is a controller $K(s)$ that internally stabilizes the closed loop such that $\|F_l(P,K)\|_\infty < \gamma$ if and only if:

1) $\gamma > \max\left(\bar{\sigma}(D_{1111}, D_{1112}), \bar{\sigma}(D_{1111}^*, D_{1121}^*)\right)$;

2) there exists $X_\infty > 0$ and $Y_\infty \geq 0$ such that $\rho(X_\infty Y_\infty) < \gamma^2$ which are solutions to the following two Riccati equations:

$$
\begin{aligned}
&\left(A - B_P R^{-1} D_{1\bullet}^* C_1\right)^* X_\infty + X_\infty\left(A - B_P R^{-1} D_{1\bullet}^* C_1\right) - X_\infty B_P R^{-1} B_P^* X_\infty \\
&\qquad + C_1^*\left(I - D_{1\bullet} R^{-1} D_{1\bullet}^*\right) C_1 = 0 \\
&\left(A - B_1 D_{\bullet 1}^* \tilde{R}^{-1} C_P\right) Y_\infty + Y_\infty\left(A^* - C_P^* \tilde{R}^{-1} D_{\bullet 1} B_1^*\right) - Y_\infty C_P^* \tilde{R}^{-1} C_P Y_\infty \\
&\qquad + B_1\left(I - D_{\bullet 1}^* \tilde{R}^{-1} D_{\bullet 1}\right) B_1^* = 0
\end{aligned}
\qquad [2.6]
$$

with:

$$
\begin{aligned}
D_{1\bullet} &= \begin{pmatrix} D_{11} & D_{12} \end{pmatrix} \\
D_{\bullet 1} &= \begin{pmatrix} D_{11} \\ D_{21} \end{pmatrix} \\
R &= D_{1\bullet}^* D_{1\bullet} - \begin{pmatrix} \gamma^2 I_{m_1} & 0 \\ 0 & 0 \end{pmatrix} \\
\tilde{R} &= D_{\bullet 1} D_{\bullet 1}^* - \begin{pmatrix} \gamma^2 I_{p_1} & 0 \\ 0 & 0 \end{pmatrix}
\end{aligned}
\qquad [2.7]
$$

When constraints 1) and 2) are satisfied, a second result is that all the controllers internally stabilizing the closed loop and such that $\|F_l(P,K)\|_\infty < \gamma$ are given by:

$$K = F_l(K_\Phi, \Phi),\ \Phi \in \mathrm{R}H_\infty,\ \|\Phi\|_\infty < \gamma \qquad [2.8]$$

with:

$$
K_\Phi = \left[\begin{array}{c|cc} \hat{A} & \hat{B}_1 & \hat{B}_2 \\ \hline \hat{C}_1 & \hat{D}_{11} & \hat{D}_{12} \\ \hat{C}_2 & \hat{D}_{21} & 0 \end{array}\right] \qquad [2.9]
$$

$$\begin{aligned}
&\hat{D}_{11} = -D_{1121}D_{1111}^{*}\left(\gamma^2 I - D_{1111}D_{1111}^{*}\right)^{-1} D_{1112} - D_{1122}\\
&\hat{D}_{12}\hat{D}_{12}^{*} = I - D_{1121}\left(\gamma^2 I - D_{1111}D_{1111}^{*}\right)^{-1} D_{1121}^{*}\\
&\hat{D}_{21}\hat{D}_{21}^{*} = I - D_{1112}^{*}\left(\gamma^2 I - D_{1111}D_{1111}^{*}\right)^{-1} D_{1112}\\
&\hat{B}_2 = \left(B_2 + H_{12\infty}\right)\hat{D}_{12}\\
&\hat{C}_2 = -\hat{D}_{21}\left(C_2 + F_{12\infty}\right)Z\\
&\hat{B}_1 = -H_{2\infty} + \hat{B}_2\hat{D}_{12}^{-1}\hat{D}_{11}\\
&\hat{C}_1 = F_{2\infty}Z + \hat{D}_{11}\hat{D}_{21}^{-1}\hat{C}_2\\
&\hat{A} = A + H_{\infty}C_P + B_2\hat{D}_{12}^{-1}\hat{C}_1\\
&Z = \left(I - \gamma^{-2}Y_{\infty}X_{\infty}\right)^{-1}
\end{aligned} \qquad [2.10]$$

The controller's order is therefore equal to that of $P(s)$, i.e. n_p, and:

$$F_{\infty} = \begin{matrix} m_1 - p_2 \updownarrow \\ p_2 \updownarrow \\ m_2 \updownarrow \end{matrix}\begin{pmatrix} F_{11\infty} \\ F_{12\infty} \\ F_{2\infty} \end{pmatrix} = -R^{-1}\left(D_{1\bullet}^{*}C_1 + B_P^{*}X_{\infty}\right)$$

$$H_{\infty} = \begin{pmatrix} \underset{\underset{p_1 - m_2}{\leftrightarrow}}{H_{11\infty}} & \underset{\underset{m_2}{\leftrightarrow}}{H_{12\infty}} & \underset{\underset{p_2}{\leftrightarrow}}{H_{2\infty}} \end{pmatrix} = -\left(B_1 D_{\bullet 1}^{*} + Y_{\infty}C_P^{*}\right)\tilde{R}^{-1} \qquad [2.11]$$

The central controller can be obtained by assuming $\Phi = 0$.

2.2.1.2.2. Implementing the standard approach

Some remarks on implementation

– Hypotheses about the standard form considered

According to hypothesis A2:

$$D_{12} = \begin{pmatrix} 0 \\ I \end{pmatrix} \text{ and } D_{21} = \begin{pmatrix} 0 & I \end{pmatrix};$$

$D_{22} = 0.$

To satisfy these hypotheses, we use the *loop-shifting* procedure [ZHO 96], the aim of which is to employ a series of transformations to obtain a form equivalent to any standard form (in the sense of the H∞ norm) and which satisfies the previous hypotheses.

$$\left\| F_l\left(P(s), K(s)\right) \right\|_\infty < \gamma$$

$$\begin{pmatrix} \dot{x}(t) \\ e(t) \\ y(t) \end{pmatrix} = \begin{pmatrix} A & B_1 & B_2 \\ C_1 & D_{11} & D_{12} \\ C_2 & D_{21} & D_{22} \end{pmatrix} \begin{pmatrix} x(t) \\ w(t) \\ u(t) \end{pmatrix}$$

$$\Downarrow$$

$$\left\| F_l\left(P_p(s), K_p(s)\right) \right\|_\infty < \gamma \qquad [2.12]$$

$$\begin{pmatrix} \dot{x}_p(t) \\ e_p(t) \\ y_p(t) \end{pmatrix} = \begin{pmatrix} A_p & B_{p1} & B_{p2} \\ C_{p1} & D_{p11} & D_{p12} \\ C_{p2} & D_{p21} & D_{p22} \end{pmatrix} \begin{pmatrix} x_p(t) \\ w_p(t) \\ u_p(t) \end{pmatrix}$$

$$D_{p22} = 0,\ D_{p12} = \begin{pmatrix} 0 \\ I \end{pmatrix},\ D_{p21} = \begin{pmatrix} 0 & I \end{pmatrix}$$

– Loop-shifting on D_{12} and D_{21}

We seek to "normalize" matrices D_{12} and D_{21}. We assume that D_{12} is full column rank and D_{21} is full row rank. We consider the following factorizations:

$$\begin{aligned} D_{12} &= U_p \begin{pmatrix} 0 \\ I \end{pmatrix} R_p \\ D_{21} &= \bar{R}_p \begin{pmatrix} 0 & I \end{pmatrix} \bar{U}_p \end{aligned} \qquad [2.13]$$

where matrices U_p and $\bar{U}_p$ are square and unitary, i.e.:

$$\begin{aligned} U_p U_p^* &= U_p^* U_p = I \\ \bar{U}_p \bar{U}_p^* &= \bar{U}_p^* \bar{U}_p = I \end{aligned} \qquad [2.14]$$

We posit:

$$e = U_p e_p,\ w = \bar{U}_p^* w_p,\ y = \bar{R}_p y_p,\ u = R_p u_p \qquad [2.15]$$

and:

$$K_p(s) = R_p K(s)\overline{R}_p \qquad [2.16]$$

so that:

$$\begin{aligned} P_p(s) &= \begin{pmatrix} U_p^* & 0 \\ 0 & \overline{R}_p^{-1} \end{pmatrix} P(s) \begin{pmatrix} \overline{U}_p^* & 0 \\ 0 & R_p^{-1} \end{pmatrix} \\ &= \left[\begin{array}{c|cc} A & B_1\overline{U}_p^* & B_2 R_p^{-1} \\ \hline U_p^* C_1 & U_p^* D_{11}\overline{U}_p^* & U_p^* D_{12} R_p^{-1} \\ \overline{R}_p^{-1} C_2 & \overline{R}_p^{-1} D_{21}\overline{U}_p^* & \overline{R}_p^{-1} D_{22} R_p^{-1} \end{array}\right] \\ &= \left[\begin{array}{c|cc} A & B_{p1} & B_{p2} \\ \hline C_{p1} & D_{p11} & D_{p12} \\ C_{p2} & D_{p21} & D_{p22} \end{array}\right] \end{aligned} \qquad [2.17]$$

Then:

$$D_{p12} = \begin{pmatrix} 0 \\ I \end{pmatrix},\ D_{p21} = \begin{pmatrix} 0 & I \end{pmatrix} \qquad [2.18]$$

And also:

$$\begin{aligned} \left\| F_l(P,K) \right\|_\infty = \left\| U_p F_l(P_p, K_p)\overline{U}_p \right\|_\infty &= \left\| F_l(P_p, K_p) \right\|_\infty \\ P_p(s) = \begin{pmatrix} U_p^* & 0 \\ 0 & \overline{R}_p^{-1} \end{pmatrix} & P(s) \begin{pmatrix} \overline{U}_p^* & 0 \\ 0 & R_p^{-1} \end{pmatrix} \\ K_p(p) = R_p & K(s)\overline{R}_p \end{aligned} \qquad [2.19]$$

– Loop-shifting on D_{22}

To have $D_{22} = 0$, we use the *loop-shifting* procedure [ZHO 96] on D_{22}; it is easy to show that:

$$\begin{aligned} &F_l(P_p, K_p) = F_l(P, K) \\ &P = P_p + \begin{pmatrix} 0 & 0 \\ 0 & D_{22} \end{pmatrix} \\ &K = K_p\left(I + D_{22}K_p\right)^{-1} \end{aligned} \qquad [2.20]$$

It is therefore easier to look for the controller $K_p(s)$, which is the solution to the problem on $P_p(s)$ by taking $D_{22} = 0$, and then to reconstruct $K = [A_K, B_K, C_K, D_K]$ from $K_p = \left[A_{K_p}, B_{K_p}, C_{K_p}, D_{K_p}\right]$ with the following relations:

$$\begin{aligned}
A_K &= A_{K_p} + B_{K_p} M_1 D_{K_p}{}^{-1} C_{K_p} \\
B_K &= B_{K_p} M_2 \\
C_K &= M_1 B_{K_p} \\
D_K &= D_{K_p} M_1 \\
M_1 &= \left(I - D_{K_p} D_{22}\right)^{-1} \\
M_2 &= \left(I - D_{22} D_{K_p}\right)^{-1}
\end{aligned} \qquad [2.21]$$

– Resolution using the γ-iteration

In practice, when computing a controller using the standard approach, there is a preliminary phase relating to the existence of the controller conditioned by verification of 1) and 2), which requires a value of γ. The optimal value of γ (γ_{opt}) is approximated by a dichotomy that stops when i) or ii) is not satisfied.

Finally, the second phase consists of computing a controller, using γ_{opt}, with the formulae given previously.

2.2.1.3. *Approach using linear matrix inequalities (LMIs)*

Computing a full-order controller (i.e. $n_k = n_p$) is possible using an approach based on Riccati equations, but this does not permit synthesis of fixed-order controllers. Appearing more recently, H_∞ synthesis using LMI is an interesting alternative. It is less restrictive and based on the standard form. It has several versions, and here we retain the one that is broken down into two steps [GAH 94]: (1) study of the feasibility of the standard H_∞ problem, and (2) reconstruction of the optimal H_∞ controller.

It can easily be shown that the standard form in Figure 2.1 looped by $K(s)$ has the following state-space representation:

$$T(s) = F_l\left(P(s), K(s)\right)$$
$$= \left[\begin{array}{cc|c} A + B_2 M_1 D_k C_2 & B_2 M_1 C_k & B_1 + B_2 M_1 D_k D_{21} \\ B_k M_2 C_2 & A_k + B_k M_2 D_{22} C_k & B_k M_2 D_{21} \\ \hline C_1 + D_{12} M_1 D_k C_2 & D_{12} M_1 C_k & D_{11} + D_{12} M_1 D_k D_{21} \end{array}\right] \quad [2.22]$$
$$M_1 = (I - D_k D_{22})^{-1}$$
$$M_2 = (I - D_{22} D_k)^{-1}$$

As for the Riccati approach, we limit ourselves (for reasons of simplicity) to a case where $D_{22} = 0$, which is not at all restrictive by virtue of the *loop-shifting* procedure [ZHO 96]:

$$T(s) = \left[\begin{array}{cc|c} A + B_2 D_k C_2 & B_2 C_k & B_1 + B_2 D_k D_{21} \\ B_k C_2 & A_k & B_k D_{21} \\ \hline C_1 + D_{12} D_k C_2 & D_{12} C_k & D_{11} + D_{12} D_k D_{21} \end{array}\right] = \left[\begin{array}{c|c} A_{cl} & B_{cl} \\ \hline C_{cl} & D_{cl} \end{array}\right] \quad [2.23]$$

2.2.1.3.1. Feasibility of the standard problem

The feasibility of the standard problem is tested using the following theorem. Assuming (A, B_2) to be stabilizable and (C_2, A) to be detectable, the standard H_∞ control problem has a solution if and only if there are two matrices, R and S, verifying the following three LMIs:

$$\begin{pmatrix} N_R & 0 \\ 0 & I_{n_e} \end{pmatrix}^T \begin{pmatrix} AR + RA^T & RC_1^T & B_1 \\ C_1 R & -\gamma I_{n_z} & D_{11} \\ B_1^T & D_{11}^T & -\gamma I_{n_e} \end{pmatrix} \begin{pmatrix} N_R & 0 \\ 0 & I_{n_e} \end{pmatrix} < 0 \quad [2.24]$$

$$\begin{pmatrix} N_S & 0 \\ 0 & I_{n_z} \end{pmatrix}^T \begin{pmatrix} A^T S + SA & SB_1 & C_1^T \\ B_1^T S & -\gamma I_{n_e} & D_{11}^T \\ C_1 & D_{11} & -\gamma I_{n_z} \end{pmatrix} \begin{pmatrix} N_S & 0 \\ 0 & I_{n_z} \end{pmatrix} < 0 \quad [2.25]$$

$$\begin{pmatrix} R & I_{n_p} \\ I_{n_p} & S \end{pmatrix} \geq 0 \quad [2.26]$$

where N_R and N_S are forming a basis for the null spaces of $\begin{pmatrix} B_2^T & D_{12}^T \end{pmatrix}$ and $\begin{pmatrix} C_2 & D_{21} \end{pmatrix}$, respectively.

Moreover, the minimal value of γ is obtained by solving the following convex optimization problem:

$$\min_{R=R^T,\ S=S^T} \gamma$$

subject to constraints [2.24]–[2.26]. [2.27]

2.2.1.3.2. Reconstructing the optimal controller

Some $n_k \le n_p$ order controllers exist if and only if LMIs [2.24]–[2.26] are verified by matrices R and S, satisfying the condition:

$$\text{rank}\begin{pmatrix} R & I_{n_p} \\ I_{n_p} & S \end{pmatrix} \le n_p + n_k \Leftrightarrow \text{rank}\left(I_{n_p} - RS\right) \le n_k \qquad [2.28]$$

The final controller is obtained by applying the bounded real lemma [BOY 94] on system [2.23]: $\|T(s)\|_\infty \le \gamma$ if and only if there is a symmetric positive definite matrix X verifying:

$$\begin{pmatrix} A_{cl}^T X + XA_{cl} & XB_{cl} & C_{cl}^T \\ B_{cl}^T X & -\gamma I_{n_e} & D_{cl}^T \\ C_{cl} & D_{cl} & -\gamma I_{n_z} \end{pmatrix} < 0 \qquad [2.29]$$

which is a bilinear matrix inequality in X, A_k, B_k, C_k, D_k. An X matrix that works is [GAH 94]:

$$X = \begin{pmatrix} S & N \\ N^T & -M^+RN \end{pmatrix} \qquad [2.30]$$

where M^+ designates the pseudo-inverse of M (i.e. $M^+M = I$) and M and N are obtained by the following singular value decomposition:

$$MN^T = I_{n_p} - RS \qquad [2.31]$$

Inequality [2.31] then becomes an LMI in the variables A_k, B_k, C_k, D_k from which we deduce $K(s)$ after its resolution.

2.2.1.3.3. Observations

When the order of the standard form is substantial, this approach can suffer from numerical problems. However, this approach is of some interest, as:

– it is less restrictive than the Riccati approach;

– the order constraint is linked explicitly to a rank constraint: we can therefore theoretically set the controller's order through this means, although in reality, as we will see later, the rank constraint is not convex;

– this approach provides a much larger framework for solving more complex problems, such as multi-objective synthesis.

2.2.2. *Fixed-structure H_∞ synthesis*

2.2.2.1. *Fixed-order robust stabilization of a coprime factor plant description*

2.2.2.1.1. Principle

In [VIN 97], we find a fixed-order approach for stabilizing a coprime factor[1] plant description. It is based on the following result; the problem of robust stabilization is equivalent to the following particular distance problem [GLO 89]:

$$\left\| \begin{pmatrix} S_y(s) \\ K(s)S_y(s) \end{pmatrix} \overline{M}^{-1}(s) \right\|_\infty \leq \gamma,\; S_y(s) = \left(I - H(s)K(s)\right)^{-1}$$

$$\Leftrightarrow \qquad [2.32]$$

$$\left\| \begin{pmatrix} -\overline{N}^* \\ M^* \end{pmatrix} + \begin{pmatrix} U \\ V \end{pmatrix} \right\|_\infty \leq \left(1-\gamma^{-2}\right)^{1/2},\; K(s) = U(s)V(s)^{-1}$$

Here, we recall some elements of demonstration. Consider the doubly coprime realizations of a system $H(s) = N(s)M(s)^{-1} = \overline{M}(s)^{-1}\overline{N}(s)$ and a controller $K(s) = U(s)V(s)^{-1} = \overline{V}(s)^{-1}\overline{U}(s)$ stabilizing it. A right coprime factorization equivalent to $K(s)$ is $K(s) = \hat{U}(s)\hat{V}(s)^{-1}$ such that:

$$\begin{pmatrix} U \\ V \end{pmatrix} = \alpha \begin{pmatrix} \hat{U} \\ \hat{V} \end{pmatrix} \left(\overline{M}\hat{V} - \overline{N}\hat{U}\right)^{-1},\; \alpha \in \mathbf{R}^* \qquad [2.33]$$

1 In Appendix A, we will find some notions on coprime factorizations.

It is easily verified that:

$$UV^{-1} = \hat{U}\hat{V}^{-1*} \tag{2.34}$$

In this case:

$$\begin{aligned} \begin{pmatrix} KS_y \\ S_y \end{pmatrix} \overline{M}^{-1} &= \begin{pmatrix} K \\ I \end{pmatrix} (I - HK)^{-1} \overline{M}^{-1} \\ &= \begin{pmatrix} \hat{U}\hat{V}^{-1} \\ I \end{pmatrix} \left(I - \overline{M}^{-1} \overline{N} \hat{U} \hat{V}^{-1} \right)^{-1} \overline{M}^{-1} \\ &= \begin{pmatrix} \hat{U} \\ \hat{V} \end{pmatrix} \left(\overline{M}\hat{V} - \overline{N}\hat{U} \right)^{-1} \end{aligned} \tag{2.35}$$

Assuming the normalized coprime factorizations of H, it arises from the previous result:

$$\begin{gathered} \left\| \begin{pmatrix} KS_y \\ S_y \end{pmatrix} \overline{M}^{-1} \right\|_\infty \leq \gamma \\ \Leftrightarrow \\ \left\| \begin{pmatrix} M^* & N^* \\ -\overline{N} & \overline{M} \end{pmatrix} \begin{pmatrix} \hat{U} \\ \hat{V} \end{pmatrix} \left(\overline{M}\hat{V} - \overline{N}\hat{U} \right)^{-1} \right\|_\infty \leq \gamma \\ \Leftrightarrow \\ \left\| \begin{matrix} \left(M^*\hat{U} + N^*\hat{V} \right) \left(\overline{M}\hat{V} - \overline{N}\hat{U} \right)^{-1} \\ I \end{matrix} \right\|_\infty \leq \gamma \end{gathered} \tag{2.36}$$

that is:

$$\left(1 + \left\| \left(M^*\hat{U} + N^*\hat{V} \right) \left(\overline{M}\hat{V} - \overline{N}\hat{U} \right)^{-1} \right\|_\infty^2 \right)^{1/2} \leq \gamma \tag{2.37}$$

Moreover:

$$\begin{aligned}\left\|\begin{pmatrix}-\overline{N}^{*}\\ \overline{M}^{*}\end{pmatrix}+\begin{pmatrix}U\\ V\end{pmatrix}\right\|_{\infty}&=\left\|\begin{pmatrix}M^{*} & N^{*}\\ -\overline{N} & \overline{M}\end{pmatrix}\left(\begin{pmatrix}-\overline{N}^{*}\\ \overline{M}^{*}\end{pmatrix}+\begin{pmatrix}U\\ V\end{pmatrix}\right)\right\|_{\infty}\\ &=\left\|\begin{pmatrix}0\\ I\end{pmatrix}+\alpha\begin{pmatrix}M^{*} & N^{*}\\ -\overline{N} & \overline{M}\end{pmatrix}\begin{pmatrix}\hat{U}\\ \hat{V}\end{pmatrix}\left(\overline{M}\hat{V}-\overline{N}\hat{U}\right)^{-1}\right\|_{\infty}\\ &=\left\|\begin{pmatrix}0\\ I\end{pmatrix}+\alpha\begin{pmatrix}\left(M^{*}\hat{U}+N^{*}\hat{U}\right)\left(\overline{M}\hat{V}-\overline{N}\hat{U}\right)^{-1}\\ I\end{pmatrix}\right\|_{\infty}\\ &=\left\|\begin{pmatrix}\alpha\left(M^{*}\hat{U}+N^{*}\hat{U}\right)\left(\overline{M}\hat{V}-\overline{N}\hat{U}\right)^{-1}\\ I(1+\alpha)\end{pmatrix}\right\|_{\infty}\end{aligned}\qquad [2.38]$$

By virtue of [2.37], we arrive at:

$$\left\|\begin{pmatrix}-\overline{N}^{*}\\ \overline{M}^{*}\end{pmatrix}+\begin{pmatrix}U\\ V\end{pmatrix}\right\|_{\infty}^{2}\leq\alpha^{2}\left(\gamma^{2}-1\right)+(1+\alpha)^{2}=f(\alpha)\qquad [2.39]$$

Indeed, the function $f()$ is minimal for:

$$\frac{df}{d\alpha}=0\Leftrightarrow\alpha=-\gamma^{-2}\qquad [2.40]$$

Finally:

$$\left\|\begin{pmatrix}-\overline{N}^{*}\\ \overline{M}^{*}\end{pmatrix}+\begin{pmatrix}U\\ V\end{pmatrix}\right\|_{\infty}\leq\left(1-\gamma^{-2}\right)^{1/2}\qquad [2.41]$$

The problem of robust stabilization therefore means looking for the controller $K(s)=\hat{U}(s)\hat{V}(s)^{-1}$ stabilizing $H(s)=N(s)M(s)^{-1}=\overline{M}(s)^{-1}\overline{N}(s)$ with $\hat{U}(s),\ \hat{V}(s)$ solutions to the problem:

$$\inf_{\substack{Q(s)\in\hat{L}_{\infty}\\ \hat{U}(s),\ \hat{V}(s)\in RH_{\infty}}}\beta:\left\|\begin{pmatrix}\overline{N}(s)^{*}\\ -\overline{M}(s)^{*}\end{pmatrix}-\begin{pmatrix}\hat{U}(s)\\ \hat{V}(s)\end{pmatrix}Q(s)\right\|_{\infty}\leq\beta\qquad [2.42]$$

for which a particular $Q(s)$ is given by [2.33].

We recall that:

$$f(s) \in \hat{L}_\infty$$

$$\Leftrightarrow$$

$f(s) \in L_\infty$ (is bounded only on $j\omega$) [2.43]

$\det(f(j\omega)) \neq 0, \forall \omega$

$\det(f(j\omega))$ does not encapsulate 0 when s describes the Nyquist contour

From this result, we can envisage an original approach for robust stabilization subject to order constraint [VIN 97]. The approach is subdivided into the following four steps:

– Initialization: choosing any initial stabilizing controller $K(s)$ from which we calculate a normalized right coprime factorization written as:

$$K(s) = \hat{U}(s)\hat{V}(s)^{-1} \qquad [2.44]$$

and:

$$\begin{pmatrix} \hat{U} \\ \hat{V} \end{pmatrix} = \left[\begin{array}{c|c} \hat{A}_{UV} & \hat{B}_{UV} \\ \hline \hat{C}_{UV} & \hat{D}_{UV} \end{array}\right] \qquad [2.45]$$

– Step 1: optimization of $Q(s)$ to fixed $K(s)$. For fixed $K(s)$, we calculate:

$$\gamma = \left\| \begin{pmatrix} S_y(s) \\ K(s)S_y(s) \end{pmatrix} \overline{M}^{-1}(s) \right\|_\infty = \left\| \begin{pmatrix} I \\ K(s) \end{pmatrix} S_y(s)\begin{pmatrix} I & H(s) \end{pmatrix} \right\|_\infty \qquad [2.46]$$

We posit $\beta = \left(1-\gamma^{-2}\right)^{1/2}$. A $Q(s) \in \hat{L}_\infty$ ensuring β is given by [2.47]:

$$Q(s) = \gamma^{-2}\left(\begin{pmatrix} \overline{N}(s) & -\overline{M}(s) \end{pmatrix} \begin{pmatrix} \hat{U}(s) \\ \hat{V}(s) \end{pmatrix} \right)^{-1} \qquad [2.47]$$

– Step 2: optimization of $K(s)$ to fixed $Q(s)$. We now search for a controller $K(s) = U(s)V(s)^{-1}$ such that:

$$\begin{pmatrix} U \\ V \end{pmatrix} = \left[\begin{array}{c|c} \hat{A}_{UV} & \hat{B}_{UV} \\ \hline C_{UV} & D_{UV} \end{array}\right] \qquad [2.48]$$

which satisfies:

$$\left\| \begin{pmatrix} \overline{N}^* \\ -\overline{M}^* \end{pmatrix} - \begin{pmatrix} U \\ V \end{pmatrix} Q \right\|_\infty \leq \beta \qquad [2.49]$$

By fixing the controller's state-space matrix, we fix its order, which is therefore the same as that of the initial controller. The fact of keeping matrices $\hat{A}_{UV}$, $\hat{B}_{UV}$ constant is not restrictive, because nearly all controllers can be described by an arbitrary choice of $\hat{A}_{UV}$, $\hat{B}_{UV}$. Optimization therefore relies on C_{UV} and D_{UV}. We note the transfer:

$$\begin{pmatrix} \overline{N}^* \\ -\overline{M}^* \end{pmatrix} - \begin{pmatrix} U \\ V \end{pmatrix} Q = \left[\begin{array}{c|c} A_\beta & B_\beta \\ \hline C_\beta & D_\beta \end{array}\right] \qquad [2.50]$$

By virtue of the real bounded lemma [BOY 94], its H_∞ norm is less than β if and only if there is an $X = X^T$ matrix verifying the following matrix inequality[2]:

$$\begin{pmatrix} A_\beta X + X A_\beta^T & X B_\beta & C_\beta^T \\ B_\beta^T X & -\beta I & D_\beta^T \\ C_\beta & D_\beta & -\beta I \end{pmatrix} < 0 \qquad [2.51]$$

We posit:

$$\begin{pmatrix} \overline{N} & -\overline{M} \end{pmatrix} = \left[\begin{array}{c|c} A_{NM} & B_{NM} \\ \hline C_{NM} & D_{NM} \end{array}\right]$$
$$Q = \left[\begin{array}{c|c} A_Q & B_Q \\ \hline C_Q & D_Q \end{array}\right] \qquad [2.52]$$

2 We recall that the matrix inequality $Y < 0$ means that all the eigenvalues of Y are strictly negative.

In this case, it is easy to show that[3]:

$$\left[\begin{array}{c|c} A_\beta & B_\beta \\ \hline C_\beta & D_\beta \end{array}\right] = \left[\begin{array}{ccc|c} -A_{NM}{}^T & 0 & 0 & C_{NM}{}^T \\ 0 & A_Q & 0 & -B_Q \\ 0 & \hat{B}_{UV}C_Q & \hat{A}_{UV} & -\hat{B}_{UV}D_Q \\ \hline -B_{NM}{}^T & D_{UV}C_Q & C_{UV} & D_{NM}{}^T + D_{UV}D_Q \end{array}\right] \quad [2.53]$$

It is clear that A_{cl} and B_{cl} do not depend on C_{UV} nor D_{UV}. Moreover C_{cl} and D_{cl} depend linearly on C_{UV} and D_{UV}. The matrix inequality [2.51] is therefore linear in optimization variables X, C_{UV}, D_{UV} and β, which allows direct optimization using LMI techniques.

After Step 2, we return to Step 1 and so on. The stop criterion might relate to the value of γ defined by:

$$\gamma = \left(1-\beta^2\right)^{-1/2} \quad [2.54]$$

When γ is sufficiently small, the optimization ends at step 3.

– Step 3: computing the final controller. We posit:

$$C_{UV} = \begin{pmatrix} C_U \\ C_V \end{pmatrix}, \; D_{UV} = \begin{pmatrix} D_U \\ D_V \end{pmatrix} \quad [2.55]$$

The controller is reconstructed with the following state-space representation:

$$K(s) = U(s)V(s)^{-1} = \left[\begin{array}{c|c} \hat{A}_{UV} - \hat{B}_{UV}D_V C_V & \hat{B}_{UV}D_V{}^{-1} \\ \hline C_U - D_U D_V{}^{-1}C_V & D_U D_V{}^{-1} \end{array}\right] \quad [2.56]$$

2.2.2.1.2. Optimizing the approach

In [VIN 97], it is shown that another $Q(s) \in \hat{L}_\infty$ ensuring β is:

$$Q(s) = \left(\begin{pmatrix} \overline{N}(s) & -\overline{M}(s) \end{pmatrix}\begin{pmatrix} \hat{U}(s) \\ \hat{V}(s) \end{pmatrix}\right)^* = \left(\begin{pmatrix} -\overline{M}(s) & \overline{N}(s) \end{pmatrix}\begin{pmatrix} \hat{V}(s) \\ \hat{U}(s) \end{pmatrix}\right)^* \quad [2.57]$$

3 Recall that $H = \left[\begin{array}{c|c} A & B \\ \hline C & D \end{array}\right] \Rightarrow H^* = \left[\begin{array}{c|c} -A^T & -C^T \\ \hline B^T & D^T \end{array}\right]$.

It can therefore be preferable to use this parametrization in the previous algorithm. As the inversion has disappeared, the numerical conditioning is better. However, its real advantage lies in the simplification of the LMI at Step 3. By substituting the expression of $Q(s)$ in [2.50], we have:

$$\left[\begin{array}{c|c} A_\beta & B_\beta \\ \hline C_\beta & D_\beta \end{array}\right] = \begin{pmatrix} \overline{N}^* \\ -\overline{M}^* \end{pmatrix} - \begin{pmatrix} U \\ V \end{pmatrix} Q = \begin{pmatrix} \overline{N}^* \\ -\overline{M}^* \end{pmatrix} - \begin{pmatrix} U \\ V \end{pmatrix}\left(\begin{pmatrix} \overline{N} & -\overline{M} \end{pmatrix}\begin{pmatrix} \hat{U} \\ \hat{V} \end{pmatrix}\right)^* = \left(I - \begin{pmatrix} U \\ V \end{pmatrix}\begin{pmatrix} \hat{U} \\ \hat{V} \end{pmatrix}^*\right)\begin{pmatrix} \overline{N} & -\overline{M} \end{pmatrix}^* \quad [2.58]$$

A minimal representation is therefore:

$$\left[\begin{array}{c|c} A_\beta & B_\beta \\ \hline C_\beta & D_\beta \end{array}\right] = \left[\begin{array}{ccc|c} -A_{NM}{}^T & 0 & 0 & C_{NM}{}^T \\ \hat{C}_{UV}{}^T B_{NM}{}^T & -\hat{A}_{UV}{}^T & 0 & -\hat{C}_{UV}{}^T D_{NM}{}^T \\ \hat{B}_{UV}\hat{D}_{UV}B_{NM}{}^T & -\hat{B}_{UV}\hat{B}_{UV}{}^T & \hat{A}_{UV} & \hat{B}_{UV}\hat{D}_{UV}{}^T D_{NM}{}^T \\ \hline -\left(\mathrm{I} - D_{UV}\hat{D}_{UV}{}^T\right)B_{NM}{}^T & -D_{UV}\hat{B}_{UV}{}^T & C_{UV} & \left(\mathrm{I} - D_{UV}\hat{D}_{UV}{}^T\right)D_{NM}{}^T \end{array}\right] \quad [2.59]$$

Recall that the optimization relates to C_{UV} and D_{UV}. This representation has a lower order than [2.53], giving rise to fewer decision variables in the LMI [2.51]. The latter is always linear in the optimization variables X, C_{UV}, D_{UV} and β. The approach therefore consists of iteratively solving [2.51], the matrices being those of the controller from the previous iteration (Step 2 becomes obsolete in the sense that it is contained in Step 3 and matrices $\hat{A}_{UV}$, $\hat{B}_{UV}$ are those of the initial stabilizing controller).

2.2.2.1.3. Observations

This optimization is only local: the result therefore depends on the initial conditions. In the case of stabilizing a *loop-shape*, the optimization is initialized by a controller with the desired order stabilizing the latter (e.g. by reducing the full-order pre-compensator K_p (Figure 2.2) to the desired order, the only constraint being that the reduced pre-compensator is stabilizing). The final controller is obtained by recombining the optimized, reduced-order pre-compensator with the *loop-shape*'s weights (Figure 2.2). Thus, the final controller will have at least the order of the *loop-shape's* weights, which is a limitation of the method (aside from the fact that it is necessary to find a stabilizing controller of the desired order); however, it is a natural extension of the synthesis method presented in the introduction.

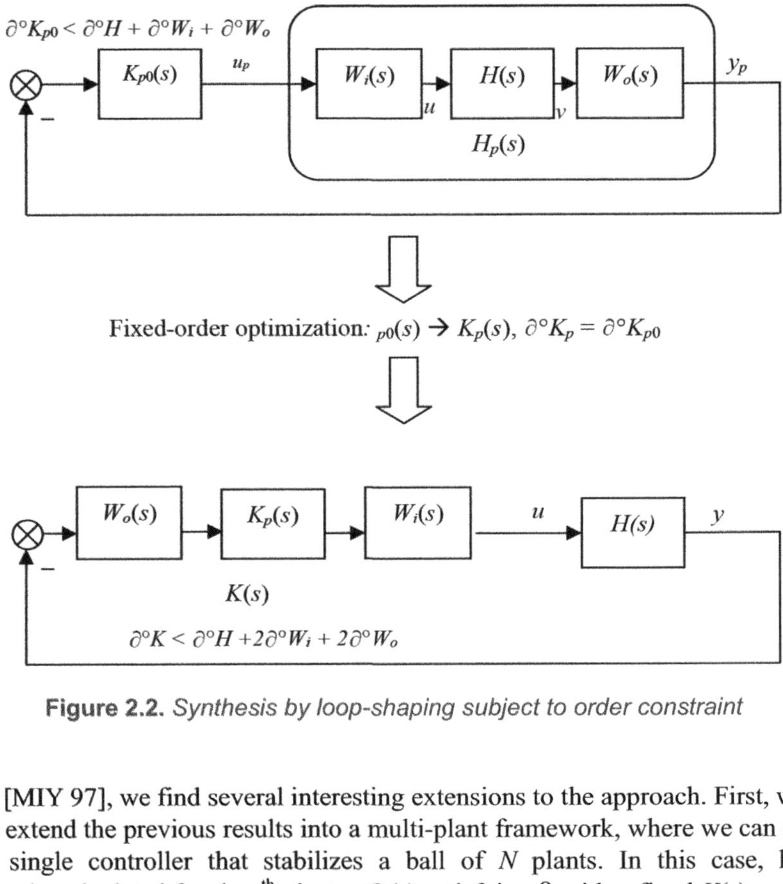

Figure 2.2. *Synthesis by loop-shaping subject to order constraint*

In [MIY 97], we find several interesting extensions to the approach. First, we can easily extend the previous results into a multi-plant framework, where we can search for a single controller that stabilizes a ball of N plants. In this case, having previously calculated for the ith plant a $Q_i(s)$ satisfying β with a fixed $K(s)$, we need only to solve the LMI [2.51] simultaneously for all the models, that is, solve the problem:

$$\min_{X,\beta,C_{UY},D_{UY}} \beta$$

$$\left(\begin{pmatrix} A_{\beta,i}X + XA_{\beta,i}^{\ T} & XB_{\beta,i} & C_{\beta}^{\ T} \\ B_{\beta,i}^{\ T}X & -\beta I & D_{\beta,i}^{\ T} \\ C_{\beta,i} & D_{\beta,i} & -\beta I \end{pmatrix} < 0,\ i = 1,\dots,N\right) \qquad [2.60]$$

Finally, another extension to the method relates to the computation of decentralized controllers. Given a system $H(s)$ of dimensions $p{\cdot}m$, we find a block diagonal controller $K(s)$ such as:

$$\left\| \begin{pmatrix} S_y(s) \\ K(s)S_y(s) \end{pmatrix} \overline{M}^{-1}(s) \right\|_\infty \leq \gamma$$
$$K(s) = diag\left(K_1(s),...,K_L(s)\right) \quad [2.61]$$

where the ith controller $K_i(s)$ is of dimension $p_i{\cdot}m_i$. Furthermore:

$$m = \sum_{i=1}^{L} m_i,\ p = \sum_{i=1}^{L} p_i \quad [2.62]$$

In this case, we need only to impose a decentralized structure on each of the matrices defining the coprime factor of the controller $K(s) = U(s)V(s)^{-1}$:

$$V = diag\left(V_1,...,V_L\right),\ V_i \in \mathbf{R}H_\infty^{p_i \times p_i}$$
$$U = diag\left(U_1,...,U_L\right),\ U_i \in \mathbf{R}H_\infty^{m_i \times p_i} \quad [2.63]$$

We then solve the LMI [2.51] in relation to the optimization variables X, C_{UV}, D_{UV} and β, where:

$$C_{VU} = diag\left(C_{VU_1},...,C_{VU_L}\right)$$
$$D_{VU} = diag\left(D_{VU_1},...,D_{VU_L}\right) \quad [2.64]$$

2.2.2.1.4. Conclusion

The drawback of the above approach is that the order constraint is only related to the pre-compensator; the final controller therefore has at least the order of the weights, which can quickly become problematic in case of strong implementation. Furthermore, the method must be initialized with a stabilizing controller with the desired order. However, it retains all the advantages related to *loop-shaping* robust synthesis. On the basis of the four-block equivalent criterion from *loop-shaping* synthesis, we can extend the framework of the above approach by directly optimizing the order of the final controller following the synthesis scheme in Figure 1.3, where the input and output weights of the plant have been externalized.

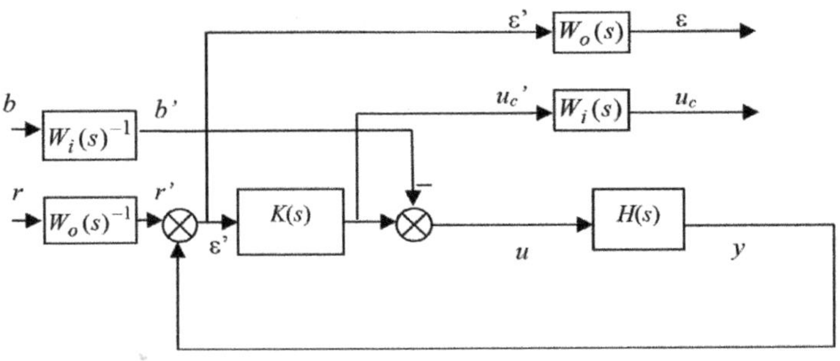

Figure 2.3. *Loop-shaping synthesis in equivalent four-block form*

Consider *H*(*s*) of dimension $p{\cdot}m$, then the synthesis scheme in Figure 2.3 is a particular case of the general standard form for control *P*(*s*) in Figure 1.4, where:

– *z* represents the signals to survey, so here $z = (\varepsilon, u_c)^T$ and dim(*z*) = $n_z = m + p$;

– *e* represents the exogenous inputs, so here $e = (b, r)^T$ and dim(*e*) = $n_e = m + p$;

– *y* represents the measurements available to the controller, so here $y = \varepsilon'$ and dim(*y*) = *p*;

– *u* represents the control signals, here $u = u_c'$ and dim(*u*) = *m*.

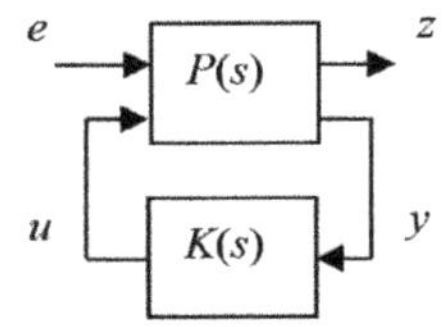

Figure 2.4. *Standard form for H∞ synthesis*

2.2.2.1.5. The notion of static output feedback

The loopback in Figure 2.4 has the following state-space representation:

$$
\begin{aligned}
T(s) &= F_l(P(s), K(s)) \\
&= \left[\begin{array}{cc|c} A + B_2 M_1 D_k C_2 & B_2 M_1 C_k & B_1 + B_2 M_1 D_k D_{21} \\ B_k M_2 C_2 & A_k + B_k M_2 D_{22} C_k & B_k M_2 D_{21} \\ \hline C_1 + D_{12} M_1 D_k C_2 & D_{12} M_1 C_k & D_{11} + D_{12} M_1 D_k D_{21} \end{array}\right] \\
M_1 &= (I - D_k D_{22})^{-1} \\
M_2 &= (I - D_{22} D_k)^{-1}
\end{aligned}
\qquad [2.65]
$$

By virtue of the *loop-shifting* procedure [ZHO 96], we can posit $D_{22} = 0$ without loss of generality so that:

$$T(s) = \left[\begin{array}{cc|c} A + B_2 D_k C_2 & B_2 C_k & B_1 + B_2 D_k D_{21} \\ B_k C_2 & A_k & B_k D_{21} \\ \hline C_1 + D_{12} D_k C_2 & D_{12} C_k & D_{11} + D_{12} D_k D_{21} \end{array}\right] \qquad [2.66]$$

In the case where the controller is static, the closed-loop response becomes:

$$T(s) = \left[\begin{array}{c|c} A + B_2 K C_2 & B_1 + B_2 K D_{21} \\ \hline C_1 + D_{12} K C_2 & D_{11} + D_{12} K D_{21} \end{array}\right] \qquad [2.67]$$

If we set:

$$\Theta_k = \begin{pmatrix} A_k & B_k \\ C_k & D_k \end{pmatrix} \qquad [2.68]$$

then, [2.66] is equivalent to:

$$T(s) = \left[\begin{array}{c|c} \overline{A} + \overline{B}_2 \Theta_k \overline{C}_2 & \overline{B}_1 + \overline{B}_2 \Theta_k \overline{D}_{21} \\ \hline \overline{C}_1 + \overline{D}_{12} \Theta_k \overline{C}_2 & \overline{D}_{11} + \overline{D}_{12} \Theta_k \overline{D}_{21} \end{array}\right] \qquad [2.69]$$

with:

$$\begin{aligned} &\overline{A} = \begin{pmatrix} A & 0_{n_p \times n_k} \\ 0_{n_k \times n_p} & 0_{n_k \times n_k} \end{pmatrix}, \overline{B}_1 = \begin{pmatrix} B_1 \\ 0_{n_k \times n_e} \end{pmatrix}, \overline{B}_2 = \begin{pmatrix} 0_{n_p \times n_k} & B_2 \\ I_{n_k} & 0_{n_k \times m} \end{pmatrix} \\ &\overline{C}_1 = \begin{pmatrix} C_1 & 0_{n_z \times n_k} \end{pmatrix}, \quad \overline{D}_{11} = D_{11}, \quad \overline{D}_{12} = \begin{pmatrix} 0_{n_z \times n_k} & D_{12} \end{pmatrix} \\ &\overline{C}_2 = \begin{pmatrix} 0_{n_k \times n_p} & I_{n_k} \\ C_2 & 0_{p \times n_k} \end{pmatrix}, \overline{D}_{21} = \begin{pmatrix} 0_{n_k \times n_e} \\ D_{21} \end{pmatrix} \end{aligned} \qquad [2.70]$$

By comparing [2.67] and [2.69], it is clear that any dynamic output feedback can formally be made into a static output feedback on the augmented system defined by [2.70]. Thus, the problem of synthesizing a dynamic output feedback of any order $K(s) = [A_k, B_k, C_k, D_k]$ can be reduced to that of synthesizing a static output feedback Θ_k defined by [2.68].

2.2.2.2. *The limits of LMI H$_\infty$ synthesis and the order constraint*

According to [2.28], to compute *a priori* a controller with an order lower than n_p, we need only to impose an additional constraint on the LMIs [2.24]–[2.26] on the rank of the matrix $I_{n_p} - RS$ during the feasibility study of the standard H$_\infty$ problem. Thus, the H∞ problem of reduced order n_k accepts a solution if and only if there are two matrices, R and S, verifying:

$$\operatorname{rank}\begin{pmatrix} R & I_{n_p} \\ I_{n_p} & S \end{pmatrix} \leq n_p + n_k \qquad [2.71]$$

subject to constraints [2.24]–[2.26].

However, the rank constraint is not convex; nevertheless, the static output feedback is a particularly interesting case, as:

– any dynamic output feedback synthesis problem can be reduced to the synthesis of a static output feedback on an augmented system;

– in this case, the rank constraint becomes:

$$\operatorname{Rank}\begin{pmatrix} R & I_{n_p} \\ I_{n_p} & S \end{pmatrix} \leq n_p \Leftrightarrow \operatorname{rank}\left(I_{n_p} - RS\right) = 0 \Leftrightarrow I_{n_p} - RS = 0 \qquad [2.72]$$

There are only extremely rare cases when the relation $R = S^{-1}$ can be taken into account explicitly in the feasibility step [PRE 04].

In [APK 03], augmented Lagrangian methods are used to solve locally the following problem:

$$\min\ f(x) := \gamma \qquad [2.73]$$

subject to the constraint $h(x) := \mathrm{I} - RS = 0$, $x \in \chi$

where:

$$x = (R, S, \gamma) \qquad [2.74]$$

and:

$$\chi := \{x \text{ such that } [2.24]\text{–}[2.26]\} \qquad [2.75]$$

The authors then suggest solving a series of relaxed problems of form [2.76] instead of [2.73]

$$\min_{R=R^T,\, S=S^T,\, \gamma} g(R,S,\gamma) \qquad [2.76]$$

subject to constraints [2.24]–[2.26]

where $g(R,S,\gamma)$ is the partially augmented Lagrangian:

$$g(R,S,\gamma) = \gamma + \text{trace}\left(\Lambda^T (RS - I)\right) + \frac{c}{2}\text{trace}\left((RS-I)^T (RS-I)\right) \qquad [2.77]$$

c: penalty factor

Λ: Lagrange matrix multiplier

This approach provides interesting results, but only for low-order standard forms.

Finally, some more conservative work has been done with the aim of tackling the rank constraint more directly, but without more guarantees of global convergence. For example, one might cite [ELG 97] where the rank constraint is "saturated" [2.71] by resolving the following optimization problem:

$$\min_{R=R^T,\, S=S^T} \text{trace}(RS) \qquad [2.78]$$

subject to constraints [2.24]–[2.26].

There is a static output feedback if and only if the problem's global minimum [2.78] is n. An iterative method using a linear approximation of the trace (RS) is then used:

$$\Phi_{\text{lin}}(R, S) = \text{constant} + \text{trace}(S_0 R + R_0 S) \qquad [2.79]$$

An implementation of this algorithm can be found in [PRE 05].

Finally, we point out the existence of some interesting sub-optimal heuristics such as the *X-K* iteration [ELG 94] where we suggest solving the LMI [2.80] of the bounded lemma alternatively following X and Θ_k.

$$\begin{pmatrix} A_{cl}^T X + XA_{cl} & XB_{cl} & C_{cl}^T \\ B_{cl}^T X & -\gamma I_{n_e} & D_{cl}^T \\ C_{cl} & D_{cl} & -\gamma I_{n_z} \end{pmatrix} < 0 \quad [2.80]$$

where

$$T(s) = \left[\begin{array}{c|c} \overline{A} + \overline{B}_2\Theta_k\overline{C}_2 & \overline{B}_1 + \overline{B}_2\Theta_k\overline{D}_{21} \\ \hline \overline{C}_1 + \overline{D}_{12}\Theta_k\overline{C}_2 & \overline{D}_{11} + \overline{D}_{12}\Theta_k\overline{D}_{21} \end{array}\right] = \left[\begin{array}{c|c} A_{cl} & B_{cl} \\ \hline C_{cl} & D_{cl} \end{array}\right] \quad [2.81]$$

In fact:

– at fixed X, [2.80] is an LMI in Θ_k and γ;

– at fixed Θ_k and γ, [2.80] is an LMI in X.

This approach is reminiscent of [VIN 97]: it is sub-optimal and needs to be initialized by a stabilizing controller, but is much more general because it is applied to any standard form.

As we can see, synthesizing the reduced-order controller with the help of the LMIs is not simple as the rank constraint is not convex. Moreover, optimization using LMI often necessitates a substantial number of decision variables (varying polynomially with the order of the system) often giving rise to lengthy calculation processes, and often affected by numerical problems.

2.2.2.3. *H_∞ synthesis using non-smooth optimization*

2.2.2.3.1. Recall on descent methods

Consider the optimization problem:

$$\min_{x\in \mathbb{R}^n} f(x), \; x^* = \arg\min_{x\in \mathbb{R}^n} f(x) \quad [2.82]$$

where the function $f(x)$ is derivative.

Descent methods consist of using a recursive algorithm to move in the search space. At each iteration, we choose to move in a direction enabling the decrease of the function *f*(*x*):

$$x_{k+1} = x_k + \lambda_k d_k \quad [2.83]$$

where:

– $d_k \in \mathbf{R}^n$ is the direction of descent;

– $\lambda_k > 0$ is called the step.

When the function *f*(*x*) is derivative, a standard method is to choose the gradient of *f*(*x*) as a direction of descent:

$$d_k = -\nabla f(x_k) \quad [2.84]$$

This direction corresponds to the direction of larger decrease in the function *f*(*x*) in the set of directions of descent of *f* in x_k:

$$\left\{ d \in \mathbf{R}^n : d^T \nabla f(x_k) < 0 \right\} \quad [2.85]$$

To choose the step, the method of deepest descent can be chosen:

– λ_k constant: a simple solution, but it rarely gives good results;

– λ_k optimized at each step:

$$\lambda_k = \arg\min_{\lambda} f(x_k + \lambda d_k) \quad [2.86]$$

Problem [2.86] is a one-dimensional optimization problem to be realized at each step, which can be understood by *Gridding* methods, the Newton–Raphson method, secant method, etc.

The drawback of the above method is that it does not involve taking into account the curvature of the function to be minimized, so that the trapping in a local minimum is very fast (in this case, the gradient will eventually be null and the optimization stops).

There are therefore other descent methods where the fitness function is approximated using a simpler expression (i.e. one we can optimize), for which it is possible to head for the point that minimizes this approximation. In the Newton method, it is customary to take for approximation an order 2 Taylor development of the function $f(x)$ in the vicinity of x_k:

$$g(x) = f(x_k) + \nabla f(x_k)(x - x_k) + \frac{1}{2}(x - x_k)^T \nabla^2 f(x_k)(x - x_k) \approx f(x) \quad [2.87]$$

We then choose to move in the direction that minimizes this approximation, that is:

$$x_{k+1} = \arg\min_x g(x) \quad [2.88]$$

By canceling the gradient $\nabla g(x)$ [2.87] of the approximation, we achieve:

$$x_{k+1} = x_k - \left(\nabla^2 f(x_k)\right)^{-1} \nabla f(x_k) \quad [2.89]$$

It is clear that the method requires the existence of the Hessian, on the one hand, and its invertibility, on the other hand. Because inverting the Hessian is time consuming and it is often poorly conditioned, a direction of descent in the following form is generally chosen:

$$x_{k+1} = x_k - M_k \nabla f(x_k) \quad [2.90]$$

where the matrix M_k is updated at each iteration and has certain properties of the Hessian:

$$\begin{aligned} & M_k \succ 0 \left(d_k^T \nabla f(x_k) < 0 \right) \\ & M_{k+1}\left(\nabla f(x_{k+1}) - \nabla f(x_k)\right) = x_{x+1} - x_k \\ & M_{k+1} = \arg\min_M \|M - M_k\| \end{aligned} \quad [2.91]$$

The most well-known method is called BFGS (Broyden–Fletcher–Goldfarb–Shanno), which consists of choosing:

$$\begin{aligned} & M_{k+1} = V_k M_k V_k^T + \alpha_k \left(d_k^T y_k\right)^{-1} d_k d_k^T \\ & y_k = \nabla f(x_{k+1}) - \nabla f(x_k) \\ & V_k = I - \left(d_k^T y_k\right)^{-1} d_k y_k^T \\ & M_0 = I \end{aligned} \quad [2.92]$$

2.2.2.3.2. Principle of the approach

The emergence of new optimization tools makes it possible to overcome the difficulty of the rank constraint encountered with the LMI approach by tackling the problem of reduced-order controller synthesis frontally.

The problem consists of finding $\Theta_k = \begin{pmatrix} A_K & B_K \\ C_K & D_K \end{pmatrix}$ with order $n_k < n_p$ such that:

– $T(s)$ is internally stable;

– $\|T(s)\|_\infty < \gamma$, with γ minimal.

with:

$$T(\Theta_k) = C_{cl}(\Theta_k)\left(sI_{n_p+n_k} - A_{cl}(\Theta_k)\right)^{-1} B_{cl}(\Theta_k) + D_{cl}(\Theta_k) \qquad [2.93]$$

and:

$$\begin{aligned} A_{cl}(\Theta_k) &= \overline{A} + \overline{B}_2\Theta_k\overline{C}_2 \\ B_{cl}(\Theta_k) &= \overline{B}_1 + \overline{B}_2\Theta_k\overline{D}_{21} \\ C_{cl}(\Theta_k) &= \overline{C}_1 + \overline{D}_{12}\Theta_k\overline{C}_2 \\ D_{cl}(\Theta_k) &= \overline{D}_{11} + \overline{D}_{12}\Theta_k\overline{D}_{21} \end{aligned} \qquad [2.94]$$

The looped system is stable if:

$$\max_i\left(\mathrm{Re}\left(\lambda_i\left(\overline{A} + \overline{B}_2\Theta_k\overline{C}_2\right)\right)\right) < 0 \qquad [2.95]$$

where Re( ) is the real part and λ_i is the ith eigenvalues.

Furthermore, by defining the H_∞ norm:

$$\|T(s)\|_\infty < \gamma \Leftrightarrow \max_{\omega[0,+\infty[}\left(\overline{\sigma}\left(T(j\omega)\right)\right) < \gamma \qquad [2.96]$$

where $\overline{\sigma}$ is the largest singular value.

The controller is synthesized in two steps. First, the stabilization step for the closed loop, which consists of solving the following optimization problem:

$$\min_{\Theta_k} f(\Theta_k),\ f(\Theta_k) = \max_i \left(\mathrm{Re}\left(\lambda_i\left(\bar{A} + \bar{B}_2\Theta_k\bar{C}_2\right)\right)\right) \qquad [2.97]$$

The H_∞ norm of the closed loop is, for its part, minimized by solving the following optimization problem:

$$\min_{\Theta_k} f(\Theta_k),\ f(\Theta_k) = \max_{\omega\in[0,+\infty[} \bar{\sigma}\left(T(j\omega)\right) \qquad [2.98]$$

Thus, the two steps are *minmax* optimization problems (not necessarily convex) of the form:

$$\min_x f(x),\ f(x) = \max_{s\in S} \varphi_s(x),\ x \in \mathbf{R}^n \qquad [2.99]$$

This type of problem is perfectly adapted to non-smooth optimization [ROC 94]: optimization algorithms exist [LEW 12], the only constraint being the Lipschitzian nature of the function to be minimized.

A $f:\mathbf{R}^n \to \mathbf{R}$ function is called Lipschitzian on $H \subset \mathbf{R}^n$ if there is a constant $L>0$ such that, for any value $y, z \in H$: w

$$\left|f(y) - f(z)\right| \le L\left\|y - z\right\| \qquad [2.100]$$

The locally Lipschitzian nature of a function in x therefore ensures that its rate of growth will be limited to the vicinity of x. A locally Lipschitzian function at one point is not necessarily differentiable in this point (e.g. the function $x \to |x|$ in 0). Moreover, any convex function $f:\mathbf{R}^n \to \mathbf{R}$ is locally Lipschitzian at any point of $\mathbf{R}^n$ (and so f is Lipschitzian on any compact of $\mathbf{R}^n$).

When the function f is Lipschitzian, the notion of gradient (established for the differentiable functions) is generalized by the notion of the sub-differential with the help of the directional derivative. For any $w \in \mathbf{R}^n$, the function f accepts a directional derivative in x in the direction w, written as $df(x)(w)$, if the following limit exists and is finite:

$$df(x)(w) = \lim_{\substack{t\to 0\\ t>0}} \frac{f(x+tw) - f(x)}{t} \qquad [2.101]$$

where $df(x)$ is called the sub-differential of f in x and when f is Lipschitzian:

$$\left|df(x)(w)\right| \leq L\|w\| \qquad [2.102]$$

It should be noted that if f is differentiable at x, then it accepts directional derivatives in all directions w, and we have:

$$df(x)(w) = (\nabla f(x))^T w \qquad [2.103]$$

It is shown that the sub-differential has the following properties:

$$\begin{aligned} &df(x)(w_1 + w_2) \leq df(x)w_1 + df(x)w_2 \\ &df(x)(\lambda w) \leq \lambda df(x)w, \lambda > 0 \end{aligned} \qquad [2.104]$$

The set of sub-gradients of f at x is defined by:

$$\partial f(x) = \left\{v \in \mathbf{R}^n \middle| v.w \leq df(x)(w), \forall w \in \mathbf{R}^n\right\} \qquad [2.105]$$

where $w \in \mathbf{R}^n$ is a direction of descent of f at x if one of the following conditions is met:

– $df(x)(w) < 0$;

– $w^T g < 0, \forall g \in \partial f(x)$.

Non-smooth optimization thus proves appropriate for the minimization of non-regular functions, because, in contrast to other usual descent methods, where the gradient of f in x is used to decide the direction of descent, we will use a sub-gradient of f in x in the context of non-regular functions, which can be defined as a linear combination of directional derivatives of f in x.

On the basis of these notions and on the standard approach for control, an H_∞ synthesis using non-smooth optimization is suggested in [APK 02].

We posit:

$$f(\Theta_k) = \|T(s)\|_\infty \qquad [2.106]$$

Consider the singular value decomposition[4] of $T(j\omega)$:

$$T(j\omega) = U\Sigma V^H \qquad [2.107]$$

where u and v are, respectively, the left and right vectors corresponding to the largest singular value resulting from the singular value decomposition of T.

It is therefore shown that [APK 02]:

$$\left.\begin{aligned} \frac{\partial f}{\partial A_{cl}} &= \phi^* C_{cl}^{\ *} uv^* B_{cl}^{\ *} \phi^* \\ \frac{\partial f}{\partial B_{cl}} &= \phi^* C_{cl}^{\ *} uv^* \\ \frac{\partial f}{\partial C_{cl}} &= uv^* B_{cl}^{\ *} \phi^* \\ \frac{\partial f}{\partial D_{cl}} &= uv^* \end{aligned}\right\} := \partial \| \ \|_\infty (T(\Theta_k)) \qquad [2.108]$$

where $\phi = (j\omega I - A_{cl})^{-1}$, $M = (C_{cl}\phi B_{cl} + D_{cl})$.

The sub-differential of the function $f(\Theta_k) = \|T(\Theta_k)\|_\infty$ is then determined in relation to Θ_k using the following composition of relations:

$$\partial f(\Theta_k) = T'(\Theta_k)^* \left(\partial \|\cdot\|_\infty (T(\Theta_k))\right) \qquad [2.109]$$

where $\partial\|\cdot\|_\infty$ is the sub-differential of the H_∞ norm determined previously, and $T'(\Theta_k)^*$ is the adjoint[5] of $T'(\Theta_k)$ defined by:

$$T'(\Theta_k)\delta\Theta_k := G_{12}\delta\Theta_k G_{21} \qquad [2.110]$$

4 In this section, *MH* indicates the conjugate transpose of *M*.

5 Note that a^* is called the adjoint of a if $\langle a(x), y\rangle = \langle x, a^*(y)\rangle$.

with:

$$\begin{pmatrix} T(\Theta_k,s) & G_{12}(\Theta_k,s) \\ G_{21}(\Theta_k,s) & * \end{pmatrix} = \begin{pmatrix} C_{cl}(\Theta_k) \\ C_2 \end{pmatrix} (sI - A_{cl}(\Theta_k))^{-1} \begin{pmatrix} B_{cl}(\Theta_k) & B_2 \end{pmatrix} + \begin{pmatrix} D_{cl}(\Theta_k) & D_{12} \\ D_{21} & * \end{pmatrix} \quad [2.111]$$

Remember, if A, B, X and C are matrices with compatible dimensions, then it is shown that the derivative of $f(A+BXC)$ in relation to X is B^*GC^* with $G = \left[\frac{\partial}{\partial x_{ij}} f(X)\right]$.

A descent method can therefore be used as an optimization.

It is important to note that an initial point has to be chosen during the optimization. In general, it is randomly chosen so that several *runs* of optimization will not be able to give the same result due to trappings in local optima. It therefore makes sense to run several.

In [GAH 11] we can find an implementation of H_∞ synthesis using non-smooth optimization with the *Hinfstruct* tool in *Matlab*. We also note the existence of another free and efficient tool based on the same optimization principles, called *Hifoo* [BUR 06].

2.2.2.3.3. Decentralized controllers

The search for Θ_k coefficients of the controller's state-space representation absolutely does not prevent us from structuring it as we see fit, especially in MIMO cases. Therefore, rather than remaining on a centralized controller (i.e. a controller where the state-space matrix is common to all its inputs and outputs), it is preferable [HIR 12] to search for a decentralized controller, that is, a controller structured according to Figure 1.5, but with a higher computing time (since the number of unknowns to be optimized increases).

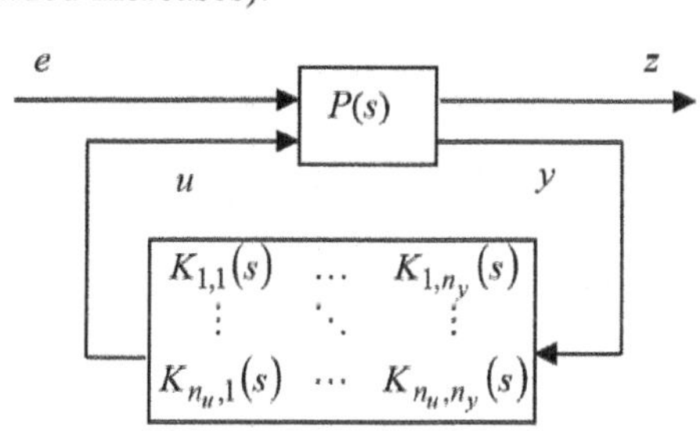

Figure 2.5. *Structured and decentralized H∞ synthesis*

2.2.2.3.3.1. Extension to the multiple-plant framework

H_∞ synthesis tools using the non-smooth optimization previously cited provide a framework for use that can be extended well beyond H_∞ synthesis and particularly enable the case of structured H_∞ synthesis for multiple plants to be tackled. As shown above, a solution to this problem has been suggested using the LMI formalism in the problem of stabilizing a coprime factor plant description subject to order constraint, but this approach was limited to the order constraint and structuration of the pre-compensator (and not of the final controller). With non-smooth optimization, one can generalize the standard H_∞ synthesis control problem to this structured multiple-plant framework.

Consider the following set of m models:

$$\wp = \{P_i(s), i = 1,\ldots,m\} \qquad [2.112]$$

According to Figure 2.6, the fixed structure H_∞ synthesis problem for multiple plants consists of finding a controller $K(s)$ with order n_k that:

– internally stabilizes each closed loop $F_l(P_i,K)$;

– minimizes γ such that:

$$\max_{i=1,\ldots,m} \left\| F_l\left(P_i, K\right) \right\|_\infty < \gamma \qquad [2.113]$$

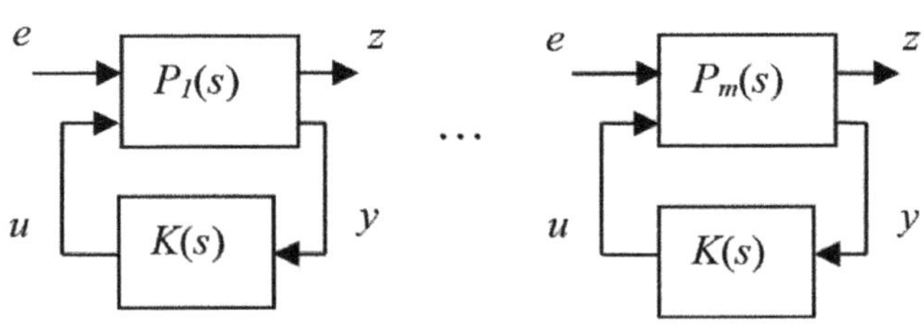

Figure 2.6. *H_∞ synthesis for multiple plants*

We find an example of implementation of this approach in [APK 12]. The implementation of this technique is possible by building the standard form depicted in Figure 1.7 and enforcing a particular structuration on the controller.

Thus, [2.113] is solved by computing a controller $K'(s)$ with order n_k that:

– internally stabilizes each closed loop $F_l(P', K')$;

– minimizes γ such that:

$$\begin{aligned} &\left\| F_l(P',K') \right\|_\infty < \gamma \\ &P'(s) = \text{diag}\left(P_i(s), i = 1,\dots,m\right) \\ &K'(s) = \text{diag}\left(K_i(s), K_i(s) = K(s), i = 1,\dots,m\right) \end{aligned} \quad [2.114]$$

It is important to note that the augmented standard form thus constructed has an order equal to the sum of the orders of the models considered, which can be high if the number of models is substantial, and so consequently slow the controller's optimization process (we will address this point again when implementing H∞ synthesis using stochastic optimization in the multiple-plant framework).

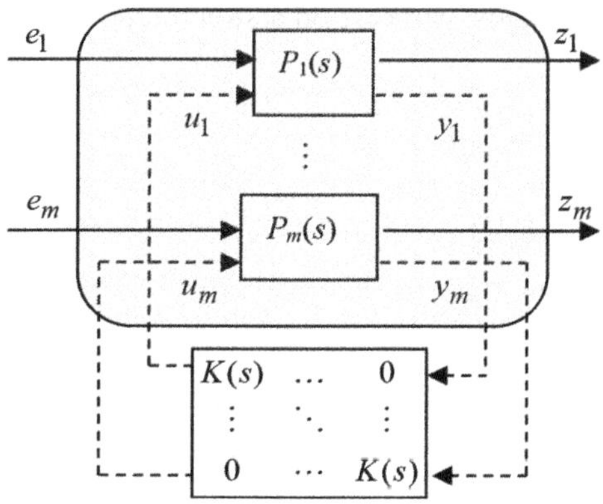

Figure 2.7. *H∞ synthesis for multiple plants using non-smooth optimization*

2.2.3. *Formulating H∞ synthesis for stochastic optimization*

2.2.3.1. *Principle*

Stochastic optimization using metaheuristics is adapted to solving problems for which it is difficult to find a global optimum or good local optima using classic methods. We recall that this type of optimization has three characteristics often decisive in global optimization:

– optimization using metaheuristics does not require us to know the gradient of the function to be minimized, the only constraint being the capability of evaluating the latter, which can therefore have any form;

– it is not necessary to have a "good" initial point, the initialization is randomly done in the search space;

– finally, this type of optimization is stochastic, which makes it possible to overcome the combinatorial explosion of possibilities and limits the trapping in the local optima.

In [DRÉ 06], we will find a detailed description of the main metaheuristics inspired by analogies with physics, ethology, biology, etc.

Despite the main drawback concerning the pertinence of the solution found and which can be evaluated statistically by several successive runs (which gives rise to a generally high computing time), this type of optimization proves interesting for difficult optimization problems and appears appropriate for reduced-order controller synthesis.

Therefore, here we suggest working on a structured H_∞ synthesis approach using stochastic optimization.

By positing:

$$\bar{\lambda} = \max_i \left(\mathrm{Re}\left(\lambda_i \left(A_{cl} \right) \right) \right) \quad [2.115]$$

where $\lambda_i(\ )$ is the ith eigenvalue.

Then, we synthesize an H_∞-optimal stabilizing controller by considering the following optimization problem:

$$\min_{\Theta_k} f(\Theta_k)$$

If $\bar{\lambda}_{cl} \geq 0$

$$f(\Theta_k) = \bar{\lambda}_{cl}$$

Unless

$$f(\Theta_k) = -\frac{1}{\left\| T(s) \right\|_\infty} \quad [2.116]$$

In [MAR 08], we find a similar idea for the definition of the fitness function envisaged here. The fitness function involves carrying out two steps: first, the stabilization of the closed loop, which involves making the spectral abscissa $\bar{\lambda}_{cl}$ strictly negative. When the closed loop becomes stable, its poles are then, in all probability, close to the imaginary axis such that the value of γ obtained with the corresponding controller is very large, and such that $-\gamma^{-1}$ is negative but close to 0. Thus, there is a pseudo-continuity in the fitness function between the stabilization step and the properly termed optimization step, from which we can benefit here.

We also note that the fitness function so defined does not depend on the stochastic optimization algorithm (or even more generally on optimization by direct search). Later on, we test our approach with several metaheuristics.

In literature, we find other formulations of the fitness function, for example in [POP 10] where the use of an evolutionary algorithm for fixed-order H_∞ synthesis is envisaged with the following fitness function:

$$\min_{\Theta_k} f(\Theta_k)$$

If $\bar{\lambda}_{cl} \geq 0$

$$f(\Theta_k) = \beta + \bar{\lambda}_{cl}$$

Unless

$$f(\Theta_k) = \|T(s)\|_\infty \qquad [2.117]$$

where β is a penalty that should be high, for example, $\beta = 10^{10}$.

For the reasons cited above, we will stay with formulation [2.116].

2.2.3.2. *Digression: formulating H_2 synthesis*

H_2 synthesis is a direct extension of formulation [2.116]. In fact, an H_2-optimal stabilizing controller is synthesized considering the following optimization problem:

$$\min_{\Theta_k} f(\Theta_k)$$

If $\bar{\lambda}_{cl} \geq 0$

$$f(\Theta_k) = \bar{\lambda}_{cl}$$

Unless

$$f(\Theta_k) = -\frac{1}{\|T(s)\|_2} \qquad [2.118]$$

It is important to note that the H_2 norm is only defined for strictly proper systems. It is therefore necessary here that:

$$\begin{aligned} & D_{11} + D_{12} M_1 D_k D_{21} = 0 \\ & M_1 = (I - D_k D_{22})^{-1} \end{aligned} \qquad [2.119]$$

If $D_{11} = 0$, a simple way to enforce this condition is to search for a controller $K(s)$ structured such that $D_k = 0$.

2.2.3.3. *Improving the rate of convergence for controller synthesis*

Below, significant improvements to the rate of convergence of the optimization process are suggested.

2.2.3.3.1. Using the tridiagonal form

For unknowns, we have posited the controller's state-space representation:

$$\Theta_k = \left[\begin{array}{c|c} A_k & B_k \\ \hline C_k & D_k \end{array}\right] \qquad [2.120]$$

Indeed, for a single transfer function, there is an infinity of possible state-space representations. Assuming all the matrices of Θ_k to be full, the number of unknowns for a SISO controller is established at $(n_k + 1)^2$.

One idea for decreasing the number of unknowns would be to work on a companion form such as the following:

$$\begin{aligned} & A_k = \begin{pmatrix} -a_1 & -a_2 & \dots & -a_{k-1} & -a_k \\ 1 & 0 & \dots & 0 & 0 \\ 0 & 1 & \dots & 0 & 0 \\ \vdots & \vdots & & \vdots & \vdots \\ 0 & 0 & \dots & 1 & 0 \end{pmatrix}, \quad B_k = \begin{pmatrix} 1 \\ 0 \\ 0 \\ \vdots \\ 0 \end{pmatrix} \\ & C_k = \begin{pmatrix} b_1 & b_2 & \dots & b_{k-1} & b_k \end{pmatrix}, \qquad D_k = d \\ & \Leftrightarrow \\ & K(s) = \frac{b_1 s^{k-1} + b_2 s^{k-2} + \dots + b_{k-1} s + b_k}{s^k + a_1 s^{k-1} + \dots + a_{k-1} s + a_k} + d \end{aligned} \qquad [2.121]$$

In fact, the number of unknowns is established in this case at $2n_k+1$. However, the values of coefficients are often too different despite their low number, which can make optimization more delicate.

A better approach consists of using the tridiagonal form suggested in [MCK 96]. By definition, a tridiagonal matrix is a real square matrix that contains non-null elements only on its main, upper and lower diagonals:

$$A=\begin{pmatrix} a_{1,1} & a_{1,2} & 0 & 0 & 0 \\ a_{2,1} & \ddots & \ddots & 0 & 0 \\ 0 & \ddots & \ddots & \ddots & 0 \\ 0 & 0 & \ddots & \ddots & a_{n-1,n} \\ 0 & 0 & 0 & a_{1,n-1} & a_{n,n} \end{pmatrix} \qquad [2.122]$$

In [MCK 96], it is shown that for any MIMO system with state-space representation $[A,B,C,D]$, there is an equivalent representation $[A',B',C',D']$ where A' is in tridiagonal form. One idea therefore involves searching for the state-space matrix of the controller in the tridiagonal form:

$$A_k=\begin{pmatrix} a_{1,1} & a_{1,2} & 0 & 0 & 0 \\ a_{2,1} & \ddots & \ddots & 0 & 0 \\ 0 & \ddots & \ddots & \ddots & 0 \\ 0 & 0 & \ddots & \ddots & a_{n_k-1,n_k} \\ 0 & 0 & 0 & a_{n_k,n_k-1} & a_{n_k,n_k} \end{pmatrix} \qquad [2.123]$$

Assuming matrices B_k, C_k, D_k to be full, the number of unknowns for a SISO controller is then established at $5n_k-1$, which is of the same order of magnitude as for the companion form.

A particular case is that of the static controller where the number of unknowns is established at 1 (this comes down to seeking only the direct transfer matrix).

2.2.3.3.2. Transforming the search space

Suppose that all the unknown parameters are strictly positive. In this case, the initial search space can be transformed as follows:

$$\begin{aligned} x\in[\underline{x},\overline{x}] \underset{x'=\log_{10}(x)}{\longrightarrow} x'\in[\underline{x}',\overline{x}'] \\ x'\in[\underline{x}',\overline{x}'] \underset{x=10^{x'}}{\longrightarrow} x\in[\underline{x},\overline{x}] \end{aligned} \qquad [2.124]$$

This transformation improves the sensitivity of the algorithm: all values of x' have the same weight, which is not the case with the initial space. Thus, it is possible to choose a very broad initial search interval, for example:

$$x \in [10^{-3}, 10^{3}] \underset{x'=\log_{10}(x)}{\longrightarrow} x' \in [-3, 3] \qquad [2.125]$$

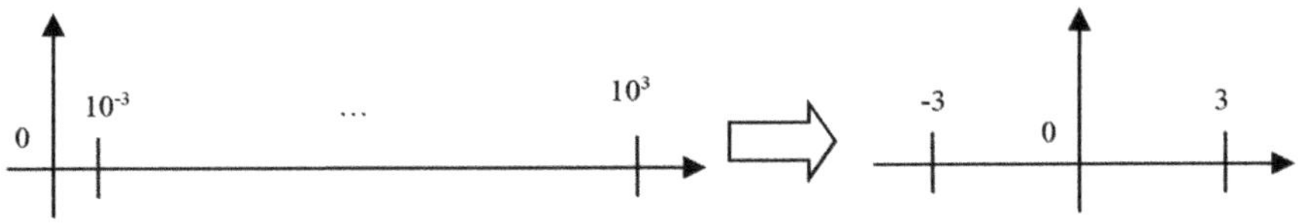

Figure 2.8. *Transforming the search space*

In reality, all the unknown parameters may be signed. As before, we can seek to improve the sensitivity of the algorithm via a transformation inspired by the previous one:

$$\begin{aligned} &x \in [\underline{x}, \overline{x}] \underset{x'=\text{ash}_{10}(Mx)}{\longrightarrow} x' \in [\underline{x}', \overline{x}'] \\ &x' \in [\underline{x}', \overline{x}'] \underset{x=\frac{1}{M}\text{sh}_{10}(x')}{\longrightarrow} x \in [\underline{x}, \overline{x}] \end{aligned} \qquad [2.126]$$

with:

$$\text{ash}_{10}(x) = \log_{10}\left(x + \sqrt{x^2 + 1}\right) \; ; \; \text{sh}_{10}(x) = \frac{10^x - 10^{-x}}{2} \qquad [2.127]$$

When $|x|$ is large, the $\log_{10}(\ )$ and $\text{ash}_{10}(\ )$ functions are very similar. They are however very different when $|x|$ is close to 0. In order for the minimum and maximum value of $|x|$ to have a similar weight during the optimization process, a coefficient M is used in [2.126] such that $M >> 1$.

In the context of this study, we will choose (Figure 2.9):

$$\underline{x} = 1000, \; \overline{x} = -1000, \; M = 1000 \qquad [2.128]$$

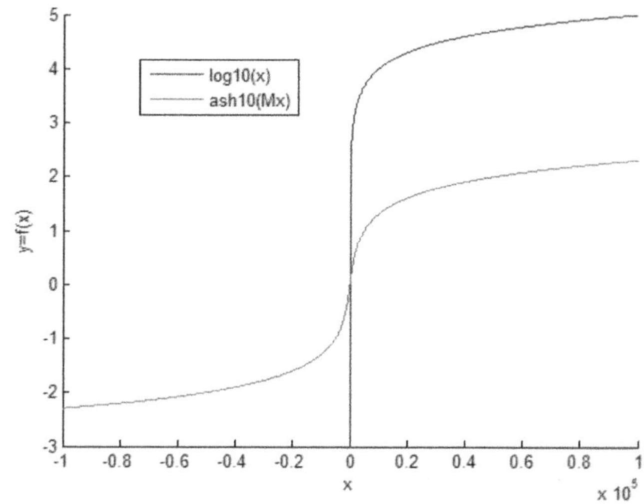

Figure 2.9. *Transforming the search space using the $ash_{10}(Mx)$ function. For a color version of this figure, see www.iste.co.uk/feyel/optimization.zip*

With M = 1,000, the rate of log10(x) is found in the vicinity of $x = 0^+$, and its symmetric in $x = 0^-$ (uneven function).

2.2.3.4. *Centralized and decentralized controller*

2.2.3.4.1. Centralized controller

For a centralized n_k order MIMO controller with n_y inputs and n_u outputs in tridiagonal form [2.129], the number of unknowns increases to $3n_k - 2 + n_k n_y + n_k n_u + n_u n_y$

$$\left[\begin{array}{ccccc|ccc} a_{1,1} & a_{1,2} & 0 & 0 & 0 & b_{1,1} & \cdots & b_{1,n_y} \\ a_{2,1} & \ddots & \ddots & 0 & 0 & b_{2,1} & \vdots & b_{2,n_y} \\ 0 & \ddots & \ddots & \ddots & 0 & \vdots & \vdots & \vdots \\ 0 & 0 & \ddots & \ddots & a_{n_k-1,n_k} & \vdots & \vdots & \vdots \\ 0 & 0 & 0 & a_{n_k,n_k-1} & a_{n_k,n_k} & b_{n_k,1} & \cdots & b_{n_k,n_y} \\ \hline c_{1,1} & c_{1,2} & \cdots & \cdots & c_{1,n_k} & d_{1,1} & \cdots & d_{1,n_y} \\ \vdots & \vdots & \vdots & \vdots & \vdots & \vdots & \ddots & \vdots \\ c_{n_u,1} & c_{n_u,1} & \cdots & \cdots & c_{n_u,n_k} & d_{n_u,1} & \cdots & d_{n_u,n_y} \end{array}\right] \qquad [2.129]$$

(A special case is that of a static controller where the number of unknowns is $n_u n_y$).

In this case, we suggest the following decision vector:

$$\begin{aligned} X &= (x_A, x_B, x_C, x_D)^T \\ x_A &= \left(a_{1,1}, \ldots, a_{n_k,n_k}, a_{1,2}, \ldots, a_{n_k-1,n_k}, a_{2,1}, \ldots, a_{n_k,n_k-1}\right) \\ x_B &= \left(b_{1,1}, \ldots, b_{n_k,1}, b_{1,2}, \ldots, b_{n_k,2}, \ldots, b_{1,n_u}, \ldots, b_{n_k,n_u}\right) \\ x_C &= \left(c_{1,1}, \ldots, c_{1,n_k}, c_{2,1}, \ldots, c_{2,n_k}, \ldots, c_{n_y,1}, \ldots, c_{n_y,n_k}\right) \\ x_D &= \left(d_{1,1}, \ldots, d_{1,n_u}, d_{2,1}, \ldots, d_{2,n_u}, \ldots, d_{n_y,1}, \ldots, d_{n_y,n_u}\right) \end{aligned} \quad [2.130]$$

2.2.3.4.2. Decentralized controller

For a decentralized MIMO controller [2.131], the number of unknowns depends on the order of each of the sub-transfers considered.

$$K(s) = \begin{bmatrix} K_{1,1}(s) & \ldots & K_{1,n_y}(s) \\ \vdots & \ddots & \vdots \\ K_{n_u,1}(s) & \cdots & K_{n_u,n_y}(s) \end{bmatrix} \quad [2.131]$$

Considering that each sub-transfer is a SISO system (in tridiagonal form), and that the order of the different transfers corresponds to matrix [2.132], the number of unknowns is set at $\sum_{j=1}^{n_y}\sum_{i=1}^{n_u}\left(5n_{ki,j}-1\right)$. This relation is only valid if none of the terms $K_{i,j}(s)$ is static, in which case $5n_{ki,j}-1$ must be replaced by 1 and by 0 if the nullity of $K_{i,j}(s)$ is imposed.

$$\partial^\circ K = \begin{bmatrix} n_{k1,1} & \ldots & n_{k1,n_y} \\ \vdots & \ddots & \vdots \\ n_{kn_u,1} & \cdots & n_{kn_u,n_y} \end{bmatrix}, \ n_{ki,j} \geq 1 \quad [2.132]$$

(Naturally, nothing prevents us from enforcing that one or several sub-transfers be null in the decentralized controller).

Each sub-term is represented in its tridiagonal form, which brings us to the previous case. In this case, we suggest as decision vector (each sub-term is assumed SISO for the sake of clarity):

$$\begin{aligned}
&X=\left(X_{k_{1,1}},\dots,X_{k_{1,n_y}},X_{k_{2,1}},\dots,X_{k_{2,n_y}},\dots,X_{k_{n_u,1}},\dots,X_{k_{n_u,n_y}}\right)^T\\
&X_{k_{i,j}}=\left(x_A,x_B,x_C,x_D\right)_{i,j}\\
&\left.\begin{aligned}
&x_A=\left(a_{1,1},\dots,a_{n,n},a_{1,2},\dots,a_{n-1,n},a_{2,1},\dots,a_{n,n-1}\right)\\
&x_B=\left(b_{1,1},\dots,b_{n,1}\right)\\
&x_C=\left(c_{1,1},\dots,c_{1,n}\right)\\
&x_D=d
\end{aligned}\right\}n=n_{ki,j}
\end{aligned}\qquad [2.133]$$

Noting that if:

$$H_1(s):=\left[\begin{array}{c|c}A_1 & B_1\\ \hline C_1 & D_1\end{array}\right],\ H_2(s):=\left[\begin{array}{c|c}A_2 & B_2\\ \hline C_2 & D_2\end{array}\right]\qquad [2.134]$$

then:

$$\begin{aligned}
&\begin{pmatrix}H_1(s)\\ H_2(s)\end{pmatrix}:=\left[\begin{array}{cc|c}A_1 & 0 & B_1\\ 0 & A_2 & B_2\\ \hline C_1 & 0 & D_1\\ 0 & C_2 & D_2\end{array}\right]\\
&\begin{pmatrix}H_1(s) & H_2(s)\end{pmatrix}:=\left[\begin{array}{cc|cc}A_1 & 0 & B_1 & 0\\ 0 & A_2 & 0 & B_2\\ \hline C_1 & C_2 & D_1 & D_2\end{array}\right]
\end{aligned}\qquad [2.135]$$

The state-space representation of the sub-transfers $K_{i,j}(s)$ being known, it is therefore easy to reconstruct the state-space representation of the controller $K(s)$ through algorithm [2.136].

For $j = 1$ to n_y columns of $K(s)$

For $i = 1$ to n_u rows of $K(s)$

If $i = 1$,

$$A_H = A_{k_{1,1}},\ B_H = B_{k_{1,1}}$$
$$C_H = C_{k_{1,1}},\ D_H = D_{k_{1,1}}$$

Else

$$A_H \leftarrow \begin{pmatrix} A_H & 0 \\ 0 & A_{k_{i,j}} \end{pmatrix},\ B_H \leftarrow \begin{pmatrix} B_H & 0 \\ 0 & B_{k_{i,j}} \end{pmatrix}$$
$$C_H \leftarrow \begin{pmatrix} C_H & C_{k_{i,j}} \end{pmatrix},\ D_H \leftarrow \begin{pmatrix} D_H & D_{k_{i,j}} \end{pmatrix}$$

End If

End For i

If $j = 1$,

$$A_G = A_H,\ B_G = B_H$$
$$C_G = C_H,\ D_G = D_H$$

Else

$$A_G \leftarrow \begin{pmatrix} A_G & 0 \\ 0 & A_H \end{pmatrix},\ B_G \leftarrow \begin{pmatrix} B_G & B_H \end{pmatrix}^T$$
$$C_G \leftarrow \begin{pmatrix} C_G & 0 \\ 0 & C_H \end{pmatrix}, D_G \leftarrow \begin{pmatrix} D_G & 0 \\ 0 & D_H \end{pmatrix}$$

End If

End For j

$$K(s) := \left[\begin{array}{c|c} A_G & B_G \\ \hline C_G & D_G \end{array}\right] \quad [2.136]$$

2.2.3.4.3. Decentralized and cascade controller

Instead of a feedback implementation of the controller, it is usual to wish to implement it in cascade form. This means implementing the correction as seen in Figure 2.10.

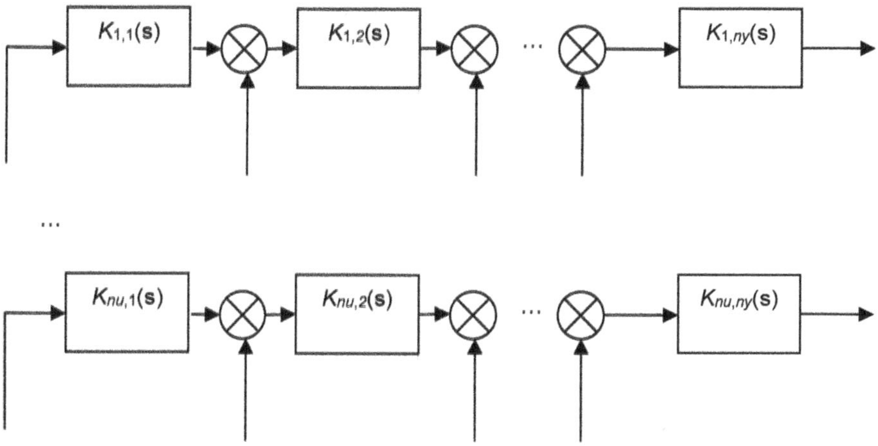

Figure 2.10. *Implementation in cascade form*

By convention, the measurement indices entering the controller go from the outermost loop (measurement 1) toward the innermost loop (measurement n_y).

For example, consider a controller with three inputs and one output. The cascade control law can be written as:

$$\begin{aligned} u &= K_{1,3}(s)\big(y_3 + K_{1,2}(s)\big(y_2 + K_{1,1}(s)y_1\big)\big) \\ &= K_{1,3}(s)y_3 + K_{1,3}(s)K_{1,2}(s)y_2 + K_{1,3}(s)K_{1,2}(s)K_{1,1}(s)y_1 \\ &= K_{1,3}(s)K_{1,2}(s)K_{1,1}(s)y_1 + K_{1,3}(s)K_{1,2}(s)y_2 + K_{1,3}(s)y_3 \\ &= \big(K_{1,3}(s)K_{1,2}(s)K_{1,1}(s) \quad K_{1,3}(s)K_{1,2}(s) \quad K_{1,3}(s)\big)y^T \\ &= K_{1,3}(s)\big(K_{1,2}(s)K_{1,1}(s) \quad K_{1,2}(s) \quad I\big)y^T \\ &= K_{1,3}(s)\big(K_{1,2}(s)\big(K_{1,1}(s) \quad I\big) \quad I\big)y^T,\ y = \big(y_1 \quad y_2 \quad y_3\big) \end{aligned} \qquad [2.137]$$

Thus, using the fact that:

$$H_2(s)\big(H_1(s) \quad I\big) := \left[\begin{array}{cc|cc} A_2 & B_2C_1 & B_2D_1 & B_2 \\ 0 & A_1 & B_1 & 0 \\ \hline C_2 & D_2C_1 & D_2D_1 & D_2 \end{array}\right] \qquad [2.138]$$

It is therefore easy to reconstruct the cascade controller $K(s)$ by recurrence from the $K_{i,j}(s)$ that form it, using algorithm [2.139].

For $j = 1$ to n_u rows of $K(s)$

For $i = 1$ to n_y columns of $K(s)$

If $i = 1$,

$$A_H = A_{k_{i,j}},\ B_H = B_{k_{i,j}}$$
$$C_H = C_{k_{i,j}},\ D_H = D_{k_{i,j}}$$

Else

$$A_H \leftarrow \begin{pmatrix} A_{k_{i,j}} & B_{k_{i,j}} C_H \\ 0 & A_H \end{pmatrix},\ B_H \leftarrow \begin{pmatrix} B_{k_{i,j}} D_H & B_{k_{i,j}} \\ B_H & 0 \end{pmatrix}$$
$$C_H \leftarrow \begin{pmatrix} C_{k_{i,j}} & D_{k_{i,j}} C_H \end{pmatrix},\ D_H \leftarrow \begin{pmatrix} D_{k_{i,j}} D_H & D_{k_{i,j}} \end{pmatrix}$$

End If

End For i

If $j = 1$,

$$A_G = A_H,\ B_G = B_H$$
$$C_G = C_H,\ D_G = D_H \qquad [2.139]$$

Else

$$A_G \leftarrow \begin{pmatrix} A_G & 0 \\ 0 & A_H \end{pmatrix},\ B_G \leftarrow \begin{pmatrix} B_G & B_H \end{pmatrix}^T$$
$$C_G \leftarrow \begin{pmatrix} C_G & 0 \\ 0 & C_H \end{pmatrix}, D_G \leftarrow \begin{pmatrix} D_G & 0 \\ 0 & D_H \end{pmatrix}$$

End If

End For j

$$K(s) := \left[\begin{array}{c|c} A_G & B_G \\ \hline C_G & D_G \end{array}\right]$$

2.2.3.5. *Evaluating the fitness function*

We want to solve the following optimization problem:

$$\min_{X \in [\underline{x};\overline{x}]} f(X) \qquad [2.140]$$

where the variable X is defined in section 2.2.3.4 and the function $f(X)$ consists of carrying out the following steps:

– construct $\left[\begin{array}{c|c} A_k & B_k \\ \hline C_k & D_k \end{array}\right]$ from X and the previous section;

– construct the following closed-loop state-space matrix A_{cl}:

$$\begin{aligned} A_{cl} &= \begin{bmatrix} A + B_2 M_1 D_k C_2 & B_2 M_1 C_k \\ B_k M_2 C_2 & A_k + B_k M_2 D_{22} C_k \end{bmatrix} \\ M_1 &= (I - D_k D_{22})^{-1} \qquad ; \\ M_2 &= (I - D_{22} D_k)^{-1} \end{aligned} \qquad [2.141]$$

– evaluate the closed loop's spectral abscissa:

$$\overline{\lambda} = \max_i \left(\mathrm{Re}\left(A_{cl} \right) \right) \qquad [2.142]$$

– if $\overline{\lambda} \geq 0$ the $f(X) = \overline{\lambda}$

– else:

- construct the following closed loop $F_l(P,K)$:

$$\begin{aligned} T(s) &= F_l\left(P(s), K(s)\right) \\ &= \left[\begin{array}{cc|c} A + B_2 M_1 D_k C_2 & B_2 M_1 C_k & B_1 + B_2 M_1 D_k D_{21} \\ B_k M_2 C_2 & A_k + B_k M_2 D_{22} C_k & B_k M_2 D_{21} \\ \hline C_1 + D_{12} M_1 D_k C_2 & D_{12} M_1 C_k & D_{11} + D_{12} M_1 D_k D_{21} \end{array}\right] \qquad [2.143] \\ M_1 &= (I - D_k D_{22})^{-1} \\ M_2 &= (I - D_{22} D_k)^{-1} \end{aligned}$$

- compute $\gamma = \|F_l(P,K)\|_\infty$;

- evaluate $f(X) = -\dfrac{1}{\gamma}$.

2.2.3.6. *Benchmarking with COMPl$_e$ib*

2.2.3.6.1. Introduction

The *COMPl$_e$ib* [LEI 03] library is made up of a collection of examples of systems (linear and invariant) that can be used to test and compare controller synthesis algorithms based on different optimization techniques such as *LMI*, *BMI* and non-smooth optimization. We also decided to use this library to test H_∞ synthesis with stochastic optimization formulated previously. Examples of *COMPl$_e$ib* range from the purely academic to the most industrially realistic and are ranged in sub-groups:

– airplane models (AC);

– helicopter models (HE);

– airplane engine models (JE) and reactor models (REA);

– a decentralized interconnected, system (DIS);

– Euler Bernoulli beam model (EB);

– 2D heat transfer model (HF2D);

– reduced-order control problem (ROC);

– academic problems (NN);

– various problems: wind turbine model (WEC).

Here, version 1.1 of *COMPl$_e$ib* is used.

The total number of test cases that form this *benchmarking* is 234: they are identical to those that can be found in [APK 10]. In Appendix E, we give the list of cases tested as well as the number of iterations used.

2.2.3.6.2. Comparing metaheuristics

Measuring performances

As the optimization algorithms tested are stochastic, a statistical evaluation of the results is needed for each test.

There are numerous ways of measuring algorithm performance, but what interests us here, from a practical point of view, is the mean behavior of the H_∞ synthesis results (minimal value for γ and associated repeatability) on all the *COMPl$_e$ib* test cases with the algorithms (and their setting) introduced in the first section.

The rule for performance measurement is the following.

For the set of 234 test cases, each test is subject to 10 successive *runs:* we therefore reveal the minimal value of γ obtained (γ_{min}), the mean value of γ (γ_{mean}), its standard deviation (γ_{std}) and finally the global computing time (t_{cpu}) for these 10 *runs*:

$$\left.\begin{aligned} \gamma_{min} &= \min_{i=1,\ldots,n \text{ runs}} \gamma_n \\ \gamma_{mean} &= \frac{1}{n}\sum_{i=1}^{n \text{ runs}} \gamma_i \\ \gamma_{std} &= \sqrt{\frac{1}{n}\sum_{i=1}^{n \text{ runs}} \left(\gamma_i - \gamma_{mean}\right)^2} \end{aligned}\right\} i\text{th test case} \qquad [2.144]$$

Therefore, for each test (e.g. AC1 and AC2), each algorithm can be classed by rank according to values [2.144].

When all the tests have been carried out, a mean rank is then established for each algorithm according to values [2.144] from the ranks obtained for each test (Figure 2.11).

We consider that the algorithm with the lowest mean rank is eventually the most efficient with regard to the value considered [2.144] for the H_∞ synthesis problem.

Measuring the computing time is interesting here, as:

– some algorithms use a local search phase linked to a global search phase (ABC, CS); therefore, during iteration, the fitness function can be evaluated more times than for more direct approaches (PSO, QPSO, DE);

– the fitness function [2.116] is broken down into two parts: stabilization and performance. Indeed, these two phases do not have the same running time (the second takes more time).

Thus, with the same number of iterations, the computing times for each algorithm to evaluate each test case can be very disparate.

For each test, the number of iterations chosen (empirically) is identical for all the algorithms tested: it is given in Appendix E for all the tests.

The whole test has been carried out on a machine equipped with Intel Core i7-3740QM (2.7 GHz) processor and 8 Gb of RAM with Matlab2010b.

Global performance

Here, we show the performances obtained on the test overall.

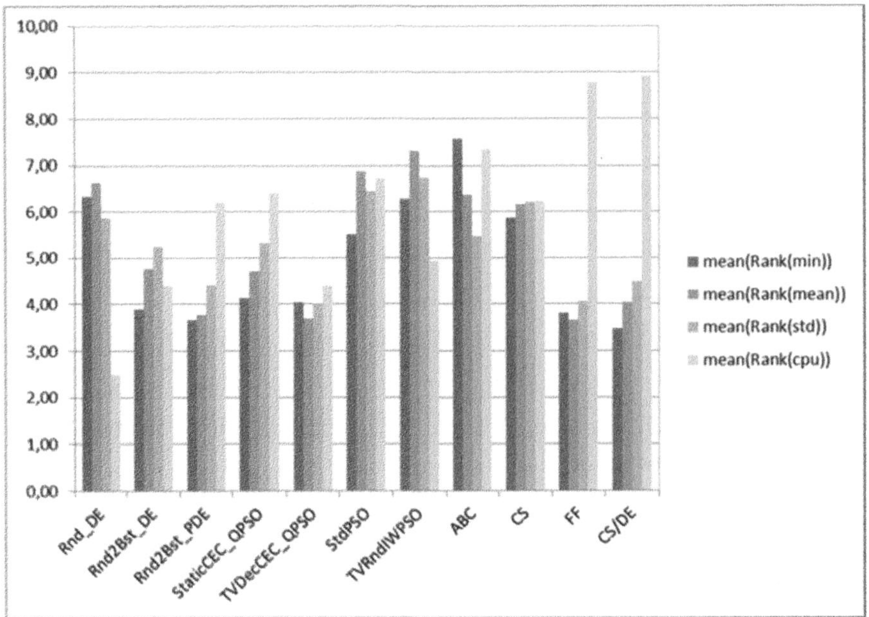

Figure 2.11. *Mean ranks of different algorithms on the test overall (234 tests)*

Discussion

We observe that the best algorithm in terms of minimal value reached is *CS/DE*; however, this is at the price of a computing time that is prohibitive for the latter, due to the algorithm's own structure: a local exploration phase and a global exploration phase give rise to a consequent number of evaluations of the fitness function.

This is very closely followed by *Rnd2Bst_PDE*, which has a better rank in terms of mean value reached for a similar standard deviation, and above all for a much more reasonable computing time. Thus, we can say that the best algorithm (for the problem posed and with this scale) is *Rnd2Bst_PDE*, and it is therefore this algorithm that we keep for the structured H_∞ problem. Moreover, it is interesting to observe that *CS/DE*'s local search phase is exactly *DE*.

FF gives interesting results but with a computing time similar to that of *CS/DE*.

Then comes *Rnd2Bst_DE*. The deviations with its perturbed version are interesting to note: the perturbation mechanism of the differential evolution enables the algorithm's repeatability from one *run* to the other to be substantially improved, because the mean rank of the mean values reached and the corresponding standard deviation(s) are much smaller for the perturbed version.

Quantum particle swarm optimization algorithms such as QPSO give interesting but poorer results for the minimal value reached than the previous ones. It is important to note that version *TVDecCEC_QPSO* is much more repeatable than version *StaticCEC_QPSO* for similar minimal values reached.

Finally, the other algorithms are less competitive for the fixed-order H_∞ synthesis problem as regards the criteria [2.144].

2.2.3.6.3. Comparison with non-smooth optimization

In [APK 10], we find a similar comparison to the previous one between *Hifoo* and *Hinfstruct*. We also decided to compare *Rnd2Bst_PDE* with the non-smooth optimization tools mentioned above on an exhaustive sample of test cases. The comparison rests on the minimal value reached and the computing time (having no access to the mean and standard deviation of the results from *Hifoo* and *Hinfstruct*). As for stochastic optimization, 10 runs are launched for each test with non-smooth optimization tools.

Test	∂	HinfStruct		HIfoo		Evolutionary	
	(P,K)	γ	cpu (min)	γ	cpu (min)	γ	cpu (min)
AC1	(5.0)	0.00	0.16	0.00	1.99	0.00	2.40
AC1	(5.2)	0.00	0.22	0.00	3.23	0.00	1.90
AC2	(5.0)	0.11	0.04	0.11	0.34	0.11	1.03
AC2	(5.2)	0.11	0.02	0.11	0.30	0.11	1.70
AC3	(5.0)	3.42	0.05	3.66	0.48	3.57	3.20
AC3	(5.2)	2.98	0.10	3.12	1.54	3.51	12,92
AC4	(4.0)	0.94	0.04	0.94	0.20	0.95	0.18
AC4	(4.2)	0.56	0,05	0.56	0.87	0.56	8.52

AC5	(4.0)	665	0.03	669	0.05	2,366	0.04
AC5	(4.2)	665	0.05	665	0.70	682	6.82
AC6	(7.0)	4.11	0.04	4.11	1.35	4.11	3.64
AC6	(7.4)	3.46	0.24	3.55	7.32	3.83	26.79
AC7	(9.0)	0.07	0,02	0.07	0.06	0.07	0.26
AC7	(9.5)	0.04	0.18	0.04	4.33	0.04	23.15
AC8	(9.0)	2.01	0.03	2.01	1.08	2.02	1.25
AC8	(9.5)	1.62	0.13	1.62	6.61	1.66	24.25
AC9	(10.0)	1.00	0.12	1.00	2.57	1.01	12.04
AC9	(10.5)	1.00	0.24	1.01	5.20	1.19	24.86
AC9	(10.8)	1.00	0.33	1.05	3.31	1.34	50.75
AC10	(55.0)	13.24	0.34	Inf	0.58	13.6	2.46
AC10	(55.3)	7.92	2.41	Inf	4.09	14	134
AC10	(55.10)	7.90	4.57	20.6	44.20	16.0	374
AC10	(55.20)	7.69	5.81	Inf	17.03	19.2	163.3
AC12	(4.2)	0.00	0.18	0.29	3.19	0.32	9.16
AC13	(28.0)	163	0.20	163	15.03	164	26.99
AC13	(28.7)	156	0.46	156	28.96	156	28.28
AC14	(40.0)	102	0.16	102	5.19	102	17.63
AC14	(40.2)	100	0.19	100	3.02	100	8.40
AC15	(4.0)	15.2	0.03	15.2	1.00	15.2	1.58
AC15	(4.2)	14.9	0.05	14.9	1.20	15.2	4.18
AC16	(4.0)	14.9	0.03	14.9	0.39	**14.8**	1.54
AC16	(4.2)	14.9	0.04	14.9	0.74	**14.8**	2.08
AC17	(4.0)	6.61	0.01	6.61	0.01	6.61	0.67
AC17	(4.2)	6.61	0.01	6.61	0.02	6.61	1.99
AC18	(10.0)	10.70	0.06	12.7	0.75	11.7	1.03
AC18	(10.2)	7.12	0.11	7.20	1.73	7.66	8.85
HE1	(4.0)	0.15	2.64	0.15	3.98	0.15	0.29
HE1	(4.2)	0.08	4.31	0.08	95.30	0.12	10.06
HE2	(4.0)	3.91	1.90	4.00	8.88	4.40	0.84
HE2	(4.2)	2.42	4.45	2.43	76.44	2.83	33.25
HE3	(8.0)	0.81	4.10	0.81	146.96	0.82	40.95
HE3	(8.4)	0.80	9.11	0.81	270.37	0.83	55.74

HE4	(8.0)	22.8	5.58	22.8	122.48	23.9	45.28
HE4	(8.4)	22.8	5.26	22.8	209.73	27.9	49.09
HE5	(8.0)	8.90	1.80	8.90	88.43	**8.86**	4.20
HE5	(8.2)	2.16	7.90	2.17	195.25	2.27	46.30
HE6	(20.0)	192.4	10/4	192.4	294.61	193.9	109.3
HE6	(20.2)	15.5	48.6	16.4	724.23	85.3	139.8
HE7	(20.0)	192.4	18.7	192.4	251.34	200.6	113.7
HE7	(20.8)	2.62	79.6	10.1	594.84	28.7	109.1
JE1	(30.0)	10.13	0.52	23.49	7.24	23.32	19.04
JE1	(30.2)	4.10	1.09	19.65	16.26	5.41	142.43
JE1	(30.8)	3.92	1.84	111.4	1.50	5.62	118.74
JE2	(21.0)	183.4	0.32	183.4	14.02	**182**	14.19
JE2	(21.2)	72.86	0.81	97.91	20.29	139.6	82.53
JE2	(21.8)	58.58	1.38	80.47	20.88	134.1	57.58
JE3	(24.0)	5.10	0.15	5.10	7.99	5.26	5.74
JE3	(24.2)	2.90	0.47	2.90	11.61	3.41	32.97
JE3	(24.8)	2.89	0.43	3.20	7.86	3.25	64.21
REA1	(4.0)	0.87	0.03	0.89	0.07	0.87	1.43
REA1	(4.2)	0.86	0.02	0.86	0.15	0.87	1.88
REA2	(4.0)	1.15	0.03	1.15	1.35	1.16	0.62
REA2	(4.2)	1.13	0.02	1.13	0.16	1.13	1.77
REA3	(12.0)	74.25	0.02	74.25	0.07	74.25	1.85
REA3	(12.2)	74.25	0.02	74.25	0.11	74.25	4.65
REA3	(12.5)	74.25	0.03	74.25	0.06	74.25	4.44
DIS1	(8.0)	4.16	0.06	4.16	0.86	4.19	5.15
DIS1	(8.4)	4.16	0.07	4.16	1.82	4.18	29.64
DIS2	(3.0)	1.06	0.02	1.03	0.23	1.14	1.29
DIS2	(3.2)	0.95	0.04	0.95	0.71	1.03	18.46
DIS3	(6.0)	1.06	0.08	1.06	1.27	1.07	11.10
DIS3	(6.2)	1.05	0.08	1.05	1.69	1.06	24.65
DIS4	(6.0)	0.74	0.04	0.74	0.81	0.74	5.32
DIS4	(6.2)	0.73	0.05	0.74	1.27	0.74	11.76
DIS5	(4.0)	1,035	0.04	1,035	0.26	1,119	0.43
DIS5	(4,2)	667	0.088	668	1,73	745	8,86

EB1	(10.0)	3.12	0.04	3.12	0.04	**3.10**	0.10
EB1	(10.4)	3.10	0.04	3.10	0.15	**3.09**	2.79
EB2	(10.0)	2.02	0.03	2.02	0.01	2.02	0.09
EB2	(10.4)	1.78	0.06	1.78	1.47	1.81	2.76
EB3	(10.0)	2.06	0.03	2.06	0.02	**2.05**	0.08
EB3	(10.2)	1.87	0.04	1.87	0.65	1.87	3.19
NN1	(3.0)	Inf	0.05	14.89	0.05	**14.76**	0.11
NN1	(3.2)	13.13	0.06	13.47	2.19	13.99	1.62
NN2	(2.0)	2.22	0.02	2.22	0.01	**2.21**	0.07
NN2	(2.2)	1.76	0.03	1.76	0.29	1.76	1.73
NN3	(4.0)	Inf	0.02	Inf	0.00	Inf	0.00
NN3	(4.2)	20.5	0.08	20.5	0.57	Inf	0.20
NN4	(4.0)	1.36	0.04	1.36	0.49	1.40	0.98
NN4	(4.2)	1.29	0.05	1.29	0.62	1.34	1.87
NN5	(7.0)	266.5	0.02	266.5	0.11	267	0.12
NN5	(7.4)	238.5	0.09	238.8	1.80	241.6	22.20
NN6	(9.0)	5,602	0.04	5,602	2.00	5,640	1.03
NN6	(9.2)	171.2	0.32	221.5	6.89	220.6	11.20
NN7	(9.0)	74.1	0.03	74.1	0.96	78.1	0.91
NN7	(9.5)	37.4	0.42	39.7	6.77	41.9	29.43
NN8	(3.0)	2.89	0.03	2.89	0.27	2.94	1.13
NN8	(3.3)	2.36	0.03	2.36	0.28	2.42	1.30
NN9	(5.0)	28.7	0.05	2.89	0.73	28.7	0.98
NN9	(5.3)	13.7	0.13	14.1	3.07	21.8	17.41
NN11	(16.0)	0.09	0.06	0.10	0.04	0.09	3.26
NN11	(16.2)	0.02	0.28	0.05	1.87	0.06	13.14
NN12	(6.0)	17.3	0.06	Inf	0.19	21.7	0.10
NN12	(6.2)	7.43	0.20	16.2	2.08	16.1	16.07
NN13	6.0)	14.1	0.05	14.1	0.83	18.9	0.73
NN13	(6.2)	10.5	0.08	11	2.07	12.9	8.52
NN14	6.0)	17.5	0.04	17.5	0.69	18.3	0.60
NN14	(6.2)	9.47	0.12	9.54	2.22	10.5	15.76
NN15	(3.0)	0.10	0.02	0.10	0.01	0.10	1.10
NN15	(3.2)	0.10	0.03	0.10	0.13	0.10	1.98

NN16	(8.0)	0.96	0.07	0.96	2.26	0.96	2.24
NN16	(8.2)	0.96	0.08	0.96	5.77	0.96	12.88
NN17	(3.0)	11.2	0.01	11.2	0.13	11.2	0.37
NN17	(3.2)	2.76	0.04	5.04	0.87	5.16	1.36
HF2D10	(5.0)	8.02E+04	0.06	8.00E+04	2.08	8.11E+04	1.55
HF2D10	(5.2)	7.96E+04	0.06	7.96E+04	1.85	7.98E+04	4.90
HF2D11	(5.0)	7.73E+04	0.02	7.72E+04	1.01	7.72**E+04**	2.31
HF2D11	(5.2)	7.65E+04	0.04	7.65E+04	2.30	7.66E+04	4.86
HF2D12	(5.0)	1.04E+06	0.01	1.04E+06	0.13	1.04E+06	2.26
HF2D12	(5.2)	1.04E+06	0.02	1.04E+06	0.12	1.04E+06	5.03
HF2D13	(5.0)	1.02E+05	0.01	1.02E+05	0.07	1.02E+05	2.41
HF2D13	(5.2)	1.02E+05	0.02	1.02E+05	0.10	1.02E+05	5.03
ROC1	(9.1)	0.24	0.07	1.24	1,63	1.24	3.64
ROC1	(9.5)	1.14	0.17	1.14	4.79	1.29	14.45
ROC2	(10.1)	0.05	0.15	0.05	5.09	0.61	3.62
ROC2	(10.5)	0.04	0.33	0.04	5.45	0.15	11.32
ROC3	(11.1)	228.2	0.57	4,281	0.11	Inf	7.04
ROC3	(11.3)	185.8	0.73	5,730	2.74	Inf	8.83
ROC4	(9.1)	302.2	0.04	302	1.69	302.7	2.60
ROC4	(9.3)	155.3	0.24	266	4.19	304.3	6.87
ROC5	(7.1)	0.00	0.34	0.00	0.78	0.01	3.40
ROC5	(7.3)	0.00	0.44	0.00	2.58	0.01	6.65
WEC1	(10.0)	4.05	0.07	4.05	1.82	6.07	6.07
WEC1	(10.2)	3.64	0.05	3.64	1.85	3.64	15.42
WEC1	(10.4)	3.64	0.04	3.64	1.65	3.64	37.51
WEC1	(10,9)	3.64	0.08	3.64	2.80	3.64	67.85
WEC2	(10.0)	4.25	0.08	4.25	2.28	5.98	6.16
WEC2	(10.2)	3.60	0.04	3.60	1.99	3.60	15.42
WEC2	(10.4)	3.60	0.06	3.60	2.96	3.60	31.96
TG1	(10.0)	12.85	0.04	12.85	1.00	12.86	1.71
TG1	(10.2)	3.47	0.09	3.47	4,.6	5.46	5.31
TG1	(10.4)	3.47	0.09	3.47	6.28	3.47	8.98
AGS	(12.0)	8.17	0.03	8.17	0.70	8.17	2.75
AGS	(12.2)	8.17	0.04	8.17	1.07	8.17	4.59

AGS	(12.4)	8.17	0.07	8.17	1.50	8.17	3.35
TMD	(6.0)	2.23	0.07	2.53	1.12	2.70	1.74
PAS	(5.0)	0.00	0.09	40.1	0.02	1e6	0.13
PAS	(5.2)	0.00	0.18	0.01	0.26	0.13	3.69
IH	(21.8)	1.01	2.25	1.81	5.89	8.59	325.5
IH	(21.15)	1.03	9.63	1.81	5.70	14.24	237.8

Table 2.1. *Non-smooth/evolutionary optimization comparison*

It can be observed that:

– *Hinfstruct* is the fastest tool; this was already the conclusion of article [APK 10]; evolutionary optimization is characterized by computing times comparable to those of *Hifoo*; note that this is not the primary objective of the evolutionary approach as we will see below, what interests us here is the flexibility of the approach.

– Evolutionary optimization and non-smooth optimization give comparable results for the value of γ reached. In some tests, evolutionary optimization even gives better results (in bold in Table 2.1) and even manages to find a solution where non-smooth optimization does not converge (NN1 (3.0)). In this sense, this could be an interesting alternative.

2.2.3.7. *Taking account of additional constraints*

This approach to structured synthesis (much simpler than non-smooth optimization as it only involves evaluating the fitness function and not the gradient) simply enables the framework of the synthesis to be extended. There are multiple, numerous possibilities as shown hereafter.

2.2.3.7.1. Stability constraint

We can try to solve the following problem: find a controller $K(s)$ with the state-space representation $\Theta_k = \begin{pmatrix} A_K & B_K \\ C_K & D_K \end{pmatrix}$, and order $n_k < n_p$ such that:

– $T(s)$ is internally stable;

– $\|T(s)\|_\infty$ is minimal;

– the controller $K(s)$ is stable.

By positing:

$$\bar{\lambda}_k = \max_i \left(\mathrm{Re}\left(\lambda_i \left(A_k \right) \right) \right) \quad [2.145]$$

where $\lambda_i(\ )$ is the ith eigenvalue.

Then, the controller is synthesized considering the following optimization problem:

$$\min_{\Theta_k} f(\Theta_k)$$

If $\max\left(\bar{\lambda}_{cl}, \bar{\lambda}_k\right) \geq 0$

$$f(\Theta_k) = \max\left(\bar{\lambda}_{cl}, \bar{\lambda}_k\right)$$

Else

$$f(\Theta_k) = -\frac{1}{\left\| T(s) \right\|_\infty} \quad [2.146]$$

2.2.3.7.2. Closed-loop pole placement constraint

We can try to solve the following problem: find a controller $K(s)$ with state-space representation $\Theta_k = \begin{pmatrix} A_K & B_K \\ C_K & D_K \end{pmatrix}$, and order $n_k < n_p$ such that:

– $T(s)$ is internally stable;

– $\left\| T(s) \right\|_\infty$ is minimal;

– the poles of $T(s)$ all have a real part less than a strictly negative number α.

Then, the H_∞ optimal controller is synthesized considering the following optimization problem:

$$\min_{\Theta_k} f(\Theta_k)$$

If $\bar{\lambda}_{cl} \geq \alpha$

$$f(\Theta_k) = \bar{\lambda}_{cl} - \alpha$$

Unless

$$f(\Theta_k) = -\frac{1}{\|T(s)\|_\infty} \qquad [2.147]$$

2.2.3.8. *Extension to other complex problems*

2.2.3.8.1. Structured H∞ synthesis for multiple plants

Principle

As we suggest in [FEY 14a], nothing prevent us from generalizing the structured H∞ shown previously to a multiple-plant framework. To do this, consider the following set of m plants:

$$\wp = \{P_i(s), i = 1, \ldots, m\} \qquad [2.148]$$

As in Figure 2.12, the structured H∞ synthesis for multiple-plant problems consists of computing an n_k order controller $K(s)$ which:

– internally stabilizes each closed loop $F_l(P_i, K)$;

– minimizes γ such that:

$$\max_{i=1,\ldots,m} \|F_l(P_i, K)\|_\infty < \gamma \qquad [2.149]$$

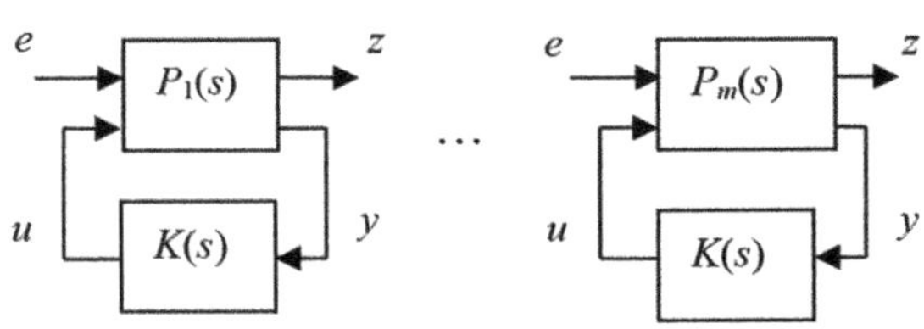

Figure 2.12. *Multiple-plant H∞ synthesis*

The ith standard form has the following state-space representation:

$$P_i(s) := \left[\begin{array}{c|cc} A_i & B_{1i} & B_{2i} \\ \hline C_{1i} & D_{11i} & D_{12i} \\ C_{2i} & D_{21i} & D_{22i} \end{array}\right] \qquad [2.150]$$

The state-space representation of the ith system looped by $K(s)$ is written as:

$$T_i(s) = F_l\left(P_i(s), K(s)\right)$$
$$= \left[\begin{array}{cc|c} A_i + B_{2i}M_{1i}D_kC_{2i} & B_{2i}M_{1i}C_k & B_{1i} + B_{2i}M_{1i}D_kD_{21i} \\ B_kM_{2i}C_{2i} & A_k + B_kM_{2i}D_{22i}C_k & B_kM_{2i}D_{21i} \\ \hline C_{1i} + D_{12i}M_{1i}D_kC_{2i} & D_{12i}M_{1i}C_k & D_{11i} + D_{12i}M_{1i}D_kD_{21i} \end{array}\right] \quad [2.151]$$
$$M_{1i} = \left(I - D_kD_{22i}\right)^{-1}$$
$$M_{2i} = \left(I - D_{22i}D_k\right)^{-1}$$

We define the spectral abscissa of the balls of plants "looped by $K(s)$":

$$\overline{\lambda}_{cl} = \max_{j=1,\ldots,m} \max_i \left(\mathrm{Re}\left(\lambda_i\left(A_{cl\,j}\right)\right)\right) \quad [2.152]$$

where $\lambda_i(\,)$ is the ith eigenvalue.

An H_∞-optimal stabilizing controller is therefore synthesized in a multiple-plant framework by considering the following optimization problem:

$$\min_{\Theta_k} f(\Theta_k)$$

If $\overline{\lambda}_{cl} \geq 0$

$$f(\Theta_k) = \overline{\lambda}_{cl}$$

Unless

$$f(\Theta_k) = -\frac{1}{\gamma},\ \gamma = \max_{j=1,\ldots,m} \left\|T_j(s)\right\|_\infty \quad [2.153]$$

Adaptation of the algorithm from section 2.2.4 of this chapter is thus immediate.

This method of controller synthesis leads us to some observations:

– the algorithm from section 2.2.4 of this chapter is found strictly by taking m = 1;

– no hypothesis has been formed on the orders of Pi(s), which can be different from one another;

– our approach does not prevent each $P_i(s)$ from having different exogenous channels z_i. It is the same for the outputs e_i;

– as for the single plant synthesis, there is nothing to stop constraints on the controller's stability from being added, as well as on the pole placement of the set of closed loops.

Example

Specifications

As an example, consider the system in Figure 2.13. It is a flexible-mode system connected to a DC motor.

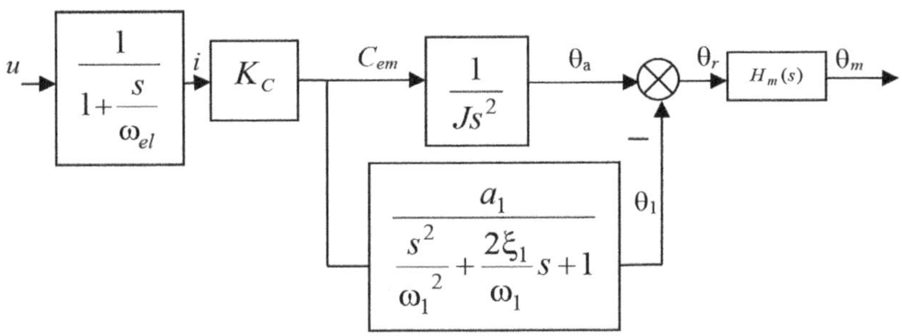

Figure 2.13. *Flexible-mode motorized system*

The values of the parameters, subject to uncertainties, are given below.

$$
\begin{aligned}
&J = 0.0014 \text{ kg.m}^2,\ K_c = 0.06 \text{ Nm.A}^{-1},\\
&a_1 = -7.10^{-4} \pm 7\%,\ \xi_1 = 0.029 \pm 70\%,\\
&\omega_1 = 1\,335 \text{ s}^{-1} \pm 10\%,\ \omega_{el} = 1\,257 \text{s}^{-1} \pm 20\%
\end{aligned}
\qquad [2.154]
$$

The angle position θ is measured using a sensor with transfer function:

$$
H_m(s) = \frac{1}{\frac{s^2}{1382^2} + \frac{s}{1382} + 1} \qquad [2.155]
$$

The nominal plant that corresponds to the nominal values of the parameters is given by $H_0(s)$.

We suggest calculating a controller that ensures the following specification:

– a bandwidth higher than 40 Hz;

– a gain margin higher than 5 dB;

– a phase margin higher than 40°;

– no steady-state error for a perturbation step at input model;

– stabilization of the system toward parametric uncertainties.

It is decided to synthesize the controller using the H_∞ loop-shaping synthesis approach. Recall that the approach involves:

1) Shaping the open loop in the form of a loop-shape $H_p(s)$ depending on the specifications, using weighting functions at the system's entrance and exit:

$$H_p(s) = W_o(s)H_0(s)W_i(s) \qquad [2.156]$$

By positing:

$$H_0(s) := \left[\begin{array}{c|c} A & B \\ \hline C & D \end{array}\right], \quad W_o(s) := \left[\begin{array}{c|c} A_o & B_o \\ \hline C_o & D_o \end{array}\right], \quad W_i(s) := \left[\begin{array}{c|c} A_i & B_i \\ \hline C_i & D_i \end{array}\right] \qquad [2.157]$$

It is easily shown that:

$$H_p(s) := \left[\begin{array}{c|c} A_p & B_p \\ \hline C_p & D_p \end{array}\right] = \left[\begin{array}{ccc|c} A & BC_i & 0 & BD_i \\ 0 & A_i & 0 & B_i \\ B_oC & B_oDC_i & A_o & B_oDD_i \\ \hline D_oC & D_oDC_i & C_o & D_oDD_i \end{array}\right] \qquad [2.158]$$

2) Computing the pre-compensator $K_p(s)$ stabilizing $H_p(s)$ by solving the four-block H_∞[6] problem from Figure 2.14 and equation [2.159]

$$K_p(s) = \arg\min \gamma, \left\| \begin{pmatrix} S_p & S_pH_p \\ K_pS_p & K_pS_pH_p \end{pmatrix} \right\|_\infty < \gamma, \; S_p = \left(I + K_pH_p\right)^{-1} \qquad [2.159]$$

6 It is also possible to solve the (equivalent) stabilization problem of a system modeled by its coprime factorizations.

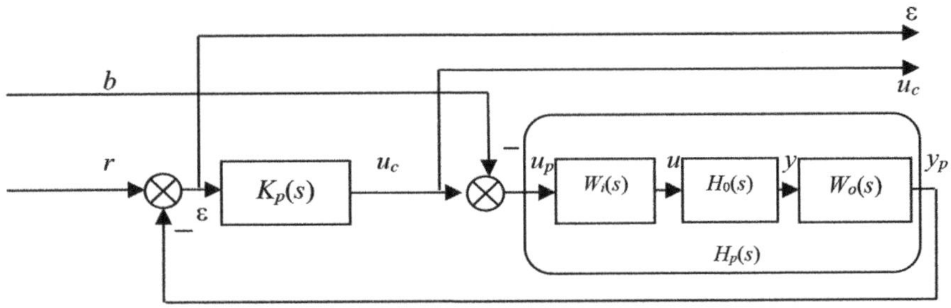

Figure 2.14. *H∞ loop-shaping synthesis*

The standard form of the problem considered is written as:

$$\begin{pmatrix} z \\ \varepsilon \end{pmatrix} = P(s)\begin{pmatrix} e \\ u_c \end{pmatrix},\ e = (r, b)^T,\ z = (\varepsilon, u_c)^T \qquad [2.160]$$

It is easily shown that:

$$P(s) := \left[\begin{array}{c|cc|c} A_p & 0 & -B_p & B_p \\ \hline -C_p & I & D_p & -D_p \\ 0 & 0 & 0 & I \\ \hline -C_p & I & D_p & -D_p \end{array}\right] \qquad [2.161]$$

3) Calculating the final controller $K(s)$ from the weighting functions and the pre-compensator:

$$K(s) = W_i(s)K_p(s)W_o(s) \qquad [2.162]$$

Setting weighting functions

The weights are set using the rules of classic automatic frequency control. Therefore, with regard to the specification, the loop-shape should:

– have an integral component in order to cancel steady-state error in response to a perturbation step;

– have a phase lead network in order to ensure the stability margins;

– have a bandwidth of at least 40 Hz;

– have a roll-off sufficient to stabilize the system when subject to parametric uncertainties: here the roll-off is chosen to satisfy the small gain theorem [ZHO 96] toward a multiplicative uncertainty Δ(s) introduced in conformity with Figure 2.15 to model the high-frequency uncertainties resulting from uncertainties on the parameters.

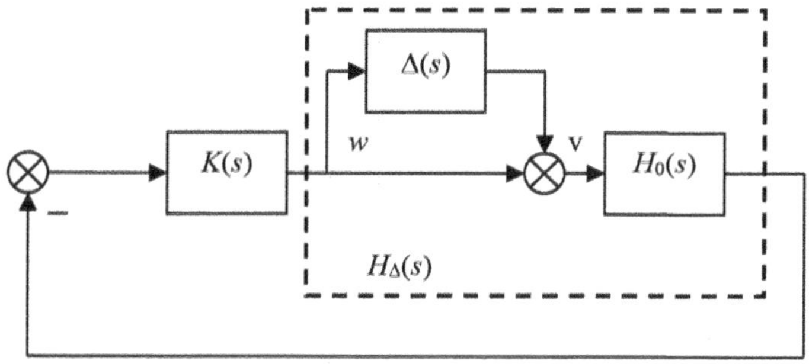

Figure 2.15. *Direct multiplicative uncertainty*

Indeed, according to Figure 2.15:

$$w(s) = T(s)v(s),\ v(s) = \Delta(s)w(s) \quad [2.163]$$

Therefore, by virtue of the small gain theorem, the closed loop is stable if and only if [2.164] is satisfied:

$$\forall\omega\ |T(j\omega)| < \left|\frac{1}{\Delta(j\omega)}\right| \quad [2.164]$$

Thus, at frequencies where $|\Delta(j\omega)|$ is high, the open loop will be of low gain to ensure stability, and in this case $|T(j\omega)| \approx |H_{BO}(j\omega)|$, which therefore makes it possible to specify the *roll-off* that must be conferred on the open loop to guarantee the aim of stability:

$$|H_{BO}(j\omega)| < \left|\frac{1}{\Delta(j\omega)}\right| \quad [2.165]$$

The aim of stability robustness can therefore be directly translated to the nominal *loop-shape* $H_p(s) = W_o(s)H_0(s)W_i(s)$ by:

$$|H_p(j\omega)| < \left|\frac{1}{\Delta(j\omega)}\right| \quad [2.166]$$

By taking the minimal, nominal and maximal values of the four uncertain parameters[7], we create a ball of $3^4 = 81$ models $H_\Delta(s)$ centered on $H_0(s)$, the frequency responses of which are given in Figure 2.16.

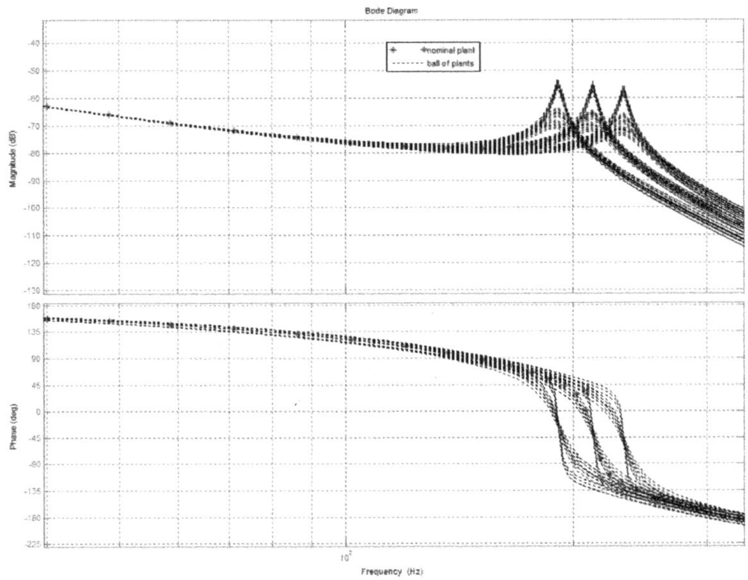

Figure 2.16. *Frequency responses from the different models. For a color version of this figure, see www.iste.co.uk/feyel/optimization.zip*

Indeed, according to Figure 2.15:

$$H_\Delta(s) = (1+\Delta(s))H_0(s) \Leftrightarrow \Delta(s) = \frac{H_\Delta(s) - H_0(s)}{H_0(s)} \quad [2.167]$$

The set of frequency responses of the corresponding $\Delta(s)$ is given in Figure 2.17.

7 $\begin{cases} a_1 = \{-7,49.10^{-4}; -7.10^{-4}; -6,51.10^{-4}\} \\ \xi_1 = \{0,0087;\, 0,029;\, 0,0493\} \\ \omega_1 = \{1201,5;\, 1\,335;\, 1468,5\} \\ \omega_{el} = \{1005,6;\, 1\,257;\, 1508,4\} \end{cases}$

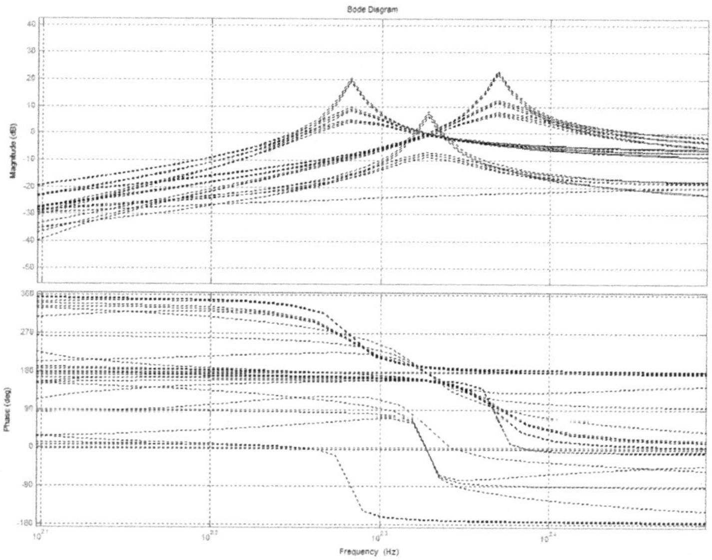

Figure 2.17. *Frequency responses of the corresponding uncertainties*

We suggest setting the weights as follows:

$$W_o(s) = \frac{128.5(s+81.24)}{s} \frac{\left(1+\dfrac{s}{5.03}\right)}{\left(1+\dfrac{s}{1257}\right)}$$

$$W_i(s) = \left(\frac{\dfrac{s^2}{942^2}+\dfrac{0.25s}{942}+1}{\dfrac{s^2}{942^2}+\dfrac{1.8s}{942}+1}\right)\left(\frac{\dfrac{s^2}{2\,200^2}+\dfrac{0.01s}{2\,200}+1}{\dfrac{s^2}{2\,200^2}+\dfrac{2.2s}{2\,200}+1}\right)\frac{1}{\left(\dfrac{s}{942}+1\right)^2} \quad [2.168]$$

In Figure 2.18, we verify that the specifications are satisfied in terms of stability and bandwidth.

Limits of the full-order H_∞ synthesis on the nominal plant

We proceed to a full-order synthesis and find $\gamma = 3.30$. In Figure 2.19, we verify that the stability margins are satisfied for the nominal model.

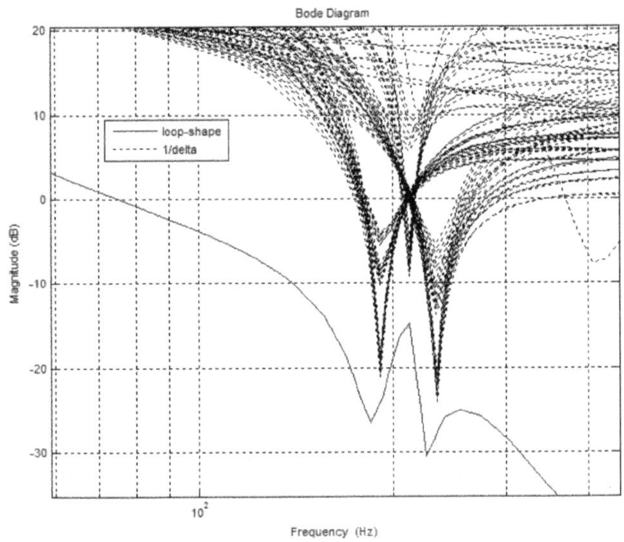

Figure 2.18. *Loop-shape and verification of the condition [2.166]. For a color version of this figure, see www.iste.co.uk/feyel/optimization.zip*

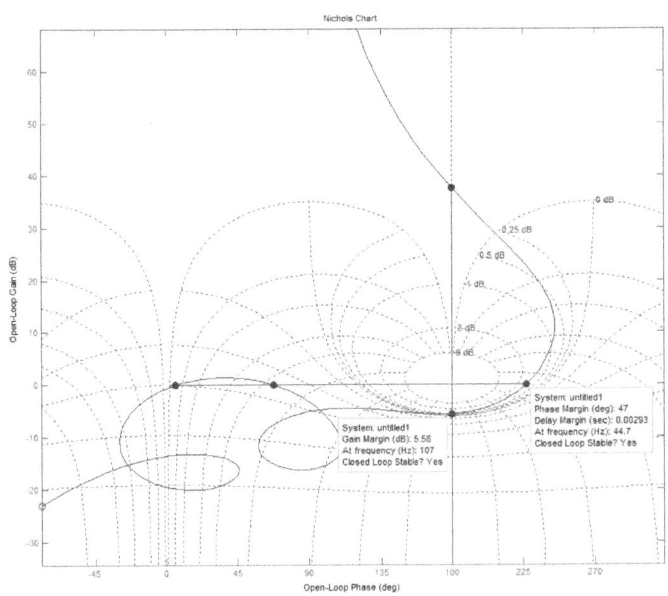

Figure 2.19. *Nominal open loop. For a color version of this figure, see www.iste.co.uk/feyel/optimization.zip*

However, this synthesis suffers from two major drawbacks:

– the set of open-loop frequency responses is traced relative to the ball of plants of Figure 2.20. It is clear that although the whole ball is stabilized, there are systems whose stability margins prove to be much reduced: the "worst case" value of γ can reach 10.5;

– the pre-compensator is order 15 and the final controller is order 23, which can be important with regard to the weakness of the stability margin robustness of the computed controller.

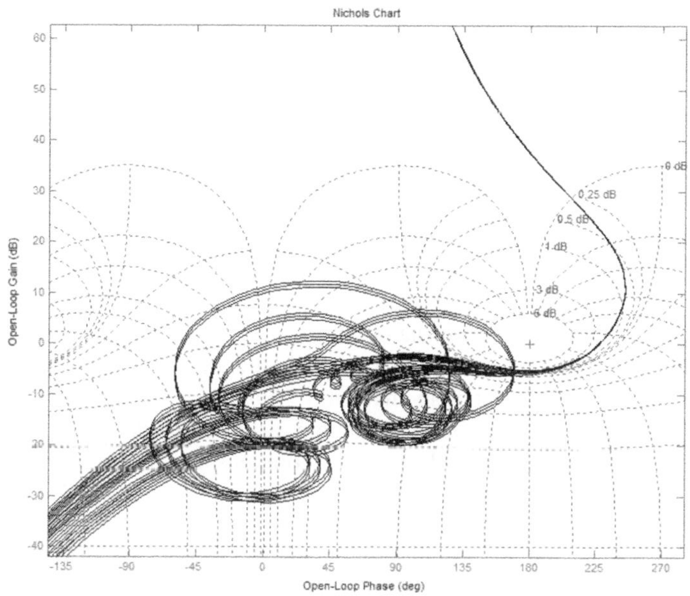

Figure 2.20. *Open loops on the set of the ball of models. For a color version of this figure, see www.iste.co.uk/feyel/optimization.zip*

Fixed-order H_∞ synthesis

In the first instance, it is decided to synthesize an order 5 pre-compensator using stochastic optimization, then to compare this with non-smooth optimization, always on the nominal model.

Because *Rnd2Bst_PDE* is a stochastic algorithm, we proceed to 10 successive *runs*. The maximum number of iterations for each *run* is 500.

For the non-smooth approach, we use *Hinfstruct*. The results provided by *Hinfstruct* depend on the initial point: so we also proceed to 10 *runs* by setting the *randomstream* option at 10. Furthermore, as for the stochastic approach, the controller is computed in tridiagonal form. The other *Hinfstruct* options are set by default.

The computations are carried out with Matlab 2011b on a machine equipped with Intel Core i7-3740QM (2.7 GHz) processor and 8 Gb of RAM.

The tests are carried out in succession with and without parallel computation (*Single Core* and *Multiple Core*, respectively). In the latter case, [MAT 10] is used with eight cores (*pool*).

A statistical evaluation of the results is needed for each test. Therefore, as in [2.144], we take the minimal value of γ obtained (γ_{min}), the mean value of γ (γ_{mean}), its standard deviation (γ_{std}) and finally the mean computing time per run ($t_{cpu/run}$).

The results are presented in Table 2.2.

It is shown that the results obtained with the two approaches in terms of γ are close, but the computing time for the evolutionary approach is 10 times greater than that for *Hinfstruct*. By contrast, the standard deviation for the γs obtained with *Rnd2Bst_PDE* is much lower than that with *Hinfstruct*, for a similar mean value: thus, *Rnd2Bst_PDE* proves to be reproducible, as much as that one or two runs can provide a very good result, which partially compensates for the increase in global computing time between the two approaches.

Finally, it is clear that the parallel computation does not significantly reduce the computing time; this is because the time gained for calculating the fitness function is lost in the time needed for the different processes to communicate. Therefore, parallel computation is only useful when the evaluation time for the fitness function is therefore much greater (see the multiple-plant example below).

	γ_{min}	γ_{mean}	γ_{std}	$t_{cpu/run}$	
				Single Core	*Multi Core*
Evolutionary approach (*Rnd2Bst_PDE*)	3.85	4.10	0.30	**28 s**	**28 s**
Non-smooth approach (*Hinfstruct*)	3.56	4.14	1.93	**2 s**	

Table 2.2. *Fixed-order synthesis – evolutionary and non-smooth approaches*

Fixed-order H_∞ synthesis for multiple plants

In the second instance, we synthesize an order 5 controller using the multiple-plant approach with the evolutionary approach, then with the non-smooth approach. The results are presented in Table 2.3.

	γ_{min}	γ_{mean}	γ_{std}	$t_{cpu/run}$	
				Single Core	*Multi Core*
Evolutionary approach (*Rnd2Bst_PDE*)	4.11	4.42	0.39	**39 min**	**10 min**
Non-smooth approach (*Hinfstruct*)	3.95	4.39	1.85	**300 min**	

Table 2.3. *Fixed-order synthesis for multiple plants – evolutionary and non-smooth approaches*

It is clear that this time, the results obtained with the two approaches in terms of γ are close, but the evolutionary approach is 10 times faster than *Hinfstruct*. Moreover, note a computing time that is four times faster with parallel computation in the evolutionary case.

We can also say that our approach to multiple-plant synthesis is a useful alternative to current, modern tools as it is much less costly in computing time, and even more so when the number of models is substantial. *Hinfstruct* searches for a controller that stabilizes a standard form covering the m sub-models [APK 12], whereas with stochastic optimization the m standard forms are stabilized independently with a controller. There again, the usefulness of the evolutionary approach lies in the flexibility it shows and in its ease of implementation.

In Figure 2.21, the set of open-loop frequency responses on the entire ball of plants is traced. It is clear that the "worst case" stability margins are better. As shown in Table 2.3, the "worst case" value of γ is then reasonable (4.11).

For clarity, the modification of the controller before and after the multiple-plant optimization is traced in Figure 2.22. It is important to note especially the modification of the roll-off needed for performance robustification.

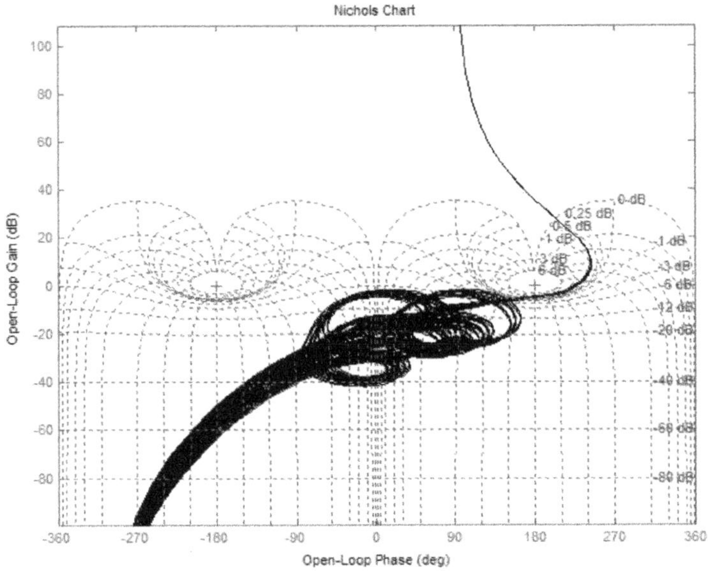

Figure 2.21. *Open loops on the whole ball of plants after fixed-order multiple-plant evolutionary synthesis*

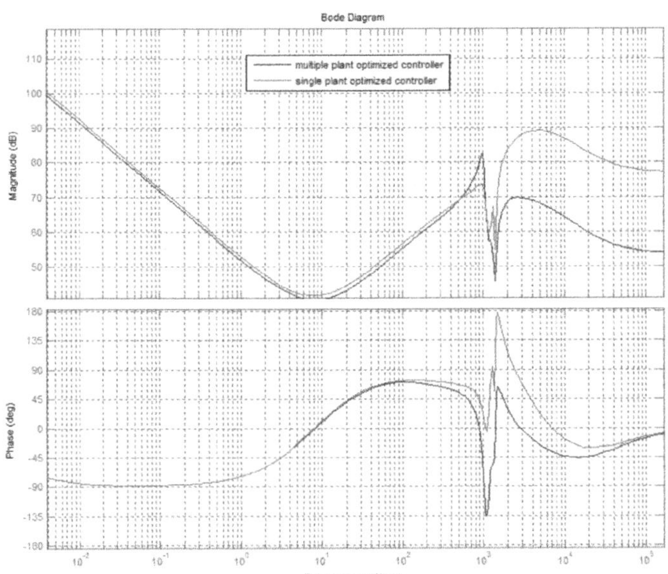

Figure 2.22. *Controllers obtained before and after the multiple-plant optimization. For a color version of this figure, see www.iste.co.uk/feyel/optimization.zip*

2.2.3.9. *General methodology for synthesizing structured stabilizing controllers with stochastic optimization*

2.2.3.9.1. General principle

We can generalize the method of synthesizing a structured stabilizing controller by using an optimization problem in which the fitness function to be minimized has the following general form:

$$\min_{\Theta_k} f(\Theta_k)$$

If $\overline{\lambda}_{cl} \geq 0$

$$f(\Theta_k) = \overline{\lambda}_{cl}$$

Else

$$f(\Theta_k) = -\frac{1}{g(\Theta_k)} \quad [2.169]$$

where $g(\Theta_k)$ is a positive function of Θ_k describing the specifications required. The benefit of this formulation is its great flexibility (easy adaptation for a multitude of complex problems, extension to the multiple-plant framework, etc.). This form will therefore be retained in the following chapter. Some examples of additional extensions are given below.

2.2.3.9.2. Example 1: structured H∞ synthesis subject to any constraint

For this, it is sufficient to choose:

$$g(\Theta_k) = \|T_\infty(s)\|_\infty + f_c(\Theta_k) \quad [2.170]$$

where the function $f_c(\Theta_k)$ is the translation of the considered constraint (in the form of a barrier function for example). This is the formulation that we have adopted for setting structured and robust controllers against high-level requirements (see the following section of the book).

2.2.3.9.3. Example 2: mixed structured H2/H∞ synthesis

Considering Figure 2.23, there are several possible formulations for the mixed H_2/H_∞ synthesis problem. We posit:

$$z_\infty(s) = T_\infty(s)e_\infty(s), \quad z_2(s) = T_2(s)e_2(s) \quad [2.171]$$

Thus, we search for an order $n_k < n_p$ controller $K(s)$ that:

– internally stabilizes the closed loop $F_l(P, K)$;

– with one of the following objectives

- $\alpha\|T_\infty(s)\|_\infty + \beta\|T_2(s)\|_2 < \gamma$, with γ minimal and ((>0, β > 0) given. In this case:

$$g(\Theta_k) = \alpha\|T_\infty(s)\|_\infty + \beta\|T_2(s)\|_2 \qquad [2.172]$$

- $\|T_\infty(s)\|_\infty < \gamma_\infty$ subject to $\|T_2(s)\|_2 < \gamma_2$, with γ_∞ minimal and γ_2 given. In this case:

$$g(\Theta_k) = \|T_\infty(s)\|_\infty + f_c(\Theta_k),\ f_c(\Theta_k) = e^{M(\|T_2(s)\|_2 - \gamma_2)} \qquad [2.173]$$

where $M >> 1$ since f_c therefore represents a barrier penalty.

- $\|T_2(s)\|_2 < \gamma_2$ subject to $\|T_\infty(s)\|_\infty < \gamma_\infty$, with γ_2 minimal and γ_∞ given. In this case:

$$g(\Theta_k) = \|T_2(s)\|_2 + f_c(\Theta_k),\ f_c(\Theta_k) = e^{M(\|T_\infty(s)\|_\infty - \gamma_\infty)} \qquad [2.174]$$

where $M >> 1$ since f_c therefore represents a barrier penalty.

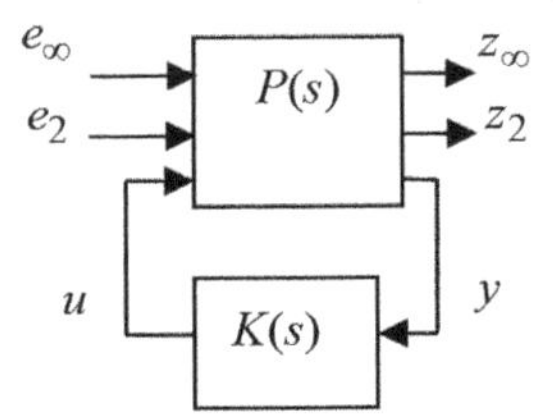

Figure 2.23. *Mixed H2/H∞ synthesis*

It is important to note that as for [2.146], there is nothing to prevent constraints on the controller's stability, or on the pole placement of closed loops being added.

Similarly, there is nothing to prevent this technique being extended to a multiple-plant framework.

2.2.4. *Conclusion*

In this section, an original method of structured controller synthesis is implemented. Although the computing time for controller synthesis using stochastic optimization is greater than that for other techniques for "simple" problems (the fitness function is evaluated a higher number of times), the approach derives its usefulness from the complexity of the criteria used, which can be any criteria, which other techniques cannot easily handle. Similarly, there is nothing to prevent exotic constraints being added on this or that transfer with penalties. The list of criteria is not remotely exhaustive: nothing prevents time criteria being added, such as bounds on the impulse or step responses and thus computing a structured controller by optimizing a criterion with any form (see section 2.3).

2.3. μ-synthesis

2.3.1. *The problem of performance robustness*

Consider the control scheme in Figure 2.24, where $P(s)$ is a linear time-invariant system, $K(s)$ is a linear time-invariant controller to be computed, $\Delta(s)$ represents the uncertainties on the system, $(z_\Delta, e_\Delta) \in \mathbf{R}^{n_\Delta \times n_\Delta}$ and $(z, e) \in \mathbf{R}^{n_z \times n_e}$ are the channels related to the uncertainties and the channels linked to performance, respectively and $y \in \mathbf{R}^{n_y}$ represents the measurement vector needed for the controller to compute the control vector $u \in \mathbf{R}^{n_u}$.

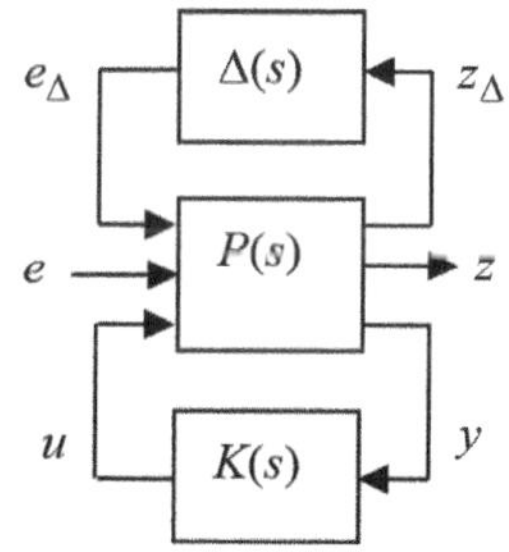

Figure 2.24. *Control scheme with uncertainty block in LFT form*

NOTE.–

$$T(s) = F_l\left(P(s), K(s)\right) = \begin{pmatrix} T_{z_\Delta e_\Delta}(s) & T_{z_\Delta e}(s) \\ T_{z e_\Delta}(s) & T_{ze}(s) \end{pmatrix} \quad [2.175]$$

We will suppose without loss of generality that $\Delta(s)$ is normalized, i.e. that $\|\Delta(s)\|_\infty < 1$.

According to the small gain theorem [ZHO 96], the system $T(s) \in \mathbf{R}H_\infty$ is internally stable for any unstructured uncertainty $\Delta(s) \in \mathbf{R}H_\infty$ such that $\|\Delta(s)\|_\infty < 1$ if and only if:

$$\|T_{z_\Delta e_\Delta}(s)\|_\infty \leq 1 \qquad [2.176]$$

Thus, the small gain theorem provides a necessary and sufficient condition for stability robustness in the unstructured framework. However, it is known that robustness analysis with the small gain theorem in this framework can lead to very conservative results in spite of the ease of implementation. Much more relevant results can be obtained by directly taking into account the block-diagonal structure obtained in [2.177] when the system and its uncertainty are placed in LFT[8] form:

$$\begin{array}{l}\Delta(s) = diag\left\{\Delta_1(s),\ldots\Delta_q(s), \delta_1 I_{r_1},\ldots,\delta_r I_{r_r}, \varepsilon_1 I_{c_1},\ldots,\varepsilon_c I_{c_c}\right\} \\ \Delta_i(s) \in \mathbf{R}H_\infty{}^{n_i x n_i},\ \delta_j \in \mathbf{R},\ \varepsilon_k \in \mathbf{C}\end{array} \qquad [2.177]$$

and verifying the normalization conditions:

$$\left(\|\Delta_i(s)\|_\infty < 1\,; \delta_j \in \left]-1;1\right[\ ; |\varepsilon_k| < 1\right) \Leftrightarrow \left(\|\Delta(s)\|_\infty < 1\right) \qquad [2.178]$$

It is important to note that according to [2.177]:

$$n_\Delta = \sum_{i=1}^{q} n_i + \sum_{i=1}^{r} r_i + \sum_{i=1}^{c} c_i \qquad [2.179]$$

Therefore, we define the set $\underline{\Delta}$ of complex matrices with the same structure as $\underline{\Delta}(s)$:

$$\underline{\Delta} := \left\{\begin{array}{c}\Delta = diag\left\{\Delta_1,\ldots\Delta_q, \delta_1 I_{r_1},\ldots,\delta_r I_{r_r}, \varepsilon_1 I_{c_1},\ldots,\varepsilon_c I_{c_c}\right\} \in \mathbf{C}^{n_\Delta x n_\Delta} \\ \Delta_i \in \mathbf{C}^{n_i x n_i},\ \delta_j \in \mathbf{R},\ \varepsilon_k \in \mathbf{C}\end{array}\right\} \qquad [2.180]$$

The hypothesis of block-diagonal structuration of the uncertainty matrix is not at all restrictive as during the LFT formatting of a system's uncertainties as in

8 Linear Fractional Transformation

Figure 2.24, it is always possible to reduce it to a structure such as [2.180]. To show this, there are some examples of setting uncertainties in LFT form in Appendix B.

To take into account the particular structure $\underline{\Delta}$, we define the structured singular value of a matrix P relative to the structure $\underline{\Delta}$ [ZHO 96]:

$$\mu_{\underline{\Delta}}(P) := \left(\inf_{\Delta\in\underline{\Delta}}\left(\overline{\sigma}(\Delta) : \det(I-\Delta P)=0\right)\right)^{-1}$$
$$\mu_{\underline{\Delta}}(P) := 0 \text{ if } \forall\Delta\in\underline{\Delta}\ \det(I-\Delta P)\neq 0 \qquad [2.181]$$

$\mu_{\underline{\Delta}}(P)^{-1}$ is therefore the smallest norm of a $\Delta \in \underline{\Delta}$ matrix for which the linear system in Figure 2.25 in z and e becomes singular.

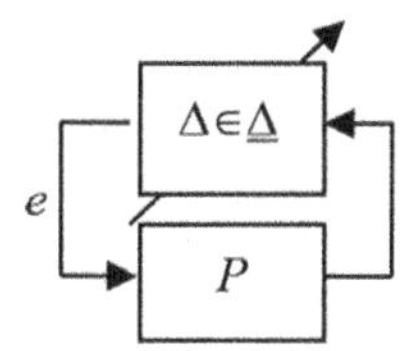

Figure 2.25. *Interpretation of the singular structured value*

$$\begin{cases} e = \Delta z \\ z = Pe \end{cases} \Leftrightarrow \begin{pmatrix} I & -\Delta \\ -P & I \end{pmatrix}\begin{pmatrix} e \\ z \end{pmatrix} = \begin{pmatrix} 0 \\ 0 \end{pmatrix} \Leftrightarrow \begin{cases} (I-\Delta P)e = 0 \\ z = Pe \end{cases} \qquad [2.182]$$

The small gain theorem can therefore be generalized [ZHO 96]. The system $T(s) \in \mathbf{R}H_\infty$ is internally stable for any structured uncertainty $\Delta(s) \in \underline{\Delta}$ if and only if:

$$\forall\omega,\ \mu_{\underline{\Delta}}\left(T_{z_\Delta w_\Delta}(j\omega)\right) \leq 1 \qquad [2.183]$$

Carrying out a robustness analysis in the structured framework with the generalized small gain theorem is called μ-analysis.

It is possible to extend analysis of stability robustness to an analysis of performance robustness. Thus, supposing that the performance is reached when $\|T_{ze}(s)\|_\infty \leq 1$, the aim is to guarantee this condition for any $\Delta(s) \in \underline{\Delta}$ matrix: this defines the problem of performance robustness. According to the small gain theorem, this is the case if and only if the system in Figure 2.26 is stable for any

$\Delta_f(s)$ such as $\left\|\Delta_f(s)\right\|_\infty < 1$, where $\Delta_f(s)$ is an unstructured fictive uncertainty introduced between the performance channel z and the exogenous inputs e.

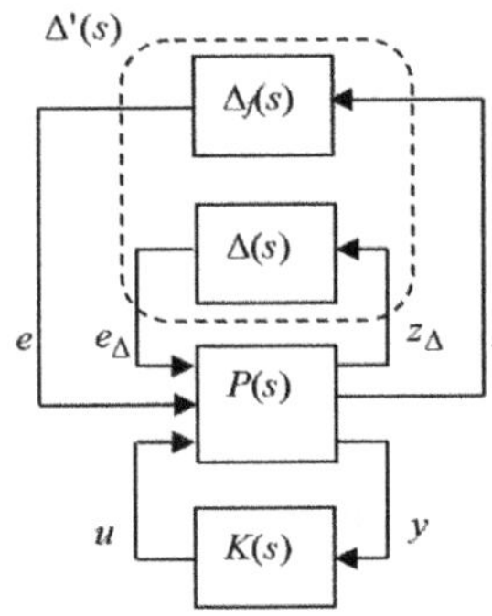

Figure 2.26. *Control scheme for performance robustness analysis*

$\Delta_f(s)$ and $\Delta(s)$ can be collected in an augmented uncertainty matrix $\Delta'(s)$:

$$\Delta'(s) = \text{diag}\left(\Delta_f(s), \Delta(s)\right) \qquad [2.184]$$

As $\Delta'(s)$ is structured, nothing prevents the generalized small gain theorem being applied. The system in Figure 2.26 is therefore internally stable for any $\Delta'(s) \in \underline{\Delta}'$ if and only if:

$$\forall\omega,\ \mu_{\underline{\Delta}'}\left(T(j\omega)\right) \leq 1 \qquad [2.185]$$

with:

$$\underline{\Delta}' := \left\{\Delta' = diag\left(\Delta, \Delta_f\right), \Delta \in \underline{\Delta},\ \Delta_f \in \mathrm{C}^{n_z \times n_w}\right\} \qquad [2.186]$$

This system is stable for any $\Delta'(s) \in \underline{\Delta}'$ such that:

$$\begin{gathered}\left\|\Delta'(s)\right\|_\infty < 1 \\ \Leftrightarrow \\ \begin{cases} \left\|\Delta(s)\right\|_\infty < 1, \forall \Delta \in \underline{\Delta} \\ \left\|\Delta_f(s)\right\|_\infty < 1 \Rightarrow \left\|T_{ze}(s)\right\|_\infty < 1 \end{cases} \\ \Leftrightarrow \\ \left\|T_{ze}(s)\right\|_\infty < 1, \forall \Delta(s) \in \underline{\Delta}\end{gathered} \qquad [2.187]$$

The problem of performance robustness analysis can therefore be reduced to an augmented stability robustness problem:

$$\|T_{ze}\|_\infty < 1,\ \forall\Delta(s)\in\underline{\Delta} \Leftrightarrow \forall\omega, \mu_{\underline{\Delta}'}\left(F_l\left(P(j\,\omega), K(j\,\omega)\right)\right) < 1 \qquad [2.188]$$

Structured analysis using $\mu_{\underline{\Delta}}(P)$ provides a less conservative framework than the unstructured framework. However, use of the structured singular value can quickly become complex, as $\mu_{\underline{\Delta}}(P)$ is very difficult to compute (in practice, except in very simple examples, calculating $\mu_{\underline{\Delta}}(P)$ is a problem of exponential complexity). Its upper bound is therefore sought. It is shown that [YOU 90]:

$$\mu_{\underline{\Delta}}(P) \le \min_{D\in\underline{D}} \overline{\sigma}\left(DPD^{-1}\right) \qquad [2.189]$$

where $D \in \underline{D}$ is a matrix that commutes with Δ (i.e. that $\Delta D = D\Delta$):

$$\underline{D} := \left\{ \begin{array}{c} D = diag\left(d_1 I_{k_1},\ldots,d_q I_{k_q}, D_1,\ldots,D_r, D_1',\ldots,D_c'\right) \\ d_i \in \mathbf{R}\ ;\ d_i > 0 \\ D_i \in \mathbf{C}^{r_i\times r_i}\ ;\ D_i = D_i^* > 0\ ;\ D_i' \in \mathbf{C}^{c_i\times c_i}\ ;\ D_i' = D_i'^* > 0 \end{array} \right\} \qquad [2.190]$$

Indeed, as D is block-diagonal and commutes with Δ, then:

$$\begin{aligned} \det\left(I - \Delta P\right) &= \det\left(I - D^{-1}\Delta DP\right) \\ &= \det\left(D^{-1}\left(D - \Delta DP\right)\right) \\ &= \det\left(\left(D - \Delta DP\right)D^{-1}\right) \\ &= \det\left(I - \Delta DPD^{-1}\right) \end{aligned} \qquad [2.191]$$

$$\Rightarrow \mu_{\underline{\Delta}}(P) = \mu_{\underline{\Delta}}(DPD^{-1}) \le \overline{\sigma}(DPD^{-1})$$

We return to the upper bound [2.189]. A search for the upper bound of $\mu_{\underline{\Delta}}(P)$ can be carried out by noting the equivalence of the following propositions:

$$\left.\begin{array}{l} \exists D\in\underline{D} : \overline{\sigma}\left(DPD^{-1}\right) \le \alpha \\ \Leftrightarrow \exists D\in\underline{D} : \overline{\lambda}\left(D^{-1}P^* DDPD^{-1}\right) \le \alpha^2 \\ \Leftrightarrow \exists D\in\underline{D} : D^{-1}P^* D^2 PD^{-1} - \alpha^2 I \le 0 \\ \Leftrightarrow \exists D\in\underline{D} : P^* D^2 P - \alpha^2 D^2 \le 0 \\ \Leftrightarrow \exists D\in\underline{D} : P^* D' P - \alpha^2 D' \le 0 \end{array}\right\} \qquad [2.192]$$

This upper bound of $\mu_{\underline{\Delta}}(P)$ can be computed by solving the following optimization problem:

$$\mu_{\underline{\Delta}}(P) \leq \alpha_{opt} \text{ with } \alpha_{opt} = \min_{\alpha,\, D \in \underline{D}} \alpha : \begin{cases} \alpha > 0 \\ P^* DP - \alpha^2 D < 0 \end{cases} \quad [2.193]$$

which is a quasi-convex optimization problem:

– at fixed α, the constraint is a linear matrix inequality in D;

– at fixed D, it is a problem of eigenvalue computation.

2.3.2. μ-Synthesis

2.3.2.1. *Introduction*

The emergence of H_∞ techniques (see the first section) has made it possible to take into account the stability robustness constraint for computing a stabilizing controller $K(s)$. In view of the previous section, a much more ambitious problem could be computing a stabilizing controller $K(s)$ satisfying $\|T_{ze}(s)\|_\infty < \gamma$ for any $\Delta(s) \in \mathbf{R}H_\infty$ such that $\|\Delta(s)\|_\infty < \gamma^{-1}$.

Assuming that $\Delta(s) \in \underline{\Delta}$, then the problem amounts to computing a stabilizing controller K(s) satisfying $\|F_l(P,K)\|_\infty < 1$ and so that $\|T_{ze}(s)\|_\infty < 1,\ \Delta(s) \in \underline{\Delta}$, that is, linking the standard problem of H_∞ synthesis with the problem of μ-analysis as shown in the previous sections: this is the μ-synthesis problem [YOU 93].

By introducing a fictive uncertainty on the performance channels as in Figure 2.26, the previous problem can be reformulated by computing a stabilizing controller $K(s)$ and satisfying:

$$\forall \omega,\ \mu_{\underline{\Delta}'}\left(F_l(P(j\omega), K(j\omega))\right) \leq 1 \quad [2.194]$$

Unfortunately, there is no exact solution to the problem of μ-synthesis as there is for the standard H_∞ problem [DOY 89]. However, it has been seen that:

$$\forall \omega,\ \mu_{\underline{\Delta}'}\left(F_l(P(j\omega), K(j\omega))\right) \leq \overline{\sigma}\left(D_\omega F_l(P(j\omega), K(j\omega)) D_\omega^{-1}\right) \quad [2.195]$$

where D_ω is any invertible matrix that commutes with the structure $\underline{\Delta}'$ (i.e. $\Delta' D_\omega = D_\omega \Delta''$).

Thus, we can approach the μ-synthesis problem via the following problem: find a stabilizing controller $K(s)$ and a transfer matrix $D(s)$ such that:

– $D(s),\ D^{-1}(s) \in \mathbf{R}H_\infty$;

– $D(s)\Delta'(s) = \Delta'(s)D(s),\ \forall \Delta'(s) \in \underline{\Delta}'$;

– $\left\| D(s) F_l(P,K) D(s)^{-1} \right\|_\infty < 1$.

It is important to note that $D(s)$ can be chosen without loss of generality in the form [2.196], where each $D_m(s)$ commutes with the corresponding block in $\Delta(s)$:

$$D(s) = \begin{pmatrix} \operatorname{diag}\left(D_m(s), m = 1,\ldots,n_\Delta\right) & 0 \\ 0 & D_{ze}(s) \end{pmatrix} \qquad [2.196]$$

It is also important to note that the scaling D_{ze}(s) can be chosen equal to the identity matrix when $n_z = n_e$. In order to simplify the following developments, we will retain this latter hypothesis.

2.3.2.2. *Full-order μ-synthesis*

2.3.2.2.1. Iterative approaches

The most standard approach is *D-K* iteration, which consists of alternatively searching for *D* and *K* [ZHO 96]:

– at fixed D(s), a controller $K(s)$ is computed by solving the standard H_∞ problem from Figure 2.27;

– at fixed $K(s)$, the matrix $D(s)$ is computed after μ-analysis by:

- calculating an upper bound of the structured singular value $\mu_{\underline{\Delta}}(P)$ for a chosen set of frequencies ω by solving [2.193], as well as the corresponding set of scalings D_ω;

- interpolating the set of scalings D_ω by a stable and inverse stable transfer $D(s)$.

The procedure is initialized by choosing the identity for $D(s)$.

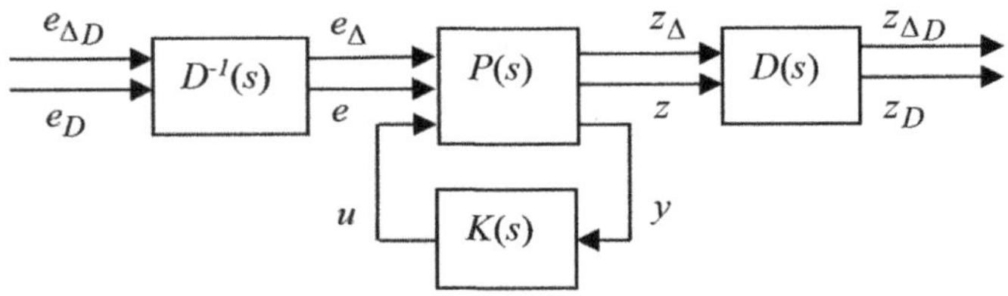

Figure 2.27. *D-K iteration: computing K(s) with fixed D(s)*

This procedure prompts the following remarks:

– it is stopped when the indicator γ no longer changes from one iteration to another or when the scalings no longer change from one iteration to another;

– note that the order of the computed controller at each iteration increases more and more since the order of the standard augmented form of the scalings $D(s)$ increases progressively with the iterations;

– the interpolation phase is very sensitive numerically: numerous possibilities resulting from the choice of frequency grid (the latter should be fine enough to take into account all the system's dynamics), choice of the method of interpolation and above all choice of the order of the scalings.

Despite all this, this approach gives very good results in many applications [DAL 08].

In reality, this approach is quite conservative as it is assumed that all the uncertainties are complex, which is not the case with real parametric uncertainties. It can be useful to take into account this distinction. Therefore, other approaches, more or less complex to implement, exist.

For example, we cite [YOU 93], which shows that an upper bound of $\mu_{\underline{\Delta}}(P)$ is:

$$\begin{gathered}\forall\omega,\ \mu_{\underline{\Delta}'}(F_l(P(j\omega),K(j\omega))) < \overline{\sigma}\left(\left(D_\omega F_l(P(j\omega),K(j\omega))D_\omega^{-1} - G_{\beta\omega}\right)G_{\alpha\omega}\right) \leq \beta_\omega \\ G_{\beta\omega} = j\beta_\omega G_\omega,\ G_{\alpha\omega} = \left(I + G_\omega^{\,2}\right)^{-1/2}\end{gathered} \qquad [2.197]$$

We then calculate the maximal bound by solving the following optimization problem:

$$\mu_{\underline{\Delta}}(P) \leq \beta_{opt} \text{ with } \beta_{opt} = \min_{\beta,\ D\in\underline{D}, G\in\underline{G}} \beta : \begin{cases} \beta \geq 0 \\ P^* DP + j\left(GP - P^* G\right) - \beta^2 D \leq 0 \end{cases} \qquad [2.198]$$

where:

$$\underline{G} := \left\{ \begin{matrix} G = diag\left(0_{k_1},\ldots,0_{k_q},G_1,\ldots,G_r,0_{c_1},\ldots,0_{c_c}\right) \\ G_i \in \mathbf{C}^{r_i \times r_i} \; ; \; G_i = G_i^* \end{matrix} \right\} \quad [2.199]$$

G contains matrices whose non-null elements correspond to the real blocks of Δ.

A procedure similar to the previous one, called *D-G*, *K* iteration, can therefore be used [YOU 93]. This approach is however very complex to implement and very sensitive numerically. We can therefore turn to the μ-*K* iteration that makes it possible to avoid interpolating matrix *G* [TOF 95].

Other approaches that make it possible to skip the scaling interpolation phase exist, but they are complex to implement and numerically sensitive. For example, we cite [TSA 02] and [LIN 05].

2.3.2.2.2. Reformulation for stochastic optimization

Introduction

As we can see, the problem of μ-synthesis remains open. Most of the approaches are iterative and suffer from two major drawbacks:

– first, these approaches have a numerically very sensitive scaling interpolation phase, so the method becomes sub-optimal among the numerous possibilities available to the engineer: the choice of interpolation method, choice of frequency grid, choice of the order of scalings, etc.;

– moreover, these methods, because they are iterative, produce controllers whose orders increases progressively with the iterations, which can quickly prove to be numerically problematic.

We thus suggest in [FEY 14b] tackling the problem of μ-synthesis using stochastic optimization.

In the literature, there are some interesting attempts, based mainly on a direct approach to the problem of μ-synthesis. Thus, for example, in [LAR 12], a structured PID controller is computed with PSO by attacking the problem frontally:

$$\min_{K(s)} \mu_{\underline{\Delta}'}\left(T(s)\right)$$

$$\text{subject to } \mu_{\underline{\Delta}}\left(T_{z_\Delta e_\Delta}(s)\right) < 1 \quad [2.200]$$

This approach seems efficient as it is "direct", but has only been tested for calculating a PID controller (four parameters to compute). Moreover, the fitness function envisaged consists of directly evaluating the structured singular value: although the scaling interpolation phase no longer exists, the question of choosing a frequency grid remains entirely and calculating the singular value is still time consuming in the case of a fine grid (necessary for obtaining a good result).

Principle of the approach

Here, we suggest solving the following global optimization problem directly:

$$\min_{X\in[\underline{x};\overline{x}]} f(X) \Leftrightarrow \min_{D(s),D(s)^{-1}\in \mathbf{R}H_\infty} \gamma \qquad [2.201]$$

with:

$$\gamma = \min_{K(s)} \left\| D(s)F_l(P,K)D(s)^{-1} \right\|_\infty \qquad [2.202]$$

This problem is not dissimilar to that of structured H_∞ synthesis and will therefore follow the same scheme.

By positing as an unknown:

$$\Theta_D = \begin{pmatrix} A_D & B_D \\ C_D & D_D \end{pmatrix} \qquad [2.203]$$

And:

$$\overline{\lambda}_D = \max\left[\max_i\left(\mathrm{Re}\left(\lambda_i\left(A_D\right)\right)\right),\ \max_i\left(\mathrm{Re}\left(\lambda_i\left(A_D - B_D D_D{}^{-1} C_D\right)\right)\right)\right] \qquad [2.204]$$

where $\lambda_i(\,)$ is the ith eigenvalue.

The μ-optimal and full-order scalings and controller are therefore found by considering the following optimization problem:

$$\min_{\Theta_D} f(\Theta_D)$$

If $\overline{\lambda}_D \geq 0$

$$f(\Theta_k) = \overline{\lambda}_D$$

Else

$$f(\Theta_k) = -\frac{1}{\gamma},\ \gamma = \min_{K(s)} \left\| D(s)F_l(P,K)D(s)^{-1} \right\|_\infty \qquad [2.205]$$

Implementation

The scaling $D(s)$ has the structure [2.196], in which, for simplicity, we will assume that $n_z = n_e$:

$$D(s) = \begin{pmatrix} \mathrm{diag}\left(D_m(s), m = 1,\ldots,n_\Delta\right) & 0 \\ 0 & I_{n_e} \end{pmatrix} \qquad [2.206]$$

Also, the unknown X represents the coefficients of the state-space representations of the $n_\Delta = q + \mathrm{r} + c$ elements $D_m(s)$ forming $D(s)$.

Each term $D_m(s)$ has the following state-space representation:

$$D_m(s) = \left[\begin{array}{c|c} A_m & B_m \\ \hline C_m & D_m \end{array}\right] \qquad [2.207]$$

As for the stochastic approach to H_∞ synthesis, the rate of convergence of the algorithm is greatly accelerated by searching for each $D_m(s)$ in tridiagonal form and by transforming the search space. Assuming that each $D_m(s)$ sought is SISO and of n_m-order, we posit for the decision vector:

$$\begin{aligned} X &= \left(X_{D_1},\ldots,X_{D_m},\ldots,X_{D_{n_\Delta}}\right)^T \\ X_{D_m} &= \left(x_A, x_B, x_C, x_D\right)_m \\ &\left.\begin{aligned} x_A &= \left(a_{1,1},\ldots,a_{n,n},a_{1,2},\ldots,a_{n-1,n},a_{2,1},\ldots,a_{n,n-1}\right) \\ x_B &= \left(b_{1,1},\ldots,b_{n,1}\right) \\ x_C &= \left(c_{1,1},\ldots,c_{1,n}\right) \\ x_D &= d \end{aligned}\right\} n = n_m \end{aligned} \qquad [2.208]$$

It is important to note that structure [2.206] is a particular, decentralized structure:

$$D(s)=\left[\begin{array}{cc|cc}\text{diag}(A_m,\ m=1,\ldots,n_\Delta) & 0 & \text{diag}(B_m,\ m=1,\ldots,n_\Delta) & 0\\ 0 & -I_{n_e} & 0 & 0_{n_e}\\ \hline \text{diag}(C_m,\ m=1,\ldots,n_\Delta) & 0 & \text{diag}(D_m,\ m=1,\ldots,n_\Delta) & 0\\ 0 & 0_{n_e} & 0 & I_{n_e}\end{array}\right]$$

$$=\left[\begin{array}{c|c}A_D & B_D\\ \hline C_D & D_D\end{array}\right]$$

[2.209]

$$D(s)^{-1}=\left[\begin{array}{c|c}A_D-B_DD_D^{-1}C_D & -B_DD_D^{-1}\\ \hline D_D^{-1}C_D & D_D^{-1}\end{array}\right]$$

$$=\left[\begin{array}{c|c}\underline{A}_D & \underline{B}_D\\ \hline \underline{C}_D & \underline{D}_D\end{array}\right]$$

It is also important to note the presence of a term $-I_{ne}$ in A_D serving to ensure that the part of $D(s)$ relative to the performance channel is stable.

The stability constraint of the scalings is written as:

$$\begin{aligned}&D(s)\in \mathrm{R}H_\infty \Leftrightarrow \max_i\left(\mathrm{Re}\left(\lambda_i\left(A_D\right)\right)\right)<0\\ &D(s)^{-1}\in \mathrm{R}H_\infty \Leftrightarrow \max_i\left(\mathrm{Re}\left(\lambda_i\left(A_D-B_DD_D^{-1}C_D\right)\right)\right)<0\end{aligned}\qquad [2.210]$$

where $\lambda_i(\ )$ is the ith eigenvalue.

Finally, assuming that $P(s)$ in Figure 2.27 has the following state-space representation:

$$\begin{bmatrix}\dot{x}\\ \hline z_\Delta\\ z\\ \hdashline y\end{bmatrix}=\left[\begin{array}{c|c:c}A & B_1 & B_2\\ \hline C_1 & D_{11} & D_{12}\\ \hdashline C_2 & D_{21} & D_{22}\end{array}\right]\begin{bmatrix}x\\ \hline e_\Delta\\ e\\ \hline u\end{bmatrix}\qquad [2.211]$$

Then, it is easy to show that the state-space representation of the augmented standard form of the scalings $D(s)P(s)D(s)^{-1}$ in Figure 2.27 is written as:

$$\begin{bmatrix} \dot{x} \\ \hline z_{D_\Delta} \\ z_D \\ \hline y \end{bmatrix} = \left[\begin{array}{ccc|c:c} A & B_1\underline{C}_D & 0 & B_1\underline{D}_D & B_2 \\ 0 & \underline{A}_D & 0 & \underline{B}_D & 0 \\ B_D C_1 & B_D D_{11}\underline{C}_D & A_D & B_D D_{11}\underline{D}_D & B_D D_{12} \\ \hline D_D C_1 & D_D D_{11}\underline{C}_D & C_D & D_D D_{11}\underline{D}_D & D_D D_{12} \\ \hdashline C_2 & D_{21}\underline{C}_D & 0 & D_{21}\underline{D}_D & D_{22} \end{array}\right] \begin{bmatrix} x \\ \hline e_{D_\Delta} \\ e_D \\ \hline u \end{bmatrix} \qquad [2.212]$$

Evaluation of the fitness function

We seek to solve the following optimization problem:

$$\min_{X \in [\underline{x};\overline{x}]} f(X) \qquad [2.213]$$

where the variable X is defined in [2.208] and the function $f(X)$ consists of carrying out the following step:

– building the realizations $\left[\begin{array}{c|c} A_m & B_m \\ \hline C_m & D_m \end{array}\right], m = 1,\ldots,n_\Delta$ from X;

– building the realizations $\left[\begin{array}{c|c} A_D & B_D \\ \hline C_D & D_D \end{array}\right]$ and $\left[\begin{array}{c|c} \underline{A}_D & \underline{B}_D \\ \hline \underline{C}_D & \underline{D}_D \end{array}\right]$ with [2.209];

– computing the following spectral abscissa:

$$\begin{aligned} \lambda_u &= \max_i \left(\mathrm{Re}\left(\lambda_i\left(A_D\right)\right)\right) \\ \lambda_d &= \max_i \left(\mathrm{Re}\left(\lambda_i\left(\underline{A}_D\right)\right)\right) \\ \overline{\lambda} &= \max\left(\lambda_u, \lambda_d\right) \end{aligned} \qquad [2.214]$$

– if $\overline{\lambda} \geq 0$ then $f(X) = \overline{\lambda}$,

– else:

- building the state-space representation of $D(s)P(s)D(s)^{-1}$ using [2.212];

- solving the standard H_∞ problem [2.202], then keeping the value of γ obtained;

- evaluating $f(X) = -\frac{1}{\gamma}$.

It is important to note that this fitness function follows the same principle as for structured H_∞ synthesis.

After optimization, the μ-optimal controller is obtained directly by solving the standard H_∞ problem [2.202] with the best scalings obtained.

The particular case of static scalings

Moreover, it is important to note that the first optimization step consists of stabilizing the scalings and their inverses. In the case of static scalings, this step is no longer necessary, and evaluating the fitness function $f(X)$ is then much simpler; it consists of carrying out the following steps:

– build each D_m, and each D_m^{-1} then construct the transfer matrices:

$$D = \begin{pmatrix} \operatorname{diag}(D_m, m = 1,\ldots,n_\Delta) & 0 \\ 0 & I_{n_e} \end{pmatrix},\ D^{-1} = \begin{pmatrix} \operatorname{diag}(D_m^{-1}, m = 1,\ldots,n_\Delta) & 0 \\ 0 & I_{n_e} \end{pmatrix}$$

– build the state-space representation $DP(s)D^{-1}$:

$$\begin{bmatrix} \dot{x} \\ \hline z_{D\Delta} \\ z_D \\ \hline y \end{bmatrix} = \left[\begin{array}{c|c:c} A & B_1 D^{-1} & B_2 \\ \hline DC_1 & DD_{11}D^{-1} & DD_{12} \\ \hdashline C_2 & D_{21}D^{-1} & D_{22} \end{array}\right] \begin{bmatrix} x \\ \hline e_{D\Delta} \\ e_D \\ \hline u \end{bmatrix} \qquad [2.215]$$

– solve the standard H_∞ problem [2.202] then retain the value of γ obtained;

– evaluate $f(X) = -\frac{1}{\gamma}$.

Example of implementation

Specification summary

We take again the example of the motorized load that we used for multiple-plant H_∞ synthesis by stochastic approach.

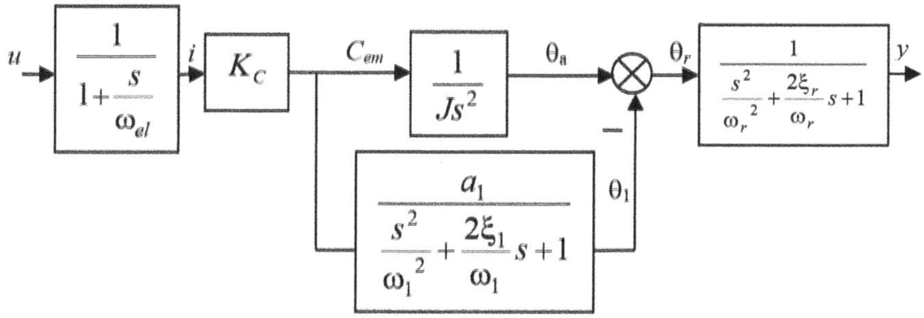

Figure 2.28. *Flexible-mode system*

We have:

$$H_0(s) = \frac{K_c}{\left(1+\frac{s}{\omega_{el}}\right)}\left(\frac{1}{Js^2} - \frac{a_1}{\frac{s^2}{\omega_1^2}+\frac{2\xi_1}{\omega_1}s+1}\right)\left(\frac{1}{\frac{s^2}{\omega_r^2}+\frac{2\xi_r}{\omega_r}s+1}\right) \quad [2.216]$$

The parameter values, subject to uncertainties, are summarized below.

$$\begin{aligned}
&J = 0.0014 \text{ kg.m}^2,\ K_c = 0.06 \text{ Nm.A}^{-1},\\
&\omega_r = 1\,382\, rad.s^{-1},\ \xi_r = 0.5\\
&a_1 = -7.10^{-4} \pm 7\%,\ \xi_1 = 0.029 \pm 70\%,\\
&\omega_1 = 1\,335 \text{ s}^{-1} \pm 10\%,\ \omega_{el} = 1\,257 \text{s}^{-1} \pm 20\%
\end{aligned} \quad [2.217]$$

These uncertainties enabled us (by considering the minimal, maximum and median values) to define a ball of 81 models and proceed to an efficient multiple-plant synthesis. We choose here to consider the system with its parametric uncertainties to proceed to a μ-synthesis.

Here is a summary of the following specification:

– bandwidth higher than 40 Hz;

– gain margin higher than 5 dB;

– phase margin higher than 40°;

– no steady-state error for a perturbation step at the model's input;

– stabilization of the system against its parametric uncertainties.

We have seen that this specification could be satisfied by *loop-shaping* ([2.219], Figure 2.29) with the following set of weights:

$$W_o(s) = \frac{128.5(s+81.24)}{s} \frac{\left(1+\dfrac{s}{5.03}\right)}{\left(1+\dfrac{s}{1257}\right)}$$

$$W_i(s) = \left(\frac{\dfrac{s^2}{942^2}+\dfrac{0.25s}{942}+1}{\dfrac{s^2}{942^2}+\dfrac{1.8s}{942}+1}\right)\left(\frac{\dfrac{s^2}{2200^2}+\dfrac{0.01s}{2200}+1}{\dfrac{s^2}{2200^2}+\dfrac{2.2s}{2200}+1}\right)\frac{1}{\left(\dfrac{s}{942}+1\right)^2} \quad [2.218]$$

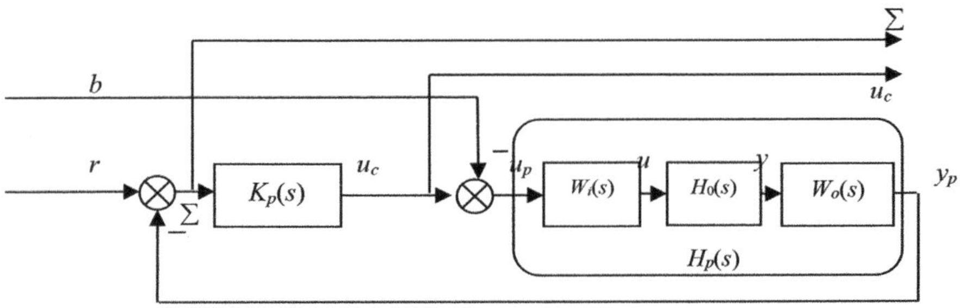

Figure 2.29. *Loop-shaping H∞ synthesis*

$$K_p(s) = \arg\min \gamma, \left\| \begin{matrix} S_p & S_p H_p \\ K_p S_p & K_p S_p H_p \end{matrix} \right\|_\infty < \gamma, \ S_p = \left(I + K_p H_p\right)^{-1} \quad [2.219]$$

The nominal synthesis leads to $\gamma = 3.30$ and the "worst case" value of γ can reach $\gamma_{wc} = 10.5$. We now decide to reduce the variation of γ in relation to the parametric uncertainties using a μ-synthesis and then posit: $\gamma_{obj} = 3.30$. The objective is therefore to synthesize a controller such that γ remains around γ_{obj} regardless of the uncertainties considered.

Formatting for μ-synthesis

LFT formatting of the uncertain system

The first step consists of placing the system [2.216] and its uncertainties [2.217] in the LFT form in Figure 2.30.

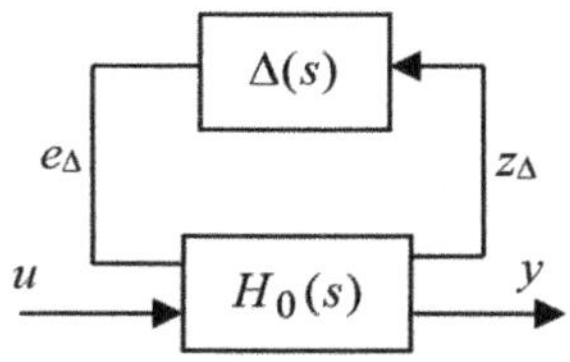

Figure 2.30. *LFT form*

This example is a particular case of a system with uncertain *n* flexible modes for which an LFT form is suggested in Appendix C.

By writing the uncertain parameters in the form:

– ω_{el} so that $\omega_{el} \in [\omega_{el\min}; \omega_{el\max}]$ and:

$$\omega_{el} = \omega_{el0} + \delta\omega_{el}\,\omega_{el1},\ |\delta\omega_{el}| \leq 1,\ \omega_{el1} = \frac{\omega_{el\max} - \omega_{el\min}}{2} \qquad [2.220]$$

– ξ_1 so that $\xi_1 \in [\xi_{1\min}; \xi_{1\max}]$ and:

$$\xi_1 = \xi_{10} + \delta\xi_1\,\xi_{i1},\ |\delta\xi_1| \leq 1,\ \xi_{11} = \frac{\xi_{1\max} - \xi_{1\min}}{2} \qquad [2.221]$$

– a_1 so that $a_1 \in [a_{1\min}; a_{1\max}]$ and:

$$a_1 = a_{10} + \delta a_1 a_{11},\ |\delta a_1| \leq 1,\ a_{11} = \frac{a_{1\max} - a_{1\min}}{2} \qquad [2.222]$$

– ω_1 so that $\omega_1 \in [\omega_{1\min}; \omega_{1\max}]$ and:

$$\omega_1 = \omega_{10} + \delta\omega_1\omega_{11},\ |\delta\omega_1| \leq 1,\ \omega_{11} = \frac{\omega_{1\max} - \omega_{1\min}}{2} \qquad [2.223]$$

An LFT form in the state-space of the plant [2.216]–[2.217] is (see Appendix C):

$$\begin{aligned} \frac{dX}{dt} &= AX + Bu + Me_\Delta \\ y &= CX + Du \\ z_\Delta &= NX \\ e_\Delta &= \Delta z_\Delta, \quad \|\Delta\|_\infty \leq 1 \end{aligned} \qquad [2.224]$$

With:

$$\begin{aligned}
&X=\begin{pmatrix} i & \theta_a & \dot{\theta}_a & \theta_1 & \dot{\theta}_1 & \theta_r & \dot{\theta}_r & \chi \end{pmatrix}^T \\
&e_\Delta=\begin{pmatrix} e_{\Delta el_i} & e_{\Delta el_u} & e_{\Delta\omega_{1\theta}} & e_{\Delta\omega_{1C}} & e_{\Delta\xi_1} & e_{\Delta a_1} \end{pmatrix}^T \\
&z_\Delta=\begin{pmatrix} z_{\Delta el_i} & z_{\Delta el_u} & z_{\Delta\omega_{1\theta}} & z_{\Delta\omega_{1C}} & z_{\Delta\xi_1} & z_{\Delta a_1} \end{pmatrix}^T
\end{aligned} \qquad [2.225]$$

And:

$$\left.\begin{aligned}
&A=\begin{pmatrix}
-\omega_{el0} & 0 & 0 & 0 & 0 & 0 & 0 & \omega_{el0} \\
0 & 0 & 1 & 0 & 0 & 0 & 0 & 0 \\
J^{-1}K_c & 0 & 0 & 0 & 0 & 0 & 0 & 0 \\
0 & 0 & 0 & 0 & 1 & 0 & 0 & 0 \\
\omega_{10}{}^2 a_{10} K_c & 0 & 0 & -\omega_{10}{}^2 & -2\xi_{10}\omega_{10} & 0 & 0 & 0 \\
0 & 0 & 0 & 0 & 0 & 0 & 1 & 0 \\
0 & \omega_r{}^2 & 0 & \omega_r{}^2 & 0 & -\omega_r{}^2 & -2\xi_r\omega_r & 0 \\
0 & 0 & 0 & 0 & 0 & 0 & 0 & -\tau_f{}^{-1}
\end{pmatrix} \\
&B=\begin{pmatrix} 0 \\ 0 \\ 0 \\ 0 \\ 0 \\ 0 \\ 0 \\ \tau_f{}^{-1} \end{pmatrix},\ M=\begin{pmatrix}
\sqrt{\omega_{el1}} & \sqrt{\omega_{el1}} & 0 & 0 & 0 & 0 \\
0 & 0 & 0 & 0 & 0 & 0 \\
0 & 0 & 0 & 0 & 0 & 0 \\
0 & 0 & \sqrt{2\omega_{11}} & \sqrt{a_{10}\omega_{10}2\omega_{11}} & \sqrt{2\xi_{11}\omega_{10}} & \sqrt{\omega_{10}a_{11}} \\
0 & 0 & 0 & 0 & 0 & 0 \\
0 & 0 & 0 & 0 & 0 & 0 \\
0 & 0 & 0 & 0 & 0 & 0 \\
0 & 0 & 0 & 0 & 0 & 0
\end{pmatrix} \\
&N=\begin{pmatrix}
-\sqrt{\omega_{el1}} & 0 & 0 & 0 & 0 & 0 & 0 & 0 \\
0 & 0 & 0 & 0 & 0 & 0 & 0 & \sqrt{\omega_{el1}} \\
0 & 0 & 0 & -\omega_{10}\sqrt{2\xi_{11}} & -\xi_{10}\sqrt{2\omega_{11}} & 0 & 0 & 0 \\
K_c\sqrt{a_{10}\omega_{10}2\omega_{11}} & 0 & 0 & 0 & 0 & 0 & 0 & 0 \\
0 & 0 & 0 & 0 & -\sqrt{2\xi_{11}\omega_{10}} & 0 & 0 & 0 \\
K_c\sqrt{\omega_{10}a_{11}} & 0 & 0 & 0 & 0 & 0 & 0 & 0
\end{pmatrix} \\
&C=\begin{pmatrix} 0 & 0 & 0 & 0 & 0 & 1 & 0 & 0 \end{pmatrix},\ D=0
\end{aligned}\right\} \qquad [2.226]$$

The uncertainty matrix is written as:

$$\Delta = diag\left(\delta\omega_{el}, \delta\omega_{el}, \delta\omega_1, \delta\omega_1, \delta\xi_1, \delta a_1\right), \ \|\Delta\|_\infty < 1 \qquad [2.227]$$

It is important to note that the nominal system is found with $\Delta = 0_{6\times 6}$:

$$H_0(s) = \left[\begin{array}{c|c} A & B \\ \hline C & D \end{array}\right] \qquad [2.228]$$

As in Appendix C, it should be noted that the state variable χ is introduced by establishing a fictive time constant (with very high frequency $\tau_f << 1$) at the model's input in order to make z_Δ strictly proper in the control signal (Figure 2.31).

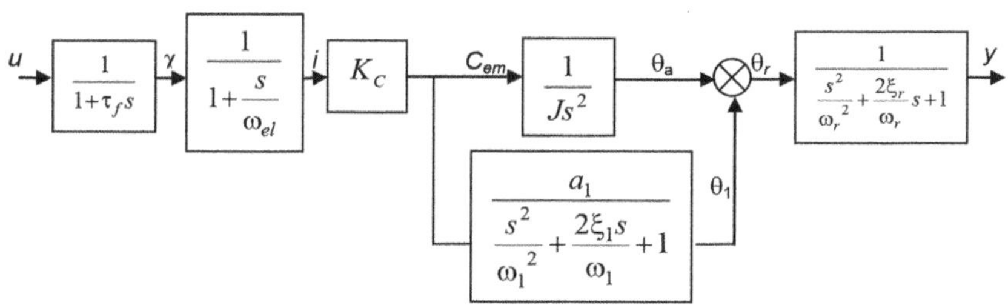

Figure 2.31. *Flexible-mode system with fictive input time constant*

Formatting for μ-analysis

The second step consists of establishing a scheme of μ-analysis for performance robustness analysis by adding a fictive complex uncertainty Δ_f between the performance channels in Figure 2.29, that is, between the vectors $(b, r)^T$ and $(\varepsilon, u_c)^T$, which enables an augmented uncertainty block to be constructed $\Delta' = \text{diag}\left(\Delta_f, \Delta\right)$:

$$\Delta_f = \begin{pmatrix} \delta_{\varepsilon r} & \delta_{ur} \\ \delta_{\varepsilon b} & \delta_{ub} \end{pmatrix} \qquad [2.229]$$

The fictive uncertainty Δ_f is normalized by the introduction of normalization coefficients $\left(\alpha_r, \alpha_e, \alpha_b, \alpha_u\right)$ as in Figure 2.32. Performance robustness is reached for

Δ such as $\|\Delta\|_\infty < 1$ if the system in Figure 2.29 remains stable for any Δ and Δ_f such as $\|\Delta\|_\infty < 1$ and $\|\Delta_f\|_\infty < 1$, that is, according to the small gain theorem if:

$$\left\| \begin{matrix} \alpha_\varepsilon S_p \alpha_r & \alpha_b S_p H_p \alpha_\varepsilon \\ \alpha_r K_p S_p \alpha_u & \alpha_b K_p S_p H_p \alpha_u \end{matrix} \right\|_\infty < 1 \qquad [2.230]$$

Indeed, the loop-shape is such that:

$$\left\| \begin{matrix} S_p & S_p H_p \\ K_p S_p & K_p S_p H_p \end{matrix} \right\|_\infty < \gamma_{obj} \qquad [2.231]$$

It is therefore enough to choose:

$$\alpha_\varepsilon = \alpha_u = \alpha_b = \alpha_r = \gamma_{obj}^{-1/2} \qquad [2.232]$$

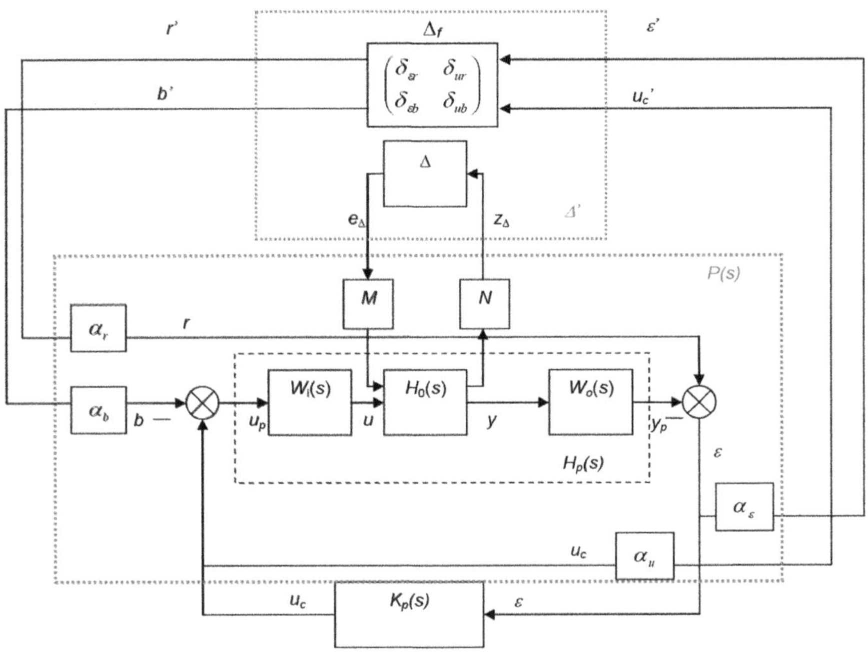

Figure 2.32. *Performance robustness analysis scheme. For a color version of this figure, see www.iste.co.uk/feyel/optimization.zip*

Formatting for μ-synthesis

The third step consists of the construction of the control scheme for μ-synthesis from Figure 2.33 by adding scalings $D(s)$ and $D(s)^{-1}$ (Figure 2.27).

The scalings will have the following structure:

$$D(s) = \begin{pmatrix} \operatorname{diag}\left(D_m(s), m = 1,\ldots,6\right) & 0 \\ 0 & I_{2\times 2} \end{pmatrix} \quad [2.233]$$

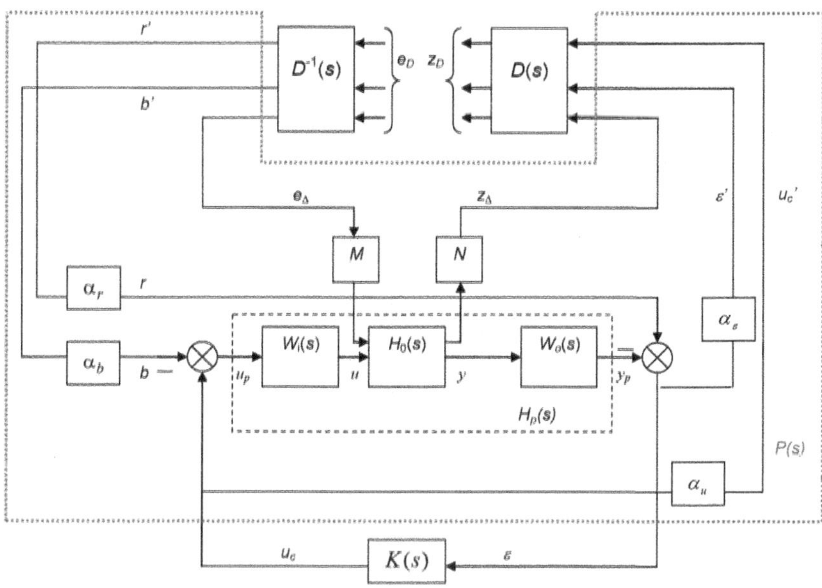

Figure 2.33. *Control scheme for μ-synthesis. For a color version of this figure, see www.iste.co.uk/feyel/optimization.zip*

It is easy to show from [2.224] and [2.158] that the loop-shape in LFT format is written as:

$$\begin{aligned} \frac{dX_p}{dt} &= A_p X_p + B_p u_p + M_p e_\Delta \\ y_p &= C_p X_p + D_p u \\ z_\Delta &= N_p X_p \\ e_\Delta &= \Delta z_\Delta, \quad \|\Delta\|_\infty \le 1 \end{aligned} \quad [2.234]$$

With:

$$A_p = \begin{bmatrix} A & BC_i & 0 \\ 0 & A_i & 0 \\ B_o C & B_o DC_i & A_o \end{bmatrix}, \; B_p = \begin{bmatrix} BD_i \\ B_i \\ B_o DD_i \end{bmatrix}, \; M_p = \begin{bmatrix} M \\ 0 \\ 0 \end{bmatrix}$$

$$C_p = \begin{bmatrix} D_o C & D_o DC_i & C_o \end{bmatrix}, \; D_p = D_o DD_i, \; N_p = \begin{bmatrix} N & 0 & 0 \end{bmatrix} \qquad [2.235]$$

According to Figure 2.33, we then have:

$$\begin{bmatrix} \dot{X} \\ \hline z_\Delta \\ \varepsilon \\ u_c \\ \hdashline \varepsilon \end{bmatrix} = \left[\begin{array}{c|ccc:c} A_p & M_p & B_p & 0 & -B_p \\ \hline N_p & 0 & 0 & 0 & 0 \\ -C_p & 0 & 0 & I & -D_p \\ 0 & 0 & 0 & 0 & I \\ \hdashline -C_p & 0 & 0 & I & -D_p \end{array}\right] \begin{bmatrix} X \\ \hline e_\Delta \\ b \\ r \\ \hline u_c \end{bmatrix} \qquad [2.236]$$

Moreover:

$$\begin{aligned} r &= \alpha_r r' \\ b &= \alpha_b b' \\ \varepsilon' &= \alpha_\varepsilon \varepsilon \\ u_c' &= \alpha_u u_c \end{aligned} \qquad [2.237]$$

We finally obtain:

$$\begin{bmatrix} \dot{X}_p \\ \hline z_\Delta \\ \varepsilon' \\ u_c' \\ \hdashline \varepsilon' \end{bmatrix} = \left[\begin{array}{c|ccc:c} A_p & M_p & \alpha_b B_p & 0 & -B_p \\ \hline N_p & 0 & 0 & 0 & 0 \\ -C_p \alpha_\varepsilon & 0 & 0 & I\alpha_r & -D_p \alpha_\varepsilon \\ 0 & 0 & 0 & 0 & \alpha_u \\ \hdashline -\alpha_\varepsilon C_p & 0 & 0 & I\alpha_\varepsilon & -D_p \alpha_\varepsilon \end{array}\right] \begin{bmatrix} X_p \\ \hline e_\Delta \\ b' \\ r' \\ \hline u_c \end{bmatrix} \qquad [2.238]$$

which is indeed as the form [2.211] with:

$$\begin{aligned} & e = (b', r')^T, \; z = (\varepsilon', u_c')^T \\ & y = \varepsilon', \; u = u_c \end{aligned} \qquad [2.239]$$

Adding scalings therefore leads to form [2.212].

Optimization results

The computations are made with Matlab 2011b on a machine equipped with Intel Core i7-3740QM (2.7 GHz) processor and 8 Gb of RAM.

The tests are successively carried out with and without parallel computation (*Single Core (SC)* and *Multiple Core (MC)*, respectively). In the latter case, [MAT 10] is used with eight cores (*pool*).

The stochastic approach to μ-synthesis is tested for different orders of scaling[9] $D(s)$: 0, 1 and 2. For each order of scaling, different metaheuristics are tested. As on [2.144], a statistical evaluation of the results is necessary. Thus, we take the minimal value of γ obtained (γ_{min}), the mean value of γ (γ_{mean}) and its standard deviation (γ_{std}). For information, we also take the average computing time per *run* ($t_{cpu/run}$) for the realization of the 10 *runs* forming each test.

The results are shown in Tables 2.4–2.6.

$\partial^\circ D = 0$, $IT/run = 200$	γ_{min}	γ_{mean}	γ_{std}	$t_{cpu/run}$ (min)	
				SC	*MC*
StaticCEC_QPSO	3.3121	3.3231	0.0130	37.4	7.59
TVDecCEC_QPSO	3.3120	3.3160	0.0050	40.50	**7.17**
Rnd_DE	3.3141	3.3448	0.055	38.6	8.54
Rnd2Bst_DE	3.3117	3.3123	0.0004	36.84	7.68
Rnd2Bst_PDE	3.3119	3.3124	**0.00037**	**33.91**	7.32
Std_PSO	3.3122	3.3256	0.027	38.52	6.96
TVRndIW_PSO	3.3124	3.3461	0.0296	72.24	7.37
ABC	3.3288	3.3588	0.0336	78.94	36.15
CS	3.3123	3.3369	0.0198	83.8	12.11
CS/DE	**3.3116**	**3.3121**	0.00051	84.52	39.60
FF	**3.3116**	3.3146	0.0029	13,697	16.11

Table 2.4. *Comparison of algorithms for full-order μ-synthesis – $\partial^\circ D = 0$*

9 Structurally, each sub-block $Dm(s)$ forming $D(s)$ is assumed to have the same order as $D(s)$.

$\partial^\circ D = 1$, $IT/run = 800$	γ_{min}	γ_{mean}	γ_{std}	$t_{cpu/run}$ (min)	
				SC	*MC*
StaticCEC_QPSO	**1.4082**	**2.6318**	0.7858	198.4	63.66
TVDecCEC_QPSO	3.0412	3.2207	**0.1217**	199.9	60.44
Rnd_DE	2.8353	20.944	29.788	155.0	**56.43**
Rnd2Bst_DE	2.4055	39.342	76.552	193.4	63.05
Rnd2Bst_PDE	**1.7035**	2.6383	0.6064	159.5	72.17
Std_PSO	1.8494	27.622	77.226	211.7	71.19
TVRndIW_PSO	2.8829	10.824	21.726	188.1	61.44
ABC	**1.4273**	2.7800	0.9077	568.98	238.49
CS	3.2460	3.5107	0.3457	313.3	79.88
CS/DE	2.7122	2.9471	0.177	331.32	198.21
FF	2.8968	3.2631	0.1312	780.15	70.55

Table 2.5. *Comparison of algorithms for full-order μ-synthesis – $\partial^\circ D = 1$*

$\partial^\circ D = 2$, $IT/run = 1200$	γ_{min}	γ_{mean}	γ_{std}	$t_{cpu/run}$	
				SC	*MC*
StaticCEC_QPSO	1.35	2.82	0.71	952.54	270.33
Rnd2Bst_PDE	**1.18**	2.39	0.72	805.34	200.18

Table 2.6. *Comparison of algorithms for full-order μ-synthesis – $\partial^\circ D = 2$*

In the case of static scalings, we can say that all the algorithms have comparable performances. We can also differentiate between the computing time measurement: the longest algorithms to optimize are *FF*, *CS*, *CS/DE* and *ABC*. The other algorithms have comparable computing times.

For the problem of full-order μ-synthesis with frequency scaling optimization, *StaticCEC_QPSO* provides the best results. *ABC* gives interesting results but at the cost of a prohibitive computing time. In terms of performances, it is followed very closely by *Rnd2Bst_PDE*.

The advantage of using parallel computation for optimization can clearly be noted.

It should be noted that the problem posed here is not a structured control problem.

As the value of γ reached in μ-synthesis is 1.35, the worst-case value of γ in H_∞ synthesis (here with loop-shaping) can reach $\gamma_{wc} = \gamma_{obj}.1.35 = 4.45$ when all the components of the uncertainty matrix vary from $1/1.35 = 74\%$.

2.3.2.3. *Structured μ-synthesis*

As we have seen in the previous sections, μ-synthesis creates high-order controllers. A post-reduction generally follows but then poses the problem of robustness of the controller obtained. Therefore, just as for H_∞ synthesis we can seek to constrain the order of the controller synthesized *a priori*. Here, we reach a field of research that is still open.

2.3.2.3.1. μ-Synthesis with non-smooth optimization

Hinfstruct enables a structured μ-synthesis to be carried out [APK 11]. We first note that the stability condition of the scalings and their inverse is only necessary for solving the problem [2.202] using "classic" computation methods (based on Riccati for example). Fundamentally, it is merely enough to satisfy [2.195]. Thus, we can seek to solve the following problem; find an n_k order stabilizing controller $K(s)$ and a transfer matrix $D(s)$ such that:

– $D(s)\Delta'(s) = \Delta'(s)D(s),\ \forall \Delta'(s) \in \underline{\Delta}'$

– $\left\| D(s)F_l(P,K)D(s)^{-1} \right\|_\infty < 1$

By positing:

$$\begin{aligned} \tilde{z}_D &= \begin{pmatrix} z_{\Delta D} & z_D \end{pmatrix}^T \\ \tilde{e}_D &= \begin{pmatrix} e_{\Delta D} & e_D \end{pmatrix}^T \end{aligned} \qquad [2.240]$$

Therefore, we note the equivalence relations in Figure 2.34. The following problems are then equivalent:

$$\begin{aligned}
&\min_{D(s),K(s)} \left\| D(s)F_l\left(P(s),K(s)\right)D(s)^{-1} \right\|_\infty \\
&\quad \text{Subject to:} \\
&\quad\quad F_l(P,K) \in \mathrm{R}H_\infty \\
&\quad\quad \Leftrightarrow \\
&\min_{K_c(s)} \left\| F_l\left(P_c(s),K_c(s)\right) \right\|_\infty \\
&\quad \text{Subject to:} \\
&\quad\quad F_l(P_c,K_c) \in \mathrm{R}H_\infty
\end{aligned} \qquad [2.241]$$

With (always assuming that $n_z = n_e$):

$$\begin{aligned}
D(s) &= \begin{pmatrix} \mathrm{diag}\left(D_m(s), m = 1,\ldots,n_\Delta\right) & 0 \\ 0 & I_{n_z} \end{pmatrix} \\
\underline{D}(s) &= D(s) - I \\
K_c(s) &= \begin{pmatrix} K(s) & & \\ & \underline{D}(s) & \\ & & \underline{D}(s) \end{pmatrix} \\
P_c(s) &:= L_c P(s) R_c + Q_c \\
&:= \left[\begin{array}{cc} I_{n_\Delta+n_z} & 0 \\ \hline 0 & I_{n_y} \\ 0 & 0 \\ I_{n_\Delta+n_z} & 0 \end{array}\right] P(s) \left[\begin{array}{c|ccc} I_{n_\Delta+n_z} & 0 & -I_{n_\Delta+n_z} & 0 \\ 0 & I_{n_u} & 0 & 0 \end{array}\right] + \left[\begin{array}{c|ccc} 0 & 0 & 0 & I \\ \hline 0 & 0 & 0 & 0 \\ I & 0 & -I & 0 \\ 0 & 0 & 0 & 0 \end{array}\right]
\end{aligned} \qquad [2.242]$$

By this means, we can then compute the controller and the scalings simultaneously by this particular decentralized structuration.

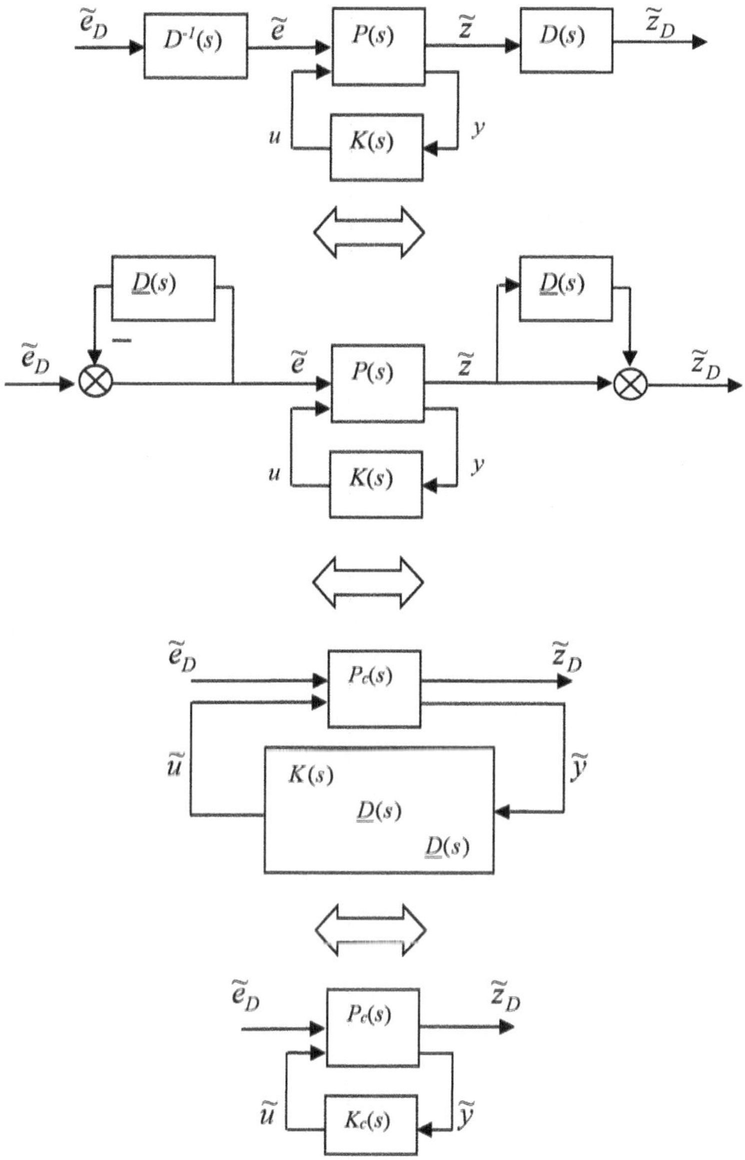

Figure 2.34. *μ-synthesis with non-smooth optimization*

2.3.2.3.2. Reformulation for stochastic optimization

By observing the reformulations of the structured H_∞ synthesis with stochastic optimization and the μ-synthesis (full-order) with dynamic scalings, the reformulation of the structured μ-synthesis with stochastic optimization becomes almost immediate as it involves unifying these two problems. This is the approach that we suggest in [FEY 14c] as an alternative to the previous μ-synthesis with non-smooth optimization, the benefit being the approach's flexibility and the possibility of taking into account numerous exotic additional constraints.

Principle of the approach

We seek to solve the following; find an n_k order stabilizing controller $K(s)$ and a transfer matrix $D(s)$ such that:

- $D(s)\Delta'(s) = \Delta'(s)D(s), \ \forall \Delta'(s) \in \underline{\Delta}'$
- $\left\| D(s)F_l(P,K)D(s)^{-1} \right\|_\infty < 1$

This problem is no more and no less than the union of problems [2.116] and [2.205], with the difference that, as seen for the non-smooth approach, the stability of the scalings is not necessary.

Assuming $D(s)$ is appropriately structured; we therefore suggest here solving the following global optimization problem directly:

$$\min_{X \in [\underline{x};\overline{x}]} f(X)$$

$$\Leftrightarrow$$

$$\min_{D(s),\, K(s)} \left\| D(s)F_l(P,K)D(s)^{-1} \right\|_\infty$$

subject to

$$F_l(P,K) \in \mathbf{R}H_\infty \qquad [2.243]$$

Positing as the unknown:

$$\Theta_D = \begin{pmatrix} A_D & B_D \\ C_D & D_D \end{pmatrix}, \ \Theta_k = \begin{pmatrix} A_k & B_k \\ C_k & D_k \end{pmatrix} \qquad [2.244]$$

and:

$$\overline{\lambda}_{cl} = \max_i \left(\mathrm{Re}\left(\lambda_i\left(A_{cl}\right)\right)\right) \qquad [2.245]$$

where $\lambda_i(\,)$ is the ith eigenvalue.

Thus, we find the μ-optimal scalings and structured controller by considering the following optimization problem:

$$\min_{\Theta_D,\Theta_k} f(\Theta_D,\Theta_k)$$

If $\overline{\lambda}_{cl} \geq 0$

Else $f(\Theta_D,\Theta_k) = \overline{\lambda}_{cl}$

$$f(\Theta_D,\Theta_k) = -\frac{1}{\gamma},\ \gamma = \left\|D(s)F_l(P,K)D(s)^{-1}\right\|_\infty \qquad [2.246]$$

It is important to note that here, $\| \; \|_\infty$ designates norm L_∞ [ZHO 96, p. 114] evaluable for an unstable system, unlike norm H_∞, which requires the system to be stable.

Implementation

The $D(s)$ scaling has exactly the same structure as for full order.

The controller $K(s)$ has the same structure as for structured H_∞ synthesis: it can be centralized or decentralized.

As for the stochastic approach to H_∞ synthesis, the rate of convergence of the algorithm is greatly accelerated by searching for the scalings and the controller in tridiagonal form and by transforming the search space.

The decision vector will take the form:

$$X = \begin{pmatrix} X_D \\ X_k \end{pmatrix} \qquad [2.247]$$

where X_D is defined in [2.208] and X_k is defined in [2.130] for a centralized controller and in [2.133] for a decentralized controller.

Using the fact that:

$$H_1 = \left[\begin{array}{c|c} A_1 & B_1 \\ \hline C_1 & D_1 \end{array}\right], \quad H_2 = \left[\begin{array}{c|c} A_2 & B_2 \\ \hline C_2 & D_2 \end{array}\right]$$

$$H_1 H_2 = \left[\begin{array}{cc|c} A_1 & B_1C_2 & B_1D_2 \\ 0 & A_2 & B_2 \\ \hline C_1 & D_1C_2 & D_1D_2 \end{array}\right] = \left[\begin{array}{cc|c} A_2 & 0 & B_2 \\ B_1C_2 & A_1 & B_1D_2 \\ \hline D_1C_2 & C_1 & D_1D_2 \end{array}\right] \qquad [2.248]$$

and positing:

$$F_l(P(s), K(s)) = \left[\begin{array}{c|c} A_{cl} & B_{cl} \\ \hline C_{cl} & D_{cl} \end{array}\right]$$

$$D(s) = \left[\begin{array}{c|c} A_D & B_D \\ \hline C_D & D_D \end{array}\right] \qquad [2.249]$$

$$D(s)^{-1} = \left[\begin{array}{c|c} A_D - B_D D_D^{-1} C_D & -B_D D_D^{-1} \\ \hline D_D^{-1} C_D & D_D^{-1} \end{array}\right] = \left[\begin{array}{c|c} \underline{A}_D & \underline{B}_D \\ \hline \underline{C}_D & \underline{D}_D \end{array}\right]$$

it is easy to show that the state-space representation of the standard augmented form with scalings $D(s)P(s)D(s)^{-1}$ in Figure 2.27 is written as:

$$\left[\begin{array}{ccc|c} A_D & B_D C_{cl} & B_D D_{cl} C_{\underline{D}} & B_D D_{cl} D_{\underline{D}} \\ 0 & A_{cl} & B_{cl} C_{\underline{D}} & B_{cl} D_{\underline{D}} \\ 0 & 0 & A_{\underline{D}} & B_{\underline{D}} \\ \hline C_D & D_D C_{cl} & D_D D_{cl} C_{\underline{D}} & D_D D_{cl} D_{\underline{D}} \end{array}\right] \qquad [2.250]$$

Evaluation of the fitness function

We seek to solve the following optimization problem:

$$\min_{X \in [\underline{x}; \overline{x}]} f(X) \qquad [2.251]$$

where the variable X is defined in [2.208] and the function $f(X)$ consists of carrying out the following steps:

– build the tridiagonal form (centralized [2.136]–[2.139] or not) $\left[\begin{array}{c|c} A_k & B_k \\ \hline C_k & D_k \end{array}\right]$ from X;

– build the matrix A_{cl} of the closed loop using [2.141];

– evaluate the spectral abscissa of the closed loop:

$$\bar{\lambda} = \max_i \left(\text{Re}\left(A_{cl} \right) \right) \quad [2.252]$$

– If $\bar{\lambda} \geq 0$, then $f(X) = \bar{\lambda}$,

– Else:

- build the closed loop $F_l(P, K)$ using [2.143];

- build realizations $\left[\begin{array}{c|c} A_m & B_m \\ \hline C_m & D_m \end{array}\right], m = 1, \ldots, n_\Delta$ from X and [2.206] to [2.209];

- build realizations $\left[\begin{array}{c|c} A_D & B_D \\ \hline C_D & D_D \end{array}\right]$ and $\left[\begin{array}{c|c} \underline{A}_D & \underline{B}_D \\ \hline \underline{C}_D & \underline{D}_D \end{array}\right]$ with [2.209];

- build the realization of $D(s)F_l(P(s), K(s))D(s)^{-1}$ using [2.250];

- evaluate $f(X) = -\dfrac{1}{\gamma}$, $\gamma = \left\| D(s)F_l(P, K)D(s)^{-1} \right\|_\infty$.

Taking account of additional constraints

As for the structured H_∞ synthesis formulated above, one benefit of this approach is its flexibility and the fact that it takes into account complex constraints.

Constraint on the stability of the controller

We can seek to solve the following problem; find an n_k order stabilizing controller $K(s)$ and a transfer matrix $D(s)$ such that:

– $D(s)\Delta'(s) = \Delta'(s)D(s), \; \forall \Delta'(s) \in \underline{\Delta}'$

– $\left\| D(s)F_l(P, K)D(s)^{-1} \right\|_\infty < 1$,

– $K(s)$ is stable.

By positing:

$$\bar{\lambda}_k = \max_i \left(\text{Re}\left(\lambda_i \left(A_k \right) \right) \right) \quad [2.253]$$

where $\lambda_i(\,)$ is the ith eigenvalue.

Then, we synthesize the μ-optimal controller by considering the following optimization problem:

$$\min_{\Theta_D,\Theta_k} f(\Theta_D,\Theta_k)$$

if $\max\left(\overline{\lambda}_{cl},\overline{\lambda}_k\right)\geq 0$

$$f(\Theta_D,\Theta_k)=\max\left(\overline{\lambda}_{cl},\overline{\lambda}_k\right)$$

Else

$$f(\Theta_D,\Theta_k)=-\frac{1}{\gamma},\ \gamma=\left\|D(s)F_l(P,K)D(s)^{-1}\right\|_\infty \qquad [2.254]$$

Closed-loop pole placement constraint

We can seek to solve the following problem; find an n_k order stabilizing controller $K(s)$ and a transfer matrix $D(s)$ such that:

- $D(s)\Delta'(s)=\Delta'(s)D(s),\ \forall\Delta'(s)\in\underline{\Delta}'$;
- $\left\|D(s)F_l(P,K)D(s)^{-1}\right\|_\infty<1$

The poles of the closed loop $F_l(P,K)$ all have a real part less than α (negative).

The μ-optimal controller is therefore synthesized by considering the following optimization problem:

$$\min_{\Theta_D,\Theta_k} f(\Theta_D,\Theta_k)$$

If $\overline{\lambda}_{cl}\geq\alpha$

$$f(\Theta_D,\Theta_k)=\overline{\lambda}_{cl}-\alpha$$

Else

$$f(\Theta_D,\Theta_k)=-\frac{1}{\gamma},\ \gamma=\left\|D(s)F_l(P,K)D(s)^{-1}\right\|_\infty \qquad [2.255]$$

Constraint on the norm of a particular transfer

We can seek to solve the following problem; find an n_k order stabilizing controller $K(s)$ and a transfer matrix $D(s)$ such that:

- $D(s)\Delta'(s) = \Delta'(s)D(s),\ \forall\Delta'(s) \in \underline{\Delta}'$;
- $\left\|D(s)F_l(P,K)D(s)^{-1}\right\|_\infty < 1$;
- $\|M(s)\|_{2 \text{ ou } \infty} < \gamma_0$;

where $M(s)$ is any looped transfer.

Therefore, the μ-optimal controller is synthesized by considering the following optimization problem:

$$\min_{\Theta_D,\Theta_k} f(\Theta_D,\Theta_k)$$

If $\overline{\lambda}_{cl} \geq 0$

$$f(\Theta_D,\Theta_k) = \overline{\lambda}_{cl}$$

Else

$$f(\Theta_D,\Theta_k) = -\frac{1}{\gamma + e^{\alpha\left(\|M(s)\|_{2 \text{ ou } \infty} - \gamma_0\right)}},$$
$$\gamma = \left\|D(s)F_l(P,K)D(s)^{-1}\right\|_\infty,\ \alpha >> 1 \quad [2.256]$$

Example

We return to the same example used for full-order μ-synthesis. It is important to recall that during multiple-plant H_∞ synthesis, we sought an order 5 controller, K_p 5. In order to compare the performances of these two structured techniques, we proceed here, too, with 0, 1 and 2 order scalings.

The computations are made with Matlab 2011b on a 2.3 GHz COREi7 with 8 Gb Ram.

The tests are carried out successively with and without parallel computation (*Single Core (SC)* and *Multiple Core (MC)*, respectively). In the latter case, [MAT 10] is used with eight cores (*pool*).

A statistical evaluation of the results is necessary. So as in [2.144], we take the minimal value of γ obtained (γ_{min}), the mean value of γ (γ_{mean}) and its standard deviation (γ_{std}). For information, we also take the average computing time per *run* ($t_{cpu/run}$) used for the realization of the 10 *runs* forming each test. The results are shown in Tables 2.7–2.9.

It is clear from Tables 2.7 and 2.8 that as for structured H_∞ synthesis, *Rnd2Bst_PDE* is most effective (of the algorithms tested) for the problem of structured μ-synthesis.

As shown in Table 2.7, the static scalings do not have enough order to have a suitable value of γ (it is clear however that *ABC* does not stabilize over 10 successive runs) and the performances obtained with order 1 are much better in Table 2.8. In Table 2.9, the order of the scaling is increased to 2: it is clear that it is therefore more effective for the value of γ obtained but at the cost of a resulting computing time. But note again (with few exceptions) the advantage of using parallel computation for optimization.

$\partial^\circ D = 0$, *IT/run* = 5000	γ_{min}	γ_{mean}	γ_{std}	$t_{cpu/run}$ (min)	
				SC	*MC*
StaticCEC_QPSO	3.67	3.92	**0.15**	11.79	5.76
TVDecCEC_QPSO	3.89	4.12	0.26	5.52	8.72
Rnd_DE	3.64	3.96	0.29	8.95	5.28
Rnd2Bst_DE	3.55	**3.88**	0.22	9.93	6.34
Rnd2Bst_PDE	**3.54**	3.93	0.20	11.51	8.10
Std_PSO	4.71	790,600	2,482,600	13.35	6.48
TVRndIW_PSO	4.05	563,540	1,782,100	11.23	6.34
ABC					
CS	3.99	28,138	87,043	3.55	3.20
CS/DE	3.57	3.92	0.17	19.10	13.36
FF	4.15	7.67	5.11	2.24	0.41

Table 2.7. *Comparison of algorithms for structured μ-synthesis – $\partial^\circ D = 0$, $\partial^\circ K_p = 5$*

$\partial^\circ D = 1$, *IT/run* = 20000	γ_{min}	γ_{mean}	γ_{std}	$t_{cpu/run}$ (min)	
				SC	*MC*
StaticCEC_QPSO	3.41	3.74	0.28	94.31	46.01
TVDecCEC_QPSO	3.41	3.65	0.17	46.23	39.95
Rnd_DE	3.22	3.38	0.16	72.14	49.67
Rnd2Bst_DE	2.94	3.14	0.15	95.34	56.36
Rnd2Bst_PDE	**2.93**	**3.13**	**0.13**	112.9	39.39
Std_PSO	3.15	4.06	1.48	96.45	55.02
TVRndIW_PSO	3.20	7.38	12.31	85.22	48.96
ABC	3.75	4.35	0.41	76.49	64.55
CS	4.15	4.51	0.21	20.49	8.57
CS/DE	3.16	3.31	0.15	163.02	179.25
FF	4.43	4.60	0.120	5.15	0.89

Table 2.8. *Comparison of algorithms for structured μ-synthesis – $\partial^\circ D = 1$, $\partial^\circ K_p = 5$*

$\partial^\circ D = 1$, *IT/run* = 40000	γ_{min}	γ_{mean}	γ_{std}	$t_{cpu/run}$ (min)	
				SC	*MC*
Rnd2Bst_PDE	**2.84**	**3.06**	**0.11**	320.23	105.57

Table 2.9. *Comparison of algorithms for structured μ-synthesis – $\partial^\circ D = 2$, $\partial^\circ K_p = 5$*

As a comparison, we proceed to the same tests with non-smooth optimization (*HinfStruct*). As for the previous tests, we proceed to 10 successive *runs* and compare the results obtained with *Rnd2Bst_PDE*. The scalings and the controller are also sought in tridiagonal form. The settings of *HinfStruct* are considered by default.

	γ_{min}	γ_{mean}	γ_{std}	$t_{cpu/run}$ (min)
$\partial^\circ D = 0$	3.28	3.67	0.14	0.14
$\partial^\circ D = 1$	3.54	4.10	0.35	0.17
$\partial^\circ D = 2$	3.63	4.21	0.32	0.24

Table 2.10. *Non-smooth approach for fixed-order μ-synthesis $-\partial^\circ K_p = 5$*

If we compare the two approaches (evolutionary and non-smooth), the evolutionary approach provides the best results but using a much longer computing time. The evolutionary approach can form an interesting alternative to the non-smooth approach with its flexibility and ease of implementation; in fact, nothing prevents the implementation of exotic constraints (controller stability, etc.) as we did for structured H_∞ synthesis. Finally, the evolutionary approach can provide a point of initialization for the non-smooth approach.

2.4. LPV/LFT synthesis

2.4.1. *Introduction*

The LPV formalism provides a powerful framework enabling the use of linear robust control tools in a broad class of nonlinear systems, provided the latter can be expressed in the form of an LPV system, that is, systems in which certain dynamics vary over time. LPV systems are linear models in which the state-space representation depends on a varying vector of parameters $\theta(t)$. If $\theta(t)$ is measurable (or estimable), then a controller with scheduled gain can be synthesized in such a way as to guarantee the stability and performance of the closed loop for any parametric trajectory evolving in the bounded domain. We will find theoretical developments when the LPV is in polytopic form in [APK 93] or when it is in LFT form in [APK 95].

Although these LPV approaches have shown great effectiveness in many applications [WIJ 06, DEV 10], one of their major drawbacks concerns the substantial order of the computed controllers, which can lead to very complex implementations.

Some work has already been undertaken to compute gain-scheduled controllers structured as in [ADE 12] or [EME 06] where convex optimization tools are used in a polytopic context. This time, we suggest extending the LFT approach to a structured framework with stochastic optimization.

2.4.2. *The LPV/LFT controller synthesis problem*

Consider an LPV system in its LFT form in Figure 2.35.

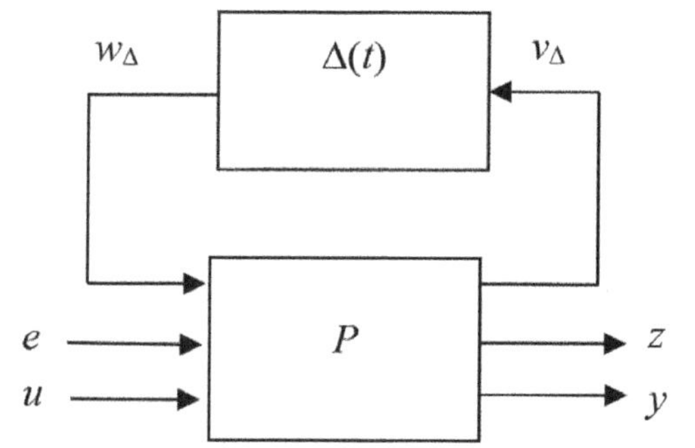

Figure 2.35. *Synthesis model of LPV/LFT form*

With:

$$A \in \mathbf{R}^{n \times n},\ z \in \mathbf{R}^{n_z},\ y \in \mathbf{R}^{n_y},\ e \in \mathbf{R}^{n_e},\ u \in \mathbf{R}^{n_u}$$

It is the same formalism as that used in the techniques of μ-synthesis and/or μ-analysis. The system P is LTI and

$$\begin{pmatrix} z \\ y \end{pmatrix} = F_u\left(P(s), \Delta(t)\right)\begin{pmatrix} e \\ u \end{pmatrix} \qquad [2.257]$$

The uncertainty matrix (which contains parameters that vary in time) is assumed to be block-diagonal:

$$\Delta(t) = blockdiag\left(\theta_1(t) I_{r_1}, \ldots, \theta_{n_\Delta}(t) I_{r_{n_\Delta}}\right) \qquad [2.258]$$

where $r_i > 1$ when the ith parameter is moreover repeated:

$$r = \sum_{i=1}^{n_\Delta} r_i \qquad [2.259]$$

In the following, the set of matrices that possess the structure of $\Delta(t)$ is described as follows:

$$\underline{\Delta} := \left\{ blockdiag\left(\theta_1(t)I_{r_1}, \ldots, \theta_{n_\Delta}(t)I_{r_{n_\Delta}}\right) : \theta_i(t) \in \mathbf{R} \right\} \qquad [2.260]$$

In the framework of LPV systems, the notion of norm H_∞ (which only applies to LTI systems) is generalized by the notion of gain L_2. We note L_2 ($\mathbf{R}^n$) the set of signals $s(t)$ of $\mathbf{R}^+$ in $\mathbf{R}^n$ whose norm L_2 is bounded:

$$\|s\|_2 := \sqrt{\int_0^\infty s(t)^T s(t) dt} < +\infty \qquad [2.261]$$

The gain L_2 of a system with varying parameters $M(\theta(t))$ with input e and output w is defined by:

$$\gamma = \|M\|_2 = \sup_{w(t) \in L_2^*} \frac{\|w\|_2}{\|e\|_2} \qquad [2.262]$$

To ensure the stability and performances of the looped LPV system, we will search for a regulator with the same parametric dependence as the system to be controlled. Thus, the controller will adapt itself to the evolution of the varying parameters (assumed to be measured). The controller is therefore itself an LPV system and the control law has the form:

$$u = F_l\left(K(s), \Delta(t)\right) y \qquad [2.263]$$

where K is LTI.

Considering Figure 2.36, the closed loop is written as:

$$T(P, K, \Delta) = F_l\left(F_u\left(P(s), \Delta(t)\right), F_l\left(K(s), \Delta(t)\right)\right) \qquad [2.264]$$

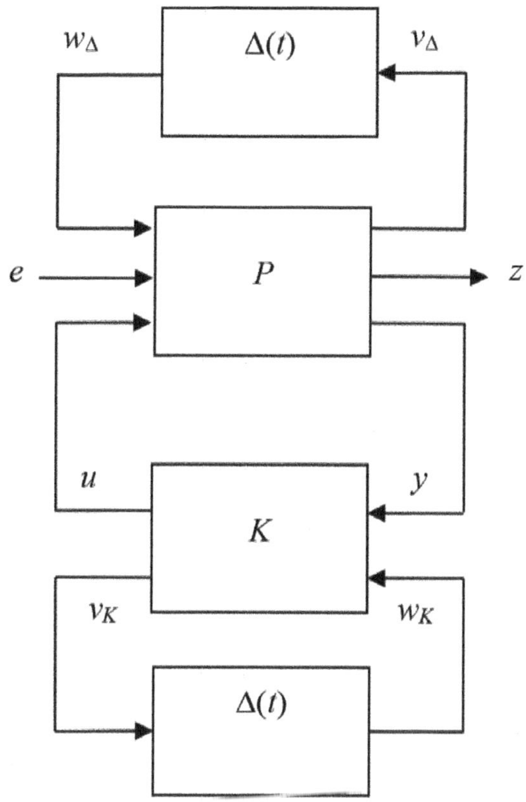

Figure 2.36. *Closed-loop LPV*

The problem is to compute an LTI controller K that:

– internally stabilizes the closed loop $T(P, K, \Delta)$ for any possible evolution of $\Delta(t)$ such that $\gamma^2\Delta^T(t)\Delta(t) \le 1$;

– which confers on the closed loop $T(P, K, \Delta)$ a gain L_2 between z and e lower than a value of γ, that is:

$$\left\|T(P,K,\Delta)\right\|_2 \le \gamma \qquad [2.265]$$

Following the same idea as in [APK 95], the schemes in Figures 2.36 and 2.37 are strictly equivalent.

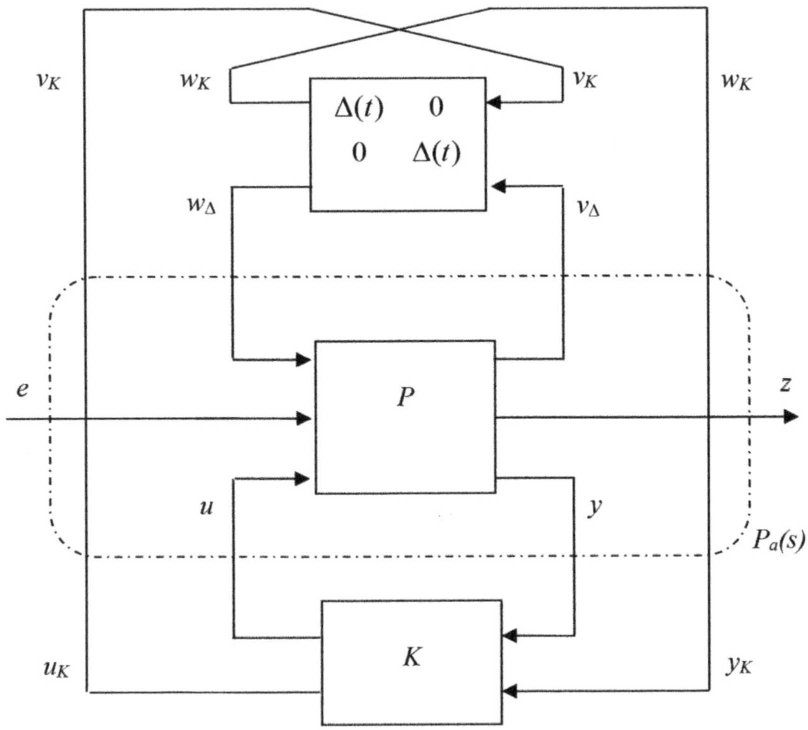

Figure 2.37. *Equivalent closed-loop LPV*

By introducing the new output to be monitored $v_K \in \mathbf{R}^r$, the new exogenous input $w_k \in \mathbf{R}^r$, the new measurement $y_K \in \mathbf{R}^r$ and the new control signal $u_K \in \mathbf{R}^r$, it becomes[10]:

$$\begin{pmatrix} v_K \\ \hline v_\Delta \\ z \\ y \\ \hline y_K \end{pmatrix} = \begin{pmatrix} 0 & 0 & I_r \\ 0 & P(s) & 0 \\ I_r & 0 & 0 \end{pmatrix} \begin{pmatrix} w_K \\ \hline w_\Delta \\ e \\ u \\ \hline u_K \end{pmatrix} = P_a(s) \begin{pmatrix} w_K \\ \hline w_\Delta \\ e \\ u \\ \hline u_K \end{pmatrix} \qquad [2.266]$$

10 It is easy to show that $T(P,K,\Delta) = F_u\left(F_l\left(P_a(s), K(s)\right), \begin{pmatrix} \Delta & 0 \\ 0 & \Delta \end{pmatrix} \right)$

The problem of LFT/LPV synthesis can therefore be seen as a classic performance robustness problem applied to the nominal model $P_a(s)$ and to the uncertainty block $\begin{pmatrix} \Delta & 0 \\ 0 & \Delta \end{pmatrix}$.

We search for the state-space representation of $P_a(s)$. We posit:

$$P(s) := \left(\begin{array}{c|cc:c} A & B_\Delta & B_e & B_u \\ \hline C_\Delta & D_{\Delta\Delta} & D_{\Delta e} & D_{\Delta u} \\ C_z & D_{z\Delta} & D_{ze} & D_{zu} \\ \hdashline C_y & D_{y\Delta} & D_{ye} & D_{yu} \end{array}\right) = \left(\begin{array}{c|c:c} A & B_1 & B_2 \\ \hline C_1 & D_{11} & D_{12} \\ \hdashline C_2 & D_{21} & D_{22} \end{array}\right) \qquad [2.267]$$

So that:

$$\left.\begin{aligned} \dot{x} &= Ax + B_\Delta w_\Delta + B_e e + B_u u \\ v_\Delta &= C_\Delta x + D_{\Delta\Delta} w_\Delta + D_{\Delta e} e + D_{\Delta u} u \\ z &= C_z x + D_{z\Delta} w_\Delta + D_{ze} e + D_{zu} u \\ y &= C_y x + D_{y\Delta} w_\Delta + D_{ye} e + D_{yu} u \end{aligned}\right\} \qquad [2.268]$$

Introducing the new output to monitor v_K, w_k, y_K and u_K, it becomes:

$$\left.\begin{aligned} \dot{x} &= Ax + B_\Delta w_\Delta + 0_{n\times r} w_K + B_e e + B_u u + 0_{n\times r} u_K \\ v_\Delta &= C_\Delta x + D_{\Delta\Delta} w_\Delta + 0_{r\times r} w_K + D_{\Delta e} e + D_{\Delta u} u + 0_{r\times r} u_K \\ v_K &= 0_{r\times n} x + 0_{r\times r} w_\Delta + 0_{r\times r} w_K + 0_{r\times n_e} e + 0_{r\times n_u} u + I_{r\times r} u_K \\ z &= C_z x + D_{z\Delta} w_\Delta + 0_{n_z\times r} w_K + D_{ze} e + D_{zu} u + 0_{n_z\times r} u_K \\ y &= C_y x + D_{y\Delta} w_\Delta + 0_{n_y\times r} w_K + D_{ye} e + D_{yu} u + 0_{n_y\times r} u_K \\ y_K &= 0_{r\times n} x + 0_{r\times r} w_\Delta + I_{r\times r} w_K + 0_{r\times n_e} e + 0_{r\times n_u} u + 0_{r\times r} u_K \end{aligned}\right\} \qquad [2.269]$$

which, matricially, is expressed as:

$$\begin{pmatrix} \dot{x} \\ \hline v_\Delta \\ v_K \\ z \\ \hdashline y \\ y_K \end{pmatrix} = \left(\begin{array}{c|cccc:cc} A & B_\Delta & 0_{n\times r} & B_e & B_u & 0_{n\times r} \\ \hline C_\Delta & D_{\Delta\Delta} & 0_{r\times r} & D_{\Delta e} & D_{\Delta u} & 0_{r\times r} \\ 0_{r\times n} & 0_{r\times r} & 0_{r\times r} & 0_{r\times n_e} & 0_{r\times n_u} & I_{r\times r} \\ C_z & D_{z\Delta} & 0_{n_z\times r} & D_{ze} & D_{zu} & 0_{n_z\times r} \\ \hdashline C_y & D_{y\Delta} & 0_{n_y\times r} & D_{ye} & D_{yu} & 0_{n_y\times r} \\ 0_{r\times n} & 0_{r\times r} & I_{r\times r} & 0_{r\times n_e} & 0_{r\times n_u} & 0_{r\times r} \end{array}\right) \begin{pmatrix} x \\ \hline w_\Delta \\ w_K \\ e \\ \hdashline u \\ u_K \end{pmatrix} \qquad [2.270]$$

which is written as:

$$P_a(s) := \begin{pmatrix} \dot{x} \\ \hline v_\Delta \\ v_K \\ z \\ \hdashline y \\ y_K \end{pmatrix} = \left(\begin{array}{c|c:c} A_a & B_{1a} & B_{2a} \\ \hline C_{1a} & D_{11a} & D_{12a} \\ \hdashline C_{2a} & D_{21a} & D_{22a} \end{array}\right) \begin{pmatrix} x \\ \hline w_\Delta \\ w_K \\ e \\ \hdashline u \\ u_K \end{pmatrix} \qquad [2.271]$$

Δ structured following $\underline{\Delta}$, we will write $\underline{\Delta} \oplus \underline{\Delta}$ the structure of the duplication $\begin{pmatrix} \Delta & 0 \\ 0 & \Delta \end{pmatrix}$.

We define the set of the symmetric positive definite scalings, which commute with the structure $\underline{\Delta}$:

$$L_{\underline{\Delta}} = \{L > 0 : L\Delta = \Delta L,\ \forall \Delta \in \underline{\Delta}\} \subset \mathbf{R}^{r \times r} \qquad [2.272]$$

Several particular cases can be imagined:

– if all the uncertainties are distinct, then the L_i matrices will be chosen to be diagonal by construction;

– if the uncertainty matrix only consists of the repetition of the same uncertainty, then the L_i matrices will be chosen to be full but symmetric by construction;

– the general case consists of uniting the two previous cases, that is, L_i block-diagonal matrices with (full) symmetric blocks for the repeated uncertainties, and pure diagonal if not (see example).

The set of symmetric positive definite scalings, which commute with the duplicated structure $\underline{\Delta} \oplus \underline{\Delta}$, is written as:

$$L_{\underline{\Delta} \oplus \underline{\Delta}} = \left\{ \begin{pmatrix} L_1 & L_2 \\ L_2^T & L_3 \end{pmatrix} > 0 : L_1, L_3 \in L_{\underline{\Delta}}\ et\ L_2\Delta = \Delta L_2,\ \forall \Delta \in \underline{\Delta} \right\} \qquad [2.273]$$

According to [APK 95], if there is a scaling $L \in L_{\underline{\Delta \oplus \Delta}}$ and an LTI controller $K(s)$ such that the closed loop $F_l\left(P_a(s), K(s)\right)$ is internally stable and satisfies:

$$K(s) := \begin{pmatrix} A_K & B_K \\ C_K & D_K \end{pmatrix} = \left(\begin{array}{c|cc} A_K & B_{K_y} & B_{K_\Delta} \\ \hline C_{K_y} & D_{K_{yy}} & D_{K_{y\Delta}} \\ C_{K_\Delta} & D_{K_{\Delta y}} & D_{K_{\Delta\Delta}} \end{array}\right) \qquad [2.274]$$

then $F_l\left(K(s), \Delta(t)\right)$ is a gain-scheduled H_∞ γ-sub-optimal controller that stabilizes the closed loop $T(P, K, \Delta)$ for any parametric trajectory such that $\gamma^2 \Delta^T(t)\Delta(t) \leq 1$.

Assuming that such a controller exists and that its state-space representation is written as:

$$K(s) := \begin{pmatrix} A_K & B_K \\ C_K & D_K \end{pmatrix} = \left(\begin{array}{c|cc} A_K & B_{K_y} & B_{K_\Delta} \\ \hline C_{K_y} & D_{K_{yy}} & D_{K_{y\Delta}} \\ C_{K_\Delta} & D_{K_{\Delta y}} & D_{K_{\Delta\Delta}} \end{array}\right) \qquad [2.275]$$

then the final gain-scheduled controller admits, for implementation, the following state form:

$$K(\Delta) = F_l\left(K(s), \Delta(t)\right) := \begin{pmatrix} A_K + B_{K_\Delta} M_\Delta C_{K_\Delta} & B_{K_y} + B_{K_\Delta} M_\Delta D_{K_{\Delta y}} \\ C_{K_y} + D_{K_{y\Delta}} M_\Delta C_{K_\Delta} & D_{K_{yy}} + D_{K_{y\Delta}} M_\Delta D_{K_{\Delta y}} \end{pmatrix} \qquad [2.276]$$

$$M_\Delta := \Delta(t)\left(I - D_{K_{\Delta\Delta}} \Delta(t)\right)^{-1}$$

We should note that if $D_{K_{\Delta\Delta}} = 0$, then $M_\Delta = \Delta(t)$, which greatly simplifies the implementation.

2.4.3. *Reformulation for stochastic optimization*

As in [APK 95], the solution to the problem posed previously can be found with the LMI formalism. The drawback is its complexity of implementation and its sometimes difficult numerical resolution, but above all the impossibility of structuring the resulting control law. So here we take the approach shown in [FEY 15] applied to structured LPV/LPT synthesis, with stochastic optimization.

2.4.3.1. *Notations*

Given a controller $K(s)$ with the state-space representation:

$$K(s) := \begin{pmatrix} A_K & B_K \\ C_K & D_K \end{pmatrix} \qquad [2.277]$$

And the augmented plant:

$$P_a(s) := \left(\begin{array}{c|c:c} A_a & B_{1a} & B_{2a} \\ \hline C_{1a} & D_{11a} & D_{12a} \\ \hdashline C_{2a} & D_{21a} & D_{22a} \end{array}\right) \qquad [2.278]$$

Then, the closed loop is written as:

$$\begin{aligned} & F_l\left(P_a(s), K(s)\right) \\ & = \begin{pmatrix} A_{cl} & B_{cl} \\ C_{cl} & D_{cl} \end{pmatrix} \\ & = \left[\begin{array}{cc|c} A_a + B_{2a} M_{1a} D_k C_{2a} & B_{2a} M_{1a} C_k & B_{1a} + B_{2a} M_{1a} D_k D_{21a} \\ B_k M_{2a} C_{2a} & A_k + B_k M_{2a} D_{22a} C_k & B_k M_{2a} D_{21a} \\ \hline C_{1a} + D_{12a} M_{1a} D_k C_{2a} & D_{12a} M_{1a} C_k & D_{11a} + D_{12a} M_{1a} D_k D_{21a} \end{array}\right] \qquad [2.279] \\ & M_{1a} = \left(I - D_k D_{22a}\right)^{-1} \\ & M_{2a} = \left(I - D_{22a} D_k\right)^{-1} \end{aligned}$$

Writing the set of the eigenvalues of the matrix A as $\lambda(A)$) then the previously defined closed loop is stable if and only if:

$$\max\left(real\left(\lambda\left(A_{cl}\right)\right)\right) < 0 \qquad [2.280]$$

2.4.3.2. *Principle approach*

Here, we suggest solving directly the following global optimization problem:

$$\begin{aligned} & \min_{L_1, L_2, L_3, K} \gamma \\ & \text{subject to} \\ & \qquad L \in L_{\underline{\Delta} \oplus \underline{\Delta}} \\ & F_l\left(P_a(s), K(s)\right) \in \mathbf{R}H_\infty \end{aligned} \qquad [2.281]$$

The problem is not dissimilar to that of structured μ-synthesis [FEY 14c] and constitutes an extension of the latter.

This problem can be rewritten as:

$$\min_{X \in [\underline{x};\overline{x}]} f(X) \tag{2.282}$$

where the unknown X covers on the one hand the coefficients of the state-space representation (in tridiagonal form) of the controller K to be sequenced and on the other hand the coefficients of the scalings L_i (i = 1, 2, 3) that are symmetric by construction and commute with Δ.

It is important to note that as the scalings are symmetric, only the upper (or lower) part is sought, which reduces the number of unknowns.

Similarly, the search space is transformed to increase the sensitivity of the optimization process.

Moreover, it should be noted that:

$$\begin{gathered}\text{Symmetric, non-positive definite } M \\ \Leftrightarrow \\ \min_i(\lambda_i(M)) \le 0 \\ \Leftrightarrow \\ -\min_i(\lambda_i(M)) \ge 0\end{gathered} \tag{2.283}$$

Thus, assuming that the L_i matrices are symmetric and that they commute with Λ, then:

$$\begin{gathered}L \notin L_{\Delta\oplus\Delta} \\ \Leftrightarrow \\ -\min\left(\min_i \lambda_i(L_1), \min_i \lambda_i(L_3), \min_i \lambda_i(L)\right) \ge 0\end{gathered} \tag{2.284}$$

2.4.3.3. *Full-order synthesis*

The unknowns are the scalings, the controller being obtained by solving (with fixed L) the following standard H_∞ problem:

$$\left\| \begin{pmatrix} L^{1/2} & 0 \\ 0 & I_{n_z \times n_z} \end{pmatrix} F_l(P_a(s), K(s)) \begin{pmatrix} L^{-1/2} & 0 \\ 0 & I_{n_e \times n_e} \end{pmatrix} \right\|_\infty < \gamma \tag{2.285}$$

By positing:

$$\Lambda = -\min\left(\min_i \lambda_i(L_1), \min_i \lambda_i(L_3), \min_i \lambda_i(L)\right) \quad [2.286]$$

then, the scaling L (and *de facto* the full-order controller $K(s)$) are obtained via the following optimization problem:

$$\min_L f(L)$$

If $\Lambda \geq 0$

$$f(L) = \Lambda$$

Else

$$f(L) = -\frac{1}{\gamma}, \quad \left\| \begin{pmatrix} L^{1/2} & 0 \\ 0 & I \end{pmatrix} F_l\left(P_a(s), K(s)\right) \begin{pmatrix} L^{-1/2} & 0 \\ 0 & I \end{pmatrix} \right\|_\infty < \gamma \quad [2.287]$$

The final LPV controller to be implemented is finally reconstructed by the relation:

$$K(\Delta) = F_l\left(K(s), \Delta(t)\right) \quad [2.288]$$

2.4.3.4. *Structured synthesis*

2.4.3.4.1. Principle of the approach

By positing:

$$\Theta_K = \begin{pmatrix} A_K & B_K \\ C_K & D_K \end{pmatrix} \quad [2.289]$$

and:

$$\left.\begin{array}{l} \underline{\lambda} = -\min\left(\min_i \lambda_i(L_1), \min_i \lambda_i(L_3), \min_i \lambda_i(L)\right) \\ \overline{\lambda} = \max_i\left(\mathrm{Re}\left[\lambda_i\left(F_l\left(P_a(s), K(s)\right)\right)\right]\right) \end{array}\right\} \Lambda = \max\left(\underline{\lambda}, \overline{\lambda}\right) \quad [2.290]$$

then, we obtain the optimal scalings and the optimal controller by solving the following problem:

$$\min_{L,\Theta_K} f(L,\Theta_K)$$

If $\Lambda \geq 0$

$$f(L,\Theta_K) = \Lambda$$

Else

$$f(L,\Theta_K) = -\frac{1}{\gamma},\ \gamma = \left\| \begin{pmatrix} L^{1/2} & 0 \\ 0 & I \end{pmatrix} F_l\left(P_a(s), K(s)\right) \begin{pmatrix} L^{-1/2} & 0 \\ 0 & I \end{pmatrix} \right\|_\infty \qquad [2.291]$$

The final LPV controller to be implemented is finally reconstructed by the relation:

$$K(\Delta) = F_l\left(K(s), \Delta(t)\right) \qquad [2.292]$$

2.4.3.4.2. Evaluation of the fitness function

We seek to solve the following optimization problem:

$$\min_{X \in \left[\underline{x};\overline{x}\right]} f(X) \qquad [2.293]$$

where the function $f(X)$ consists of carrying out the following steps:

– build the realization $\left[\begin{array}{c|c} A_k & B_k \\ \hline C_k & D_k \end{array}\right]$ from X;

– build the matrix A_{cl};

– build each scaling L_i (i=1, 2, 3) from X;

– build the scaling L from L_i;

– evaluate the spectral abscissa of the closed loop:

$$\overline{\lambda} = \max_i \left(\text{Re}\left[\lambda_i\left(A_{cl}\right)\right]\right) \qquad [2.294]$$

– evaluate:

$$\underline{\lambda} = -\min\left(\min_i \lambda_i(L_1), \min_i \lambda_i(L_3), \min_i \lambda_i(L)\right) \quad [2.295]$$

– If $\Lambda = \max\left(\underline{\lambda}, \overline{\lambda}\right) > 0$, then:

$$f(X) = \Lambda \quad [2.296]$$

– Else:

- build the state-space representation of $F_l\left(P_a(s), K(s)\right)$;

- compute:

$$\gamma = \left\| \begin{pmatrix} L^{1/2} & 0 \\ 0 & I \end{pmatrix} F_l\left(P_a(s), K(s)\right) \begin{pmatrix} L^{-1/2} & 0 \\ 0 & I \end{pmatrix} \right\|_\infty \quad [2.297]$$

- evaluate:

$$f(X) = -\frac{1}{\gamma} \quad [2.298]$$

2.4.3.5. *Example of implementation – the cart*

2.4.3.5.1. Description of the system

Consider the system in Figure 2.38.

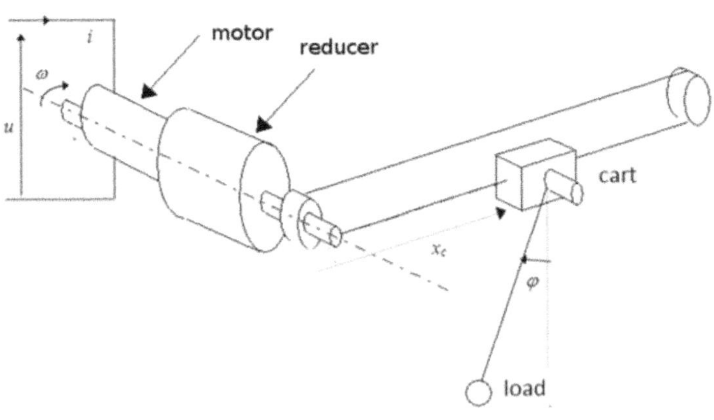

Figure 2.38. *Mobile pendulum in the cart*

When the system is running, the length of the cable and the reduced inertia on the engine can vary in the following intervals:

$$
\begin{aligned}
&0{,}1\ \text{m} < l < 0.3\ \text{m} \\
&3.10^{-6}\ \text{kg.m}^2 < J < 50.10^{-6}\ \text{kg.m}^2
\end{aligned}
\qquad [2.299]
$$

Typically in fact, the cart bears a load, the operator winds the cable and moves the cart, then unwinds the cable and deposits the load.

2.4.3.5.2. LPV modelization in LFT form

As for μ-synthesis, it makes sense in the first place to put the system and its uncertainty in LFT form.

With the previous hypotheses, the system is described by the following equations:

$$
\begin{cases}
L\dfrac{di(t)}{dt} + Ri(t) + \phi\,\omega(t) = u(t) \\
\dfrac{d\left(J(t)\,\omega(t)\right)}{dt} + f\,\omega(t) + \gamma(t) = \phi\, i(t) \\
\dfrac{dx_c(t)}{dt} = \dfrac{r}{N}\omega(t) \\
\cos\left(\phi(t)\right)\dfrac{d^2x_c(t)}{dt^2} + l(t)\dfrac{d^2\phi(t)}{dt^2} + 2\dfrac{d\,l(t)}{dt}\dfrac{d\phi(t)}{dt} + \alpha\dfrac{d\phi(t)}{dt} + g\sin\left(\phi(t)\right) = 0
\end{cases}
\qquad [2.300]
$$

Assume that the inertia is constant per piece, so that:

$$
\frac{d\left(J(t)\,\omega(t)\right)}{dt} = J(t)\frac{d\,\omega(t)}{dt} \qquad [2.301]
$$

Moreover, in the vicinity of $\phi(t) = 0$:

$$
\frac{d^2x_c(t)}{dt^2} + l(t)\frac{d^2\phi(t)}{dt^2} + 2\frac{dl(t)}{dt}\frac{d\,\phi(t)}{dt} + \alpha\frac{d\phi(t)}{dt} + g\,\phi(t) = 0 \qquad [2.302]
$$

Furthermore:

$$
\frac{d^2\left(l(t)\,\phi(t)\right)}{dt^2} = l(t)\frac{d^2\phi(t)}{dt^2} + 2\frac{dl(t)}{dt}\frac{d\,\phi(t)}{dt} + \frac{d^2l(t)}{dt^2}\phi(t) \qquad [2.303]
$$

Therefore, the fourth equation becomes:

$$\frac{d^2 x_c(t)}{dt^2} + \frac{d\, v_l(t)}{dt} + \alpha \frac{d\,\phi(t)}{dt} + g\,\phi(t) = 0 \quad [2.304]$$

with:

$$v_l(t) = \frac{d\left(l(t)\phi(t)\right)}{dt} = l(t)\frac{d\phi(t)}{dt} + \frac{dl(t)}{dt}\phi(t) \approx l(t)\frac{d\phi(t)}{dt} \quad [2.305]$$

it finally reaches the following linear state-space representation:

$$\frac{d}{dt}\begin{pmatrix} i(t) \\ \omega(t) \\ x_c(t) \\ \phi(t) \\ v_l(t) \end{pmatrix} = \begin{pmatrix} -\frac{R}{L} & -\frac{\phi}{L} & 0 & 0 & 0 \\ \frac{\phi}{J(t)} & -\frac{f}{J(t)} & 0 & 0 & 0 \\ 0 & \frac{r}{N} & 0 & 0 & 0 \\ 0 & 0 & 0 & 0 & \frac{1}{l(t)} \\ -\frac{r\phi}{NJ(t)} & \frac{rf}{NJ(t)} & 0 & -g & -\frac{\alpha}{l(t)} \end{pmatrix} \begin{pmatrix} i(t) \\ \omega(t) \\ x_c(t) \\ \phi(t) \\ v_l(t) \end{pmatrix}$$

$$+ \begin{pmatrix} 0 \\ -\frac{1}{J(t)} \\ 0 \\ 0 \\ \frac{r}{NJ(t)} \end{pmatrix} \gamma(t) + \begin{pmatrix} \frac{1}{L} \\ 0 \\ 0 \\ 0 \\ 0 \end{pmatrix} u(t) \quad [2.306]$$

$$\begin{pmatrix} x_c \\ \phi \end{pmatrix} = \begin{pmatrix} 0 & 0 & 1 & 0 & 0 \\ 0 & 0 & 0 & 1 & 0 \end{pmatrix} \begin{pmatrix} i(t) \\ \omega(t) \\ x_c(t) \\ \phi(t) \\ v_l(t) \end{pmatrix} + \begin{pmatrix} 0 \\ 0 \end{pmatrix} \gamma(t) + \begin{pmatrix} 0 \\ 0 \end{pmatrix} u(t)$$

By positing $\tilde{J}(t) = \dfrac{1}{J(t)}$ and $\tilde{l}(t) = \dfrac{1}{l(t)}$, it arrives at:

$$\frac{d}{dt}\begin{pmatrix} i(t) \\ \omega(t) \\ x_c(t) \\ \phi(t) \\ v_l(t) \end{pmatrix} = \begin{pmatrix} -\frac{R}{L} & -\frac{\phi}{L} & 0 & 0 & 0 \\ \phi\tilde{J}(t) & -f\tilde{J}(t) & 0 & 0 & 0 \\ 0 & \frac{r}{N} & 0 & 0 & 0 \\ 0 & 0 & 0 & 0 & \tilde{l}(t) \\ -\frac{r\phi}{N}\tilde{J}(t) & \frac{rf}{N}\tilde{J}(t) & 0 & -g & -\alpha\tilde{l}(t) \end{pmatrix}\begin{pmatrix} i(t) \\ \omega(t) \\ x_c(t) \\ \phi(t) \\ v_l(t) \end{pmatrix} + \begin{pmatrix} 0 \\ -\tilde{J}(t) \\ 0 \\ 0 \\ \frac{r}{N}\tilde{J}(t) \end{pmatrix}\gamma(t) + \begin{pmatrix} \frac{1}{L} \\ 0 \\ 0 \\ 0 \\ 0 \end{pmatrix}u(t)$$

$$\begin{pmatrix} x_c \\ \phi \end{pmatrix} = \begin{pmatrix} 0 & 0 & 1 & 0 & 0 \\ 0 & 0 & 0 & 1 & 0 \end{pmatrix}\begin{pmatrix} i(t) \\ \omega(t) \\ x_c(t) \\ \phi(t) \\ v_l(t) \end{pmatrix} + \begin{pmatrix} 0 \\ 0 \end{pmatrix}\gamma(t) + \begin{pmatrix} 0 \\ 0 \end{pmatrix}u(t) \qquad [2.307]$$

An LFT form can be established by positing:

$$\begin{aligned} &\tilde{J}(t) = \tilde{J}_0 + \delta_{\tilde{J}}(t)\tilde{J}_1, \ \left|\delta_{\tilde{J}}(t)\right| \le 1, \\ &\quad \tilde{J}_1 = \frac{J_{\min}^{-1} - J_{\max}^{-1}}{2}, \ \tilde{J}_0 = \frac{J_{\max}^{-1} + J_{\min}^{-1}}{2} \\ &\tilde{l}(t) = \tilde{l}_0 + \delta_{\tilde{l}}(t)\tilde{l}_1, \ \left|\delta_{\tilde{l}}(t)\right| \le 1, \\ &\quad \tilde{l}_1 = \frac{l_{\min}^{-1} - l_{\max}^{-1}}{2}, \ \tilde{l}_0 = \frac{l_{\max}^{-1} + l_{\min}^{-1}}{2} \end{aligned} \qquad [2.308]$$

Returning to the equations with varying parameters:

$$\begin{aligned}
\dot{\omega}(t) &= \left(\tilde{J}_0 + \delta_{\tilde{J}}(t)\tilde{J}_1\right)\phi i(t) - \left(\tilde{J}_0 + \delta_{\tilde{J}}(t)\tilde{J}_1\right) f\omega(t) - \left(\tilde{J}_0 + \delta_{\tilde{J}}(t)\tilde{J}_1\right)\gamma(t) \\
\dot{\phi}(t) &= \left(\tilde{l}_0 + \delta_{\tilde{l}}(t)\tilde{l}\right)\phi(t) \\
\dot{v}_l(t) &= -\left(\tilde{J}_0 + \delta_{\tilde{J}}(t)\tilde{J}_1\right)\frac{r\phi}{N} i(t) + \left(\tilde{J}_0 + \delta_{\tilde{J}}(t)\tilde{J}_1\right)\frac{rf}{N}\omega(t) \\
&\quad - g\phi(t) - \left(\tilde{l}_0 + \delta_{\tilde{l}}(t)\tilde{l}\right)\alpha v_l(t) + \left(\tilde{J}_0 + \delta_{\tilde{J}}(t)\tilde{J}_1\right)\frac{r}{N}\gamma(t)
\end{aligned} \qquad [2.309]$$

Thus:

$$\begin{aligned}
\dot{\omega}(t) &= \tilde{J}_0\phi i(t) + \delta_{\tilde{J}}(t)\tilde{J}_1\phi i(t) - \tilde{J}_0 f\omega(t) - \delta_{\tilde{J}}(t)\tilde{J}_1 f\omega(t) - \tilde{J}_0\gamma(t) \\
&\quad - \delta_{\tilde{J}}(t)\tilde{J}_1\gamma(t) \\
\dot{\phi}(t) &= \tilde{l}_0\phi(t) + \delta_{\tilde{l}}(t)\tilde{l}_1\phi(t) \\
\dot{v}_l(t) &= -\left(\tilde{J}_0\frac{r\phi}{N} i(t) + \delta_{\tilde{J}}(t)\tilde{J}_1\frac{r\phi}{N} i(t)\right) + \\
&\quad \left(\tilde{J}_0\frac{rf}{N}\omega(t) + \delta_{\tilde{J}}(t)\tilde{J}_1\frac{rf}{N}\omega(t)\right) - g\phi(t) \\
&\quad - \left(\tilde{l}_0\,\alpha v_l(t) + \delta_{\tilde{l}}(t)\tilde{l}\alpha v_l(t)\right) + \left(\tilde{J}_0\frac{r}{N}\gamma(t) + \delta_{\tilde{J}}(t)\tilde{J}_1\frac{r}{N}\gamma(t)\right)
\end{aligned} \qquad [2.310]$$

And:

$$\begin{aligned}
\dot{\omega}(t) &= \tilde{J}_0\phi i(t) - \tilde{J}_0 f\omega(t) - \tilde{J}_0\gamma(t) + \delta_{\tilde{J}}(t)\tilde{J}_1\left(\phi i(t) - f\omega(t)\right) - \delta_{\tilde{J}}(t)\tilde{J}_1\gamma(t) \\
\dot{\phi}(t) &= \tilde{l}_0\phi(t) + \delta_{\tilde{l}}(t)\tilde{l}_1\phi(t) \\
\dot{v}_l(t) &= -\tilde{J}_0\frac{r\phi}{N} i(t) + \tilde{J}_0\frac{rf}{N}\omega(t) + \tilde{J}_0\frac{r}{N}\gamma(t) - \tilde{l}_0\alpha v_l(t) - g\phi(t) + \\
&\quad \delta_{\tilde{J}}(t)\tilde{J}\left(-\frac{r\phi}{N} i(t) + \frac{rf}{N}\omega(t)\right) - \delta_{\tilde{l}}(t)\tilde{l}_1\alpha v_l(t) + \delta_{\tilde{J}}(t)\tilde{J}_1\frac{r}{N}\gamma(t)
\end{aligned} \qquad [2.311]$$

That is:

$$\begin{aligned}
&\dot{\omega}(t) = \tilde{J}_0 \phi i(t) - \tilde{J}_0 f \omega(t) - \tilde{J}_0 \gamma(t) + w_{\Delta 1}(t) + w_{\Delta 2}(t) \\
&\dot{\phi}(t) = \tilde{l}_0 \phi(t) + w_{\Delta 3}(t) \\
&\dot{v}_l(t) = -\tilde{J}_0 \frac{r\phi}{N} i(t) + \tilde{J}_0 \frac{rf}{N} \omega(t) + \tilde{J}_0 \frac{r}{N} \gamma(t) - \tilde{l}_0 \alpha v_l(t) - g\phi(t) - \frac{r}{N} w_{\Delta 1}(t) \\
&\quad + w_4(t) - \frac{r}{N} w_2(t)
\end{aligned} \quad [2.312]$$

$$\begin{aligned}
&w_{\Delta 1}(t) = \delta_{\tilde{J}}(t) \tilde{J}_1 \left(\phi i(t) - f \omega(t) \right) \\
&w_{\Delta 2}(t) = -\delta_{\tilde{J}}(t) \tilde{J}_1 \gamma(t) \\
&w_{\Delta 3}(t) = \delta_{\tilde{l}}(t) \tilde{l}_1 \phi(t) \\
&w_{\Delta 4}(t) = -\delta_{\tilde{l}}(t) \tilde{l}_1 \alpha v_l(t)
\end{aligned}$$

Or again:

$$\begin{aligned}
&\dot{\omega}(t) = \tilde{J}_0 \phi i(t) - \tilde{J}_0 f \omega(t) - \tilde{J}_0 \gamma(t) + w_{\Delta 1}(t) + w_{\Delta 2}(t) \\
&\dot{\phi}(t) = \tilde{l}_0 \phi(t) + w_{\Delta 3}(t) \\
&\dot{v}_l(t) = -\tilde{J}_0 \frac{r\phi}{N} i(t) + \tilde{J}_0 \frac{rf}{N} \omega(t) + \tilde{J}_0 \frac{r}{N} \gamma(t) - \tilde{l}_0 a v_l(t) \\
&-g\phi(t) - \frac{r}{N} w_{\Delta 1}(t) + w_4(t) - \frac{r}{N} w_2(t)
\end{aligned}$$

$$\begin{aligned}
&w_{\Delta 1}(t) = \delta_{\tilde{J}}(t) v_{\Delta 1}(t) \\
&w_{\Delta 2}(t) = \delta_{\tilde{J}}(t) v_{\Delta 2}(t) \\
&w_{\Delta 3}(t) = \delta_{\tilde{l}}(t) v_{\Delta 3}(t) \\
&w_{\Delta 4}(t) = \delta_{\tilde{l}}(t) v_{\Delta 4}(t)
\end{aligned} \quad [2.313]$$

$$\begin{aligned}
&v_{\Delta 1}(t) = \tilde{J}_1 \left(\phi i(t) - f \omega(t) \right) \\
&v_{\Delta 2}(t) = -\tilde{J}_1 \gamma(t) \\
&v_{\Delta 3}(t) = \tilde{l}_1 \phi(t) \\
&v_{\Delta 4}(t) = -\tilde{l}_1 \alpha v_l(t)
\end{aligned}$$

Thus, the LPV model is established in the following LFT form:

$$
\frac{d}{dt}\begin{pmatrix} i(t) \\ \omega(t) \\ x_c(t) \\ \phi(t) \\ v_l(t) \end{pmatrix} = \begin{pmatrix} -\frac{R}{L} & -\frac{\phi}{L} & 0 & 0 & 0 \\ \phi\tilde{J}_0 & -f\tilde{J}_0 & 0 & 0 & 0 \\ 0 & \frac{r}{N} & 0 & 0 & 0 \\ 0 & 0 & 0 & 0 & \tilde{l}_0 \\ -\frac{r\phi}{N}\tilde{J}_0 & \frac{rf}{N}\tilde{J}_0 & 0 & -g & -\alpha\tilde{l}_0 \end{pmatrix}\begin{pmatrix} i(t) \\ \omega(t) \\ x_c(t) \\ \phi(t) \\ v_l(t) \end{pmatrix}
$$

$$
+\begin{pmatrix} 0 & 0 & 0 & 0 \\ 1 & 1 & 0 & 0 \\ 0 & 0 & 0 & 0 \\ 0 & 0 & 1 & 0 \\ -\frac{r}{N} & -\frac{r}{N} & 0 & 1 \end{pmatrix} w_\Delta(t) + \begin{pmatrix} 0 \\ -\tilde{J}_0 \\ 0 \\ 0 \\ \frac{r}{N}\tilde{J}_0 \end{pmatrix}\gamma(t) + \begin{pmatrix} \frac{1}{L} \\ 0 \\ 0 \\ 0 \\ 0 \end{pmatrix} u(t)
$$

$$
v_\Delta(t) = \begin{pmatrix} \tilde{J}_1\phi & -\tilde{J}_1 f & 0 & 0 & 0 \\ 0 & 0 & 0 & 0 & 0 \\ 0 & 0 & 0 & \tilde{l}_1 & 0 \\ 0 & 0 & 0 & 0 & -\tilde{l}_1\alpha \end{pmatrix}\begin{pmatrix} i(t) \\ \omega(t) \\ x_c(t) \\ \phi(t) \\ v_l(t) \end{pmatrix} + 0w_\Delta(t) + \begin{pmatrix} 0 \\ -\tilde{J}_1 \\ 0 \\ 0 \end{pmatrix}\gamma(t) + \begin{pmatrix} 0 \\ 0 \\ 0 \\ 0 \end{pmatrix} u(t)
$$

$$
\begin{pmatrix} x_c \\ \phi \end{pmatrix} = \begin{pmatrix} 0 & 0 & 1 & 0 & 0 \\ 0 & 0 & 0 & 1 & 0 \end{pmatrix}\begin{pmatrix} i(t) \\ \omega(t) \\ x_c(t) \\ \phi(t) \\ v_l(t) \end{pmatrix} + 0w_\Delta(t) + \begin{pmatrix} 0 \\ 0 \end{pmatrix}\gamma(t) + \begin{pmatrix} 0 \\ 0 \end{pmatrix} u(t)
$$

$$
\begin{aligned}
& w_\Delta(t) = diag\left(\delta_{\tilde{J}}(t), \delta_{\tilde{J}}(t), \delta_{\tilde{l}}(t), \delta_{\tilde{l}}(t)\right) v_\Delta(t) \\
& w_\Delta(t) = \left(w_{\Delta 1}(t), w_{\Delta 2}(t), w_{\Delta 3}(t), w_{\Delta 4}(t)\right)^T \\
& v_\Delta(t) = \left(v_{\Delta 1}(t), v_{\Delta 2}(t), v_{\Delta 3}(t), v_{\Delta 4}(t)\right)^T
\end{aligned} \qquad [2.314]
$$

Finally, we note:

$$x(t) = \begin{pmatrix} i(t) & \omega(t) & x_c(t) & \phi(t) & v_l(t) \end{pmatrix}^T \qquad [2.315]$$

$$H(s) := \left(\begin{array}{c|ccc} A & B_\delta & B_\gamma & B_u \\ \hline C_\delta & D_{\delta\delta} & D_{\delta\gamma} & D_{\delta u} \\ \hdashline C_x & D_{x\delta} & D_{x\gamma} & D_{xu} \\ C_\phi & D_{\phi\delta} & D_{\phi\gamma} & D_{\phi u} \end{array} \right) \qquad [2.316]$$

Then, the following will be posited:

$$x(t) = \begin{pmatrix} i(t) & \omega(t) & x_c(t) & \phi(t) & v_l(t) \end{pmatrix}^T \qquad [2.317]$$

The numerical values retained are given below.

Letter	Meaning	Numerical value	
R	Resistance to the motor's armature	$]2{,}1\,;2{,}5[$	Ω
L	Inductance of the motor's armature	$<10^{-3}$	H
ϕ	Torque and f.c.e.m. coefficient	0.0162	U.S.I.
J	Reduced inertia on the motor's axis	$]4\,;6[\times 10^{-6}$	kg.m^{-2}
f	Friction coefficient (motor)	$]5\,;8[\times 10^{-5}$	N.m.s
r	Radius of the pulley	0.022	m
N	Reduction ratio	17	-
l	Length of the cable	0.275	m
α	Friction coefficient (cable)	$]0{,}2\,;0{,}4[$	m.s^{-1}
g	Acceleration due to gravity	9.81	m.s^{-2}
k_x	Gain of the position sensor	39.77	V.m^{-1}
k_φ	Gain of the angle sensor	4.77	V.rad^{-1}

Table 2.11. *Sizing of the pendulum in the cart*

2.4.3.5.3. Synthesis scheme

The aim is to control the position $x_c(t)$ of the cart while regulating the angle of the cable $\phi(t)$ around 0. The controller uses the tensions v_x and v_ϕ as measurements and should ensure the following performances:

– monitoring reference tracking corresponding to a motion of 0.4 m with a rate of 0.1 m/s (see Figure 2.39);

– no steady-state error (a tracking error is acceptable), overshoot less than 0.01 m, cable angle less than ± 0.1 rad, response time (cart and cable) less than 6 s;

– no steady-state error in response to a constant torque perturbation $\gamma(t)$ of 0.05 Nm;

– command limited to ±15 V and not very sensitive to measurement noise;

– modulus margin on the synthesis model higher than 0.7;

– stability robustness toward parameter uncertainties and on the neglected dynamic, with a modulus margin at least equal to 0.5.

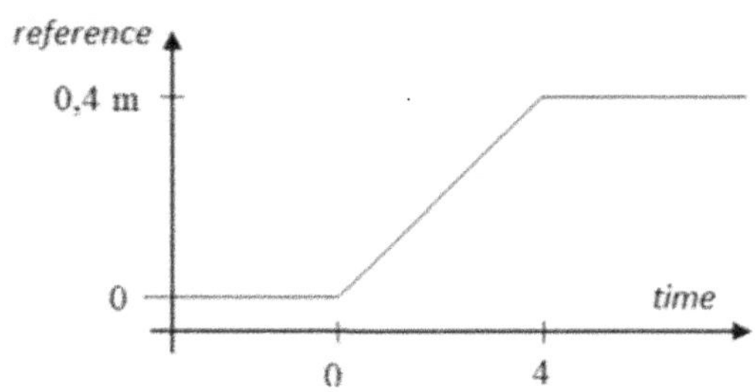

Figure 2.39. *Reference of the cart's motion*

For the control, there are two measures:

$$\begin{cases} v_x(t) = k_x \ x_c(t) \\ v_\phi(t) = k_\phi \ \phi(t) \end{cases} \qquad [2.318]$$

That is:

$$\begin{aligned} v_x &= k_x C_x x + k_x D_{x\delta} w_\Delta + k_x D_{x\gamma}\gamma + k_x D_{xu} u \\ v_\phi &= k_\phi C_\phi x + k_\phi D_{\phi\delta} w_\Delta + k_\phi D_{\phi\gamma}\gamma + k_\phi D_{\phi u} u \end{aligned} \qquad [2.319]$$

The synthesis scheme envisaged is shown in the following.

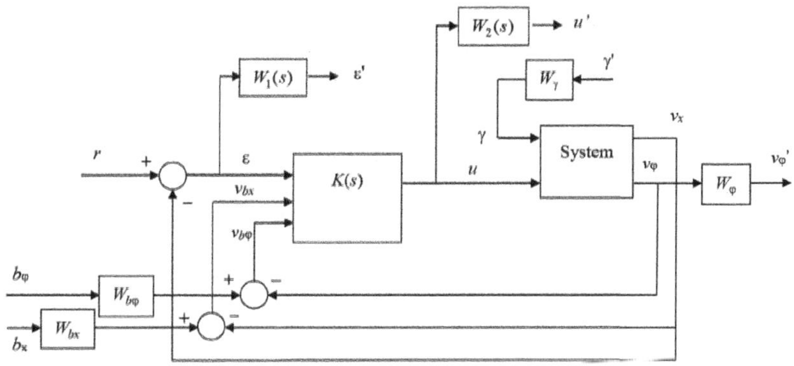

Figure 2.40. *Control synthesis scheme of the pendulum in the cart*

We have:

$$\gamma = W_\gamma \gamma' \qquad [2.320]$$

And:

$$\begin{aligned}
v_{bx} &= W_{bx} b_x - v_x \\
&= W_{bx} b_x - k_x C_x x - k_x D_{x\delta} w_\Delta - k_x D_{x\gamma} W_\gamma \gamma' - k_x D_{xu} u \\
v_{b\phi} &= W_{b\phi} b_\phi - v_\phi \\
&= W_{b\phi} b_\phi - k_\phi C_\phi x - k_\phi D_{\phi\delta} w_\Delta - k_\phi D_{\phi\gamma} W_\gamma \gamma' - k_\phi D_{\phi u} u \\
\dot{x}_1 &= A_1 x_1 + B_1 \varepsilon = A_1 x_1 + B_1 (r - v_x) \\
&= A_1 x_1 + B_1 r - B_1 k_x C_x x - B_1 k_x D_{x\delta} w_\Delta - B_1 k_x D_{x\gamma} W_\gamma \gamma' - B_1 k_x D_{xu} u \\
\varepsilon' &= C_1 x_1 + D_1 \varepsilon = C_1 x_1 + D_1 (r - v_x) \\
&= C_1 x_1 + D_1 r - D_1 k_x C_x x - D_1 k_x D_{x\delta} w_\Delta - D_1 k_x D_{x\gamma} W_\gamma \gamma' - D_1 k_x D_{xu} u \\
\dot{x}_2 &= A_2 x_2 + B_2 u \\
u' &= C_2 x_2 + D_2 u \\
v_\phi' &= W_\phi v_\phi = W_\phi k_\phi C_\phi x + W_\phi k_\phi D_{\phi\delta} w_\Delta + W_\phi k_\phi D_{\phi\gamma} W_\gamma \gamma' + W_\phi k_\phi D_{\phi u} u
\end{aligned} \qquad [2.321]$$

$$\begin{aligned}
\dot{x} &= Ax + B_\delta w_\Delta + B_\gamma W_\gamma \gamma' + B_u u \\
v_\Delta &= C_\Delta x + D_{\delta\delta} w_\Delta + D_{\delta\gamma} W_\gamma \gamma' + D_{\delta u} u
\end{aligned}$$

There is therefore:

$$\frac{d}{dt}\begin{pmatrix} x \\ x_1 \\ x_2 \end{pmatrix} = \begin{pmatrix} A & 0 & 0 \\ -B_1 k_x C_x & A_1 & 0 \\ 0 & 0 & A_2 \end{pmatrix}\begin{pmatrix} x \\ x_1 \\ x_2 \end{pmatrix} + \begin{pmatrix} B_\delta \\ -B_1 k_x D_{x\delta} \\ 0 \end{pmatrix} w_\Delta +$$

$$\begin{pmatrix} B_\gamma W_\gamma & 0 & 0 & 0 \\ -B_1 k_x D_{x\gamma} W_\gamma & B_1 & 0 & 0 \\ 0 & 0 & 0 & 0 \end{pmatrix}\begin{pmatrix} \gamma' \\ r \\ b_x \\ b_\phi \end{pmatrix} + \begin{pmatrix} B_u \\ -B_1 k_x D_{xu} \\ B_2 \end{pmatrix} u$$

$$v_\Delta = \begin{pmatrix} C_\delta & 0 & 0 \end{pmatrix}\begin{pmatrix} x \\ x_1 \\ x_2 \end{pmatrix} + D_{\delta\delta} w_\Delta + \begin{pmatrix} D_{\delta\gamma} W_\gamma & 0 & 0 & 0 \end{pmatrix}\begin{pmatrix} \gamma' \\ r \\ b_x \\ b\phi \end{pmatrix} + D_{\delta u} u$$

$$\begin{pmatrix} \varepsilon' \\ u' \\ v_\phi' \end{pmatrix} = \begin{pmatrix} -D_1 k_x C_x & C_1 & 0 \\ 0 & 0 & C_2 \\ W_\phi k_\phi C_\phi & 0 & 0 \end{pmatrix}\begin{pmatrix} x \\ x_1 \\ x_2 \end{pmatrix} + \begin{pmatrix} -D_1 k_x D_{x\delta} \\ 0 \\ W_\phi k_\phi D_{\phi\delta} \end{pmatrix} w_\Delta +$$

$$\begin{pmatrix} -D_1 k_x D_{x\gamma} W_\gamma & D_1 & 0 & 0 \\ 0 & 0 & 0 & 0 \\ W_\phi k_\phi D_{\phi\gamma} W_\gamma & 0 & 0 & 0 \end{pmatrix}\begin{pmatrix} \gamma' \\ r \\ b_x \\ b_\phi \end{pmatrix} + \begin{pmatrix} -D_1 k_x D_{xu} \\ D_2 \\ W_\phi k_\phi D_{\phi u} \end{pmatrix} u$$

$$\begin{pmatrix} \varepsilon \\ v_{bx} \\ v_{b\phi} \end{pmatrix} = \begin{pmatrix} -k_x C_x & 0 & 0 \\ -k_x C_x & 0 & 0 \\ -k_\phi C_\phi & 0 & 0 \end{pmatrix}\begin{pmatrix} x \\ x_1 \\ x_2 \end{pmatrix} + \begin{pmatrix} -k_x D_{x\delta} \\ -k_x D_{x\delta} \\ -k_\phi D_{\phi\delta} \end{pmatrix} w_\Delta + \quad [2.322]$$

$$\begin{pmatrix} -k_x D_{x\gamma} W_\gamma & 1 & 0 & 0 \\ -k_x D_{x\gamma} W_\gamma & 0 & W_{bx} & 0 \\ -k_\phi D_{\phi\gamma} W_\gamma & 0 & 0 & W_{b\phi} \end{pmatrix}\begin{pmatrix} \gamma' \\ r \\ b_x \\ b_\phi \end{pmatrix} + \begin{pmatrix} -k_x D_{xu} \\ -k_x D_{xu} \\ -k_\phi D_{\phi u} \end{pmatrix} u$$

which has the form:

$$P(s) := \left(\begin{array}{c|cc:c} A' & B_\Delta & B_e & B_u' \\ \hline C_\Delta & D_{\Delta\Delta} & D_{\Delta e} & D_{\Delta u} \\ C_z & D_{z\Delta} & D_{ze} & D_{zu} \\ \hdashline C_y & D_{y\Delta} & D_{ye} & D_{yu} \end{array}\right) \qquad [2.323]$$

We choose the following weights (enabling the specifications on the nominal model to be ensured):

$$\begin{aligned} W_\varepsilon(s) &= \frac{500s+866}{1000s+0.866} \\ W_u(s) &= \frac{99.5s+200}{0.995s+2000} \\ W_\gamma &= 0.01 \\ W_\phi &= 2 \\ W_{bx} &= 1 \\ W_{b\phi} &= 0.1 \end{aligned} \qquad [2.324]$$

2.4.3.5.4. H∞ synthesis

A full-order synthesis would lead to an order 7 controller. We therefore decide to synthesize a controller with order reduced to 4. It leads to γ = 1.07. Figure 2.41 shows a time response of the nominal system looped by the H_∞ controller.

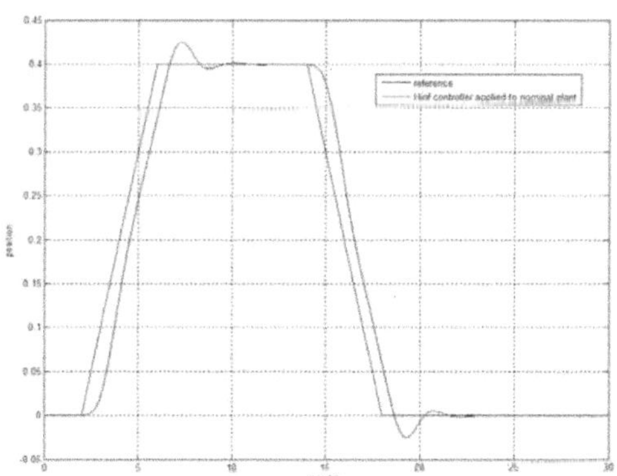

Figure 2.41. *H∞ correction on a nominal system. For a color version of this figure, see www.iste.co.uk/feyel/optimization.zip*

We now vary the parameters $J(t)$ and $l(t)$ following the profiles in Figure 2.42. The response of the varying system, looped with the H_∞ controller, is given in Figure 2.43. It is clear that the performances are much degraded. The looped system even becomes unstable at the end of the trajectory. Assuming these parameters are measurable, this clearly justifies the synthesis of an LPV controller in the following section.

2.4.3.5.5. LPV synthesis

We now proceed to synthesis of an LPV controller. The varying parameters measured are $J(t)$ and $l(t)$. Putting the uncertain system in LFT form has led to an uncertainty (normalized) matrix with the following form:

$$\Delta(t)=\begin{pmatrix}\delta_{\bar{J}}(t) & 0 & 0 & 0\\ 0 & \delta_{\bar{J}}(t) & 0 & 0\\ 0 & 0 & \delta_{\bar{l}}(t) & 0\\ 0 & 0 & 0 & \delta_{\bar{l}}(t)\end{pmatrix} \qquad [2.325]$$

This is a block-diagonal matrix with two repeated uncertainties. Thus, the (symmetric) structure of each L_i can be the following:

$$L_i=\begin{pmatrix}l_{11}^{\{i\}} & l_{12}^{\{i\}} & 0 & 0\\ l_{12}^{\{i\}} & l_{22}^{\{i\}} & 0 & 0\\ 0 & 0 & l_{33}^{\{i\}} & l_{34}^{\{i\}}\\ 0 & 0 & l_{34}^{\{i\}} & l_{44}^{\{i\}}\end{pmatrix}, i=1,2,3 \qquad [2.326]$$

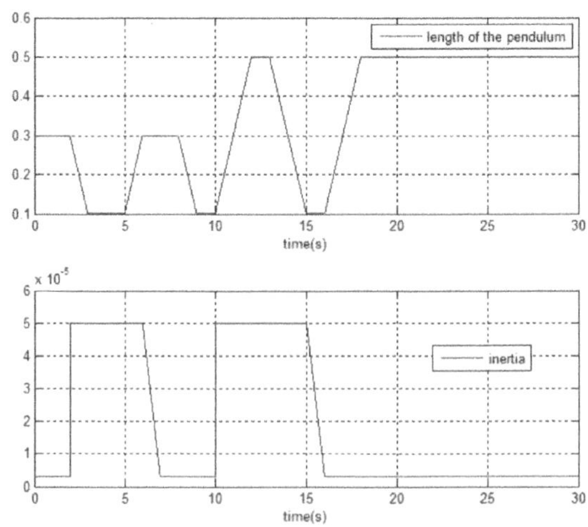

Figure 2.42. *Variation of parameters J(t) and l(t)*

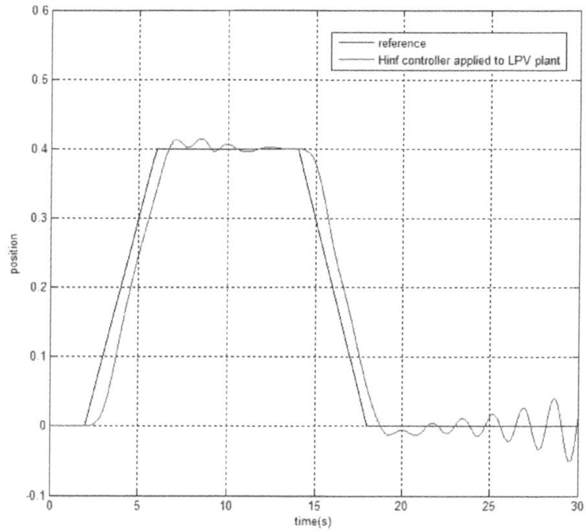

Figure 2.43. *H∞ correction on a varying system. For a color version of this figure, see www.iste.co.uk/feyel/optimization.zip*

It is verified that such a structure commutes with $\Delta(t)$:

$$
\begin{aligned}
\Delta L_i &= \begin{pmatrix} \delta_{\bar{J}} & 0 & 0 & 0 \\ 0 & \delta_{\bar{J}} & 0 & 0 \\ 0 & 0 & \delta_{\bar{l}} & 0 \\ 0 & 0 & 0 & \delta_{\bar{l}} \end{pmatrix} . \begin{pmatrix} l_{11}^{\{i\}} & l_{12}^{\{i\}} & 0 & 0 \\ l_{12}^{\{i\}} & l_{22}^{\{i\}} & 0 & 0 \\ 0 & 0 & l_{33}^{\{i\}} & l_{34}^{\{i\}} \\ 0 & 0 & l_{34}^{\{i\}} & l_{44}^{\{i\}} \end{pmatrix} \\
&= \begin{pmatrix} \delta_{\bar{J}} l_{11}^{\{i\}} & \delta_{\bar{J}} l_{12}^{\{i\}} & 0 & 0 \\ \delta_{\bar{J}} l_{12}^{\{i\}} & \delta_{\bar{J}} l_{22}^{\{i\}} & 0 & 0 \\ 0 & 0 & \delta_{\bar{l}} l_{33}^{\{i\}} & \delta_{\bar{l}} l_{34}^{\{i\}} \\ 0 & 0 & \delta_{\bar{l}} l_{34}^{\{i\}} & \delta_{\bar{l}} l_{44}^{\{i\}} \end{pmatrix} \\
&= \begin{pmatrix} l_{11}^{\{i\}} & l_{12}^{\{i\}} & 0 & 0 \\ l_{12}^{\{i\}} & l_{22}^{\{i\}} & 0 & 0 \\ 0 & 0 & l_{33}^{\{i\}} & l_{34}^{\{i\}} \\ 0 & 0 & l_{34}^{\{i\}} & l_{44}^{\{i\}} \end{pmatrix} \begin{pmatrix} \delta_{\bar{J}} & 0 & 0 & 0 \\ 0 & \delta_{\bar{J}} & 0 & 0 \\ 0 & 0 & \delta_{\bar{l}} & 0 \\ 0 & 0 & 0 & \delta_{\bar{l}} \end{pmatrix} \\
&= L_i \Delta
\end{aligned}
\qquad [2.327]
$$

Full-order LPV synthesis

The computations are carried out with Matlab 2011b on a machine equipped with Intel Core i7-3740QM (2.7 GHz) processor and 8 Gb of RAM.

The tests are carried out successively with and without parallel computation (*Single Core (SC)* and *Multiple Core (MC)*, respectively). In this latter case, [MAT 10] is used with eight cores (*pool*).

The results obtained with all the algorithms are shown in Table 2.12.

IT/run = 1000	γ_{min}	γ_{mean}	γ_{std}	$t_{cpu/run}$ (min)	
				SC	*MC*
StaticCEC_QPSO	12.13	35.02	33.72	21.93	12.13
TVDecCEC_QPSO	2.44	12.41	11.42	12.26	3.78
Rnd_DE	**1.65**	1.87	0.30	7.31	**3.21**
Rnd2Bst_DE	**1.65**	1.77	0.27	14.08	4.29
Rnd2Bst_PDE	**1.65**	**1.65**	**0.003**	11.91	4.43
Std_PSO	1.74	9,916	20,884	12.33	5.66
TVRndIW_PSO	1.86	2,316.9	7,311.23	13.91	4.16
ABC	12.92	33.92	13.86	37.06	24.72
CS	7.37	20.27	10.21	21.4	5.68
CS/DE	**1.65**	**1.65**	0.0075	26.98	18.32
FF	2.29	4.40	1.87	10.51	3.42

Table 2.12. *Comparison of algorithms for full-order LPV synthesis*

It is clear that the evolutionary and even hybrid algorithms such as *CS/DE*, provide the best results. *Rnd2Bst_PDE* provides the best results with remarkable reproducibility (γ_{std} = 0.003). We keep the value of γ = 1.65 obtained. Note there, again (with a few exceptions) the advantage of using parallel computation for optimization.

The time response obtained with the LPV system looped by the LPV controller is given in Figure 2.44, which shows the stability of the closed loop, and in fact validates our evolutionary approach. We compare this response with that obtained with an LPV correction computed using a classic approach, for example, polytopic [APK 93]. Although the two responses are close, it is clear that the evolutionary approach is still better; this is due to the polytopic representation, which is more conservative than LFT representation for systems with varying parameters.

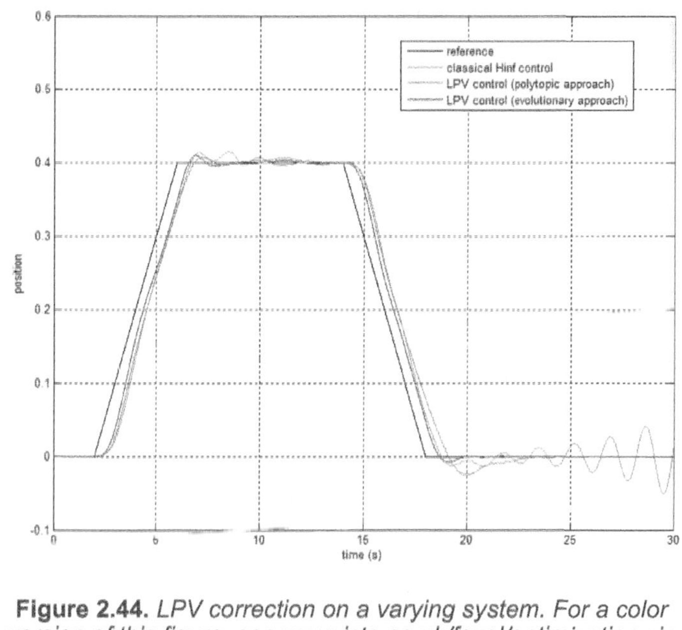

Figure 2.44. *LPV correction on a varying system. For a color version of this figure, see www.iste.co.uk/feyel/optimization.zip*

Fixed-order LPV synthesis

The computations are made with Matlab 2011b on a machine equipped with Intel Core i7-3740QM (2.7 GHz) processor and 8 Gb and RAM.

The tests are carried out successively with and without parallel computation (*Single Core (SC)* and *Multiple Core (MC)*, respectively). In this latter case, [MAT 10] is used with eight cores (*pool*).

We now search for an order 4 controller.

IT/run = 2000	γ_{min}	γ_{mean}	γ_{std}	$t_{cpu/run}$ (min)	
				SC	*MC*
StaticCEC_QPSO	3.32	49.13	42.34	84.38	46.12
TVDecCEC_QPSO	16.46	65.69	30.66	49.09	35.71
Rnd_DE	263.6	5,841	13,672	97.93	56.43
Rnd2Bst_DE	224.6	585.3	553.9	115.78	65.57
Rnd2Bst_PDE	**2.77**	**24.65**	**32.29**	137.08	50.76
Std_PSO	89.48	10,594	31,592	162.7	52.01
TVRndIW_PSO	93.12	5,564.7	15,470.8	135.8	46.68
ABC	80.98	154.70	54.19	246.8	129.12
CS	373.4	2,222.2	2,675.5	168.2	72.16
CS/DE	124.4	656.37	412.25	216.2	156.7
FF	60.49	413.03	263.97	4.68	0.73

Table 2.13. *Comparison of algorithms for structured LPV synthesis* – $\partial^o K_p = 4$

Therefore, we will retain *Rnd2Bst_PDE,* which proves to be by far the best of the algorithms for structured LPV/LFT synthesis. We will retain $\gamma = 2.77$. Because of the reduction of the controller's order, this value is higher than that obtained with full-order synthesis.

Figure 2.45 traces the closed-loop response obtained with the structured controller. It shows the controller's success in correctly stabilizing the varying system, indeed in desensitizing the closed-loop response.

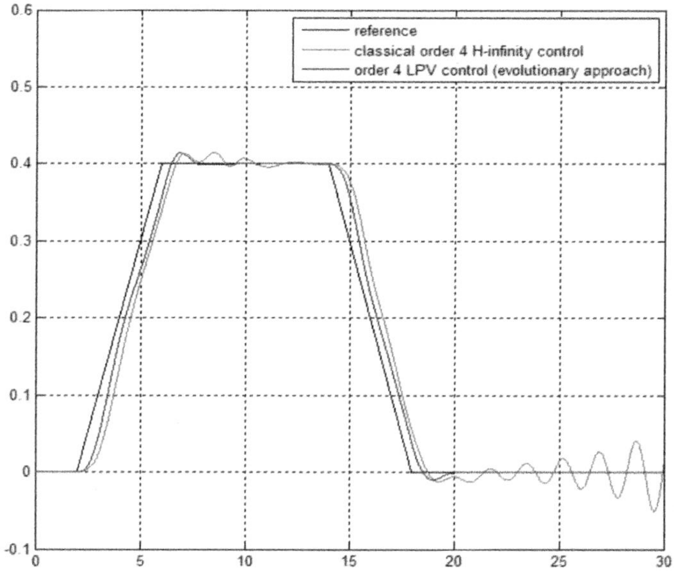

Figure 2.45. *Structured LPV correction on a varying system. For a color version of this figure, see www.iste.co.uk/feyel/optimization.zip*

3

Optimal Tuning of Structured and Robust H_∞ Controllers Against High-level Requirements

3.1. Introduction and motivations

In this chapter, we will address another major drawback inherent to modern advanced synthesis techniques: engineers must have a significant level of expertise for the weight tuning process. Indeed, starting with a frequently complex high-level specification for most of the given values, often, the first step is to simplify it (linearization procedure, approximations, etc.) in order to deduce a frequency specification for the servo-loop. This results in either more or less synthesis rounds according to the engineer's experience since the controller synthesized for this frequency specification must also crucially guarantee the high-level specification. In this way, these iterations can lead to significant additional development costs. So, in the interest of developing robust, high-performing servo-loops at a reasonable cost, this chapter will address a method for the optimal tuning of H_∞ controllers for complex industrial specifications while bypassing the particularly time-consuming frequency specification phase.

One of the main challenges that automation engineers face is in establishing a control law that loops a complex, potentially nonlinear system in order to meet a complex, potentially nonlinear specification involving the values $Z_1, Z_2, \ldots$ in response to the given inputs, disturbances, functionalities, etc. This specification often involves a finite number m of type [3.1] requirements

$$Z_1 < \overline{Z}_1,\ Z_2 < \overline{Z}_2, \ldots, Z_m < \overline{Z}_m \qquad [3.1]$$

This "high-level" industrial specification typically involves potentially mass-product systems working in an environmental range that can vary more or less (vibration, thermal environment, etc.) with variable implementation conditions. Using the adequate control law, automation engineers must consequently stabilize the system with a certain degree of robustness subject to satisfy high-level [3.1] requirements. This mission is depicted in Figure 3.1.

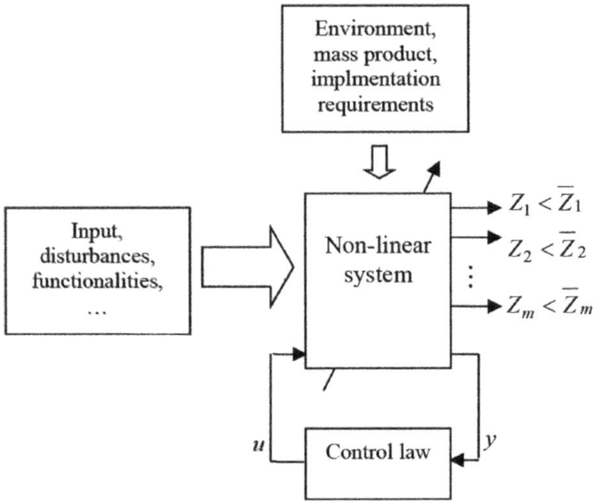

Figure 3.1. *High-level system specifications*

The complexity of this problem frequently prevents the use of more established linear tools for the controller calculation. In addition, we initially proceed via a specification simplification phase (e.g. linearization) involving another linearized system[1] whose uncertainties are seen as internal disturbances. This "simplified" specification then appears as frequency templates involving different closed loop transfers between exogenous inputs e and monitored outputs z.

In the standard form of Figure 3.2, vector e represents the exogenous inputs (references, disturbances, etc.), z represents the monitored signals (tracking errors, control signals, etc.) and y represents the measurements available to the controller to calculate the control signal u.

1 Note that the differential between the real and linearized system can also be taken into account in an uncertainty block acting on the system.

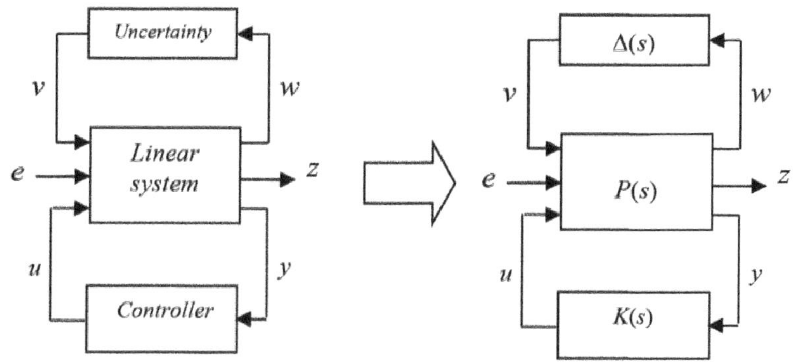

Figure 3.2. *Standard form for control*

In the 1990s, automation made significant progress with the development of controller synthesis techniques based on the H_∞ norm:

– First, the use of frequency specification was made possible by the use of frequency weights $W_o(s)$ and $W_i(s)$ applied respectively on the exogenous inputs and the monitored signals (Figure 3.3) by the minimization of the H_∞ norm concerning the transfer between these signals using a stabilizing controller for the solution to problem [3.2]. The resolution of this in the state space was explicitly formulated in [DOY 89] and derived from a dichotomy on the resolution of Riccati equations (γ-iteration).

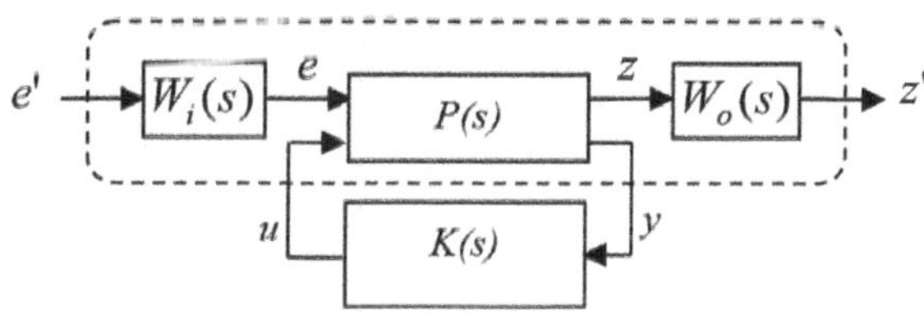

Figure 3.3. *Standard approach*

$$K(s) = \arg\min \gamma, \ \left\| W_o(s) F_l(P, K) W_i(s) \right\|_\infty < \gamma \qquad [3.2]$$

– Second, the small gain theorem provided a necessary and sufficient condition to ensure the stability of the loop for any $\Delta(s)$ uncertainty such as $\|\Delta(s)\|_\infty < \gamma^{-1}$, which is stable if and only if $\|T_{vw}(s)\|_\infty < \gamma$, thereby permitting the *a priori* integration of robustness objectives during the selection of controller synthesis weights (Figure 3.4).

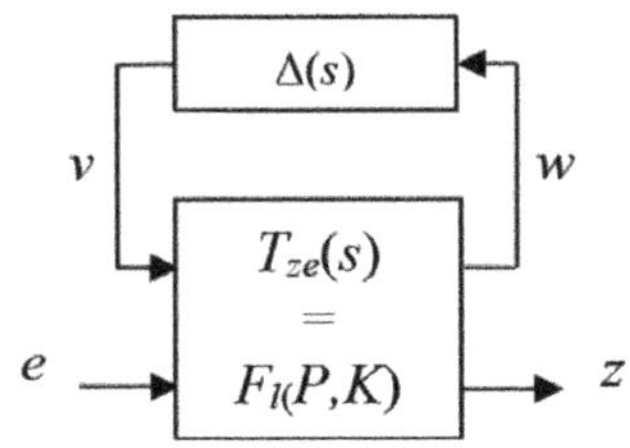

Figure 3.4. *Standard plant for robustness analysis*

In this way, with the robust control standard approach, the complexity of the controller calculation, which was thus far classically based on the examination of the open loop, is reported in the determination of the pertinent frequency weights that contribute at the first order to the final controller's performances. The problem posed by tuning these weighting functions and the conceptual divergence from the classical frequency automation rules means that this is a complex method to implement due to the expertise required. In addition, the number of degrees of freedom of tuning increases sharply with the number of measurements available, the number of exogenous signals and the number of signals to survey (at least one weight to tune per additional signal). If the specification actually does cover certain loop transfers, the total number of loop transfers defined by the standard form (and which need tuning) soon becomes significant (presence of crossed terms, etc.).

At the same time as the standard approach, the explicit solution of the robust stabilization of a plant modeled by its coprime factors [MCF 90] was also published according to the schema in Figure 3.5.

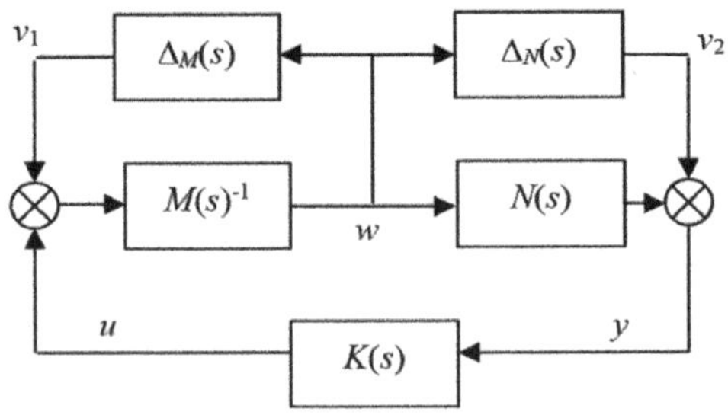

Figure 3.5. *Robust stabilization of a system modeled by its coprime factors*

Most appealing in its simplicity, this solution consists of the resolution of just two LQG-type (linear quadratic gaussian) Riccati equations. With its four-block equivalent form, it is a special instance of the standard approach of the H_∞ robust control presented above. On noting that we can model the direct and complementary sensitivity functions by an open-loop shaping and on observing that any loop transfer is proportional to these same sensitivity functions, it becomes possible to shape any loop transfer by working on a single, open-loop transfer, which sets the foundations of loop-shaping synthesis. This consists of shaping the singular values of the open loop by using weights at the system's input and output ($W_o(s)$ and $W_i(s)$ in Figure 3.6), forming a loop shape for which a stabilizing controller can be calculated, thus establishing the H_∞ loop-shaping synthesis. The controller is then calculated by resolving problem [3.3].

$$\begin{aligned} & H_p(s) = W_o(s)H(s)W_i(s) \\ & K_p(s) = \arg\min \gamma, \quad \left\| \begin{bmatrix} I \\ K_p(s) \end{bmatrix} \left(I + H_p(s)K_p(s)\right)^{-1} \begin{bmatrix} I & H_p(s) \end{bmatrix} \right\|_\infty < \gamma \\ & K(s) = W_i(s)K_p(s)W_o(s) \end{aligned} \qquad [3.3]$$

On the other hand, thanks to the gap metric concept (which mathematically announces the distance between two systems) and the small gain theorem, the stability robustness of the closed loop against uncertainties on coprime factors can be assessed before the controller is even calculated explicitly.

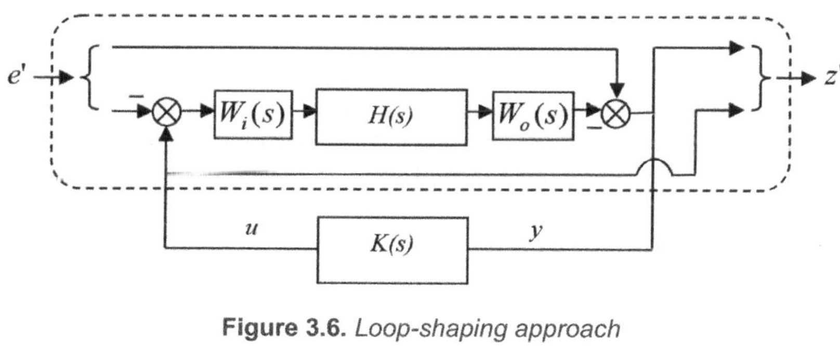

Figure 3.6. *Loop-shaping approach*

Unlike the standard approach, in the loop-shaping approach, the number of degrees of tuning freedom is limited by the system dimensions, making the approach much more straightforward. In addition, we can adjust the weighting functions according to the concepts of classical frequency automation while using the qualities of robustness inherent to H_∞ techniques.

However, for both approaches, tuning the frequency weights $W_o(s)$ and $W_i(s)$ is the hard part in terms of both difficulty and development cost. Indeed, through the controller calculation, if γ is too high, it indicates that the closed-loop frequency templates or the loop shape are too restrictive and must be adjusted[2]. This questions their fit into the high-level specification [3.1] and the robustness tradeoff. When γ is appropriate, the engineer must ensure that the controller derived from the tuning of these weights ultimately satisfies (and this is the main objective) the specification [3.1] associated with the robustness constraints as established previously. This validation phase is particularly expensive since it might require simulations and experimental validation. If the controller is not satisfactory, then the set of the weighting functions $W_o(s)$ and $W_i(s)$ has to be modified and so on.

By means of this expensive, iterative approach presented in Figure 3.7, automation engineers can choose the linear weights that best fit the high-level system specification [3.1] while ensuring the required degree of robustness. Accordingly, we must resolve the following problem:

$$\begin{array}{c} \min\limits_{W_o(s),W_i(s)} \gamma \\ \text{Subject to:} \\ K(s) \text{ ensures the required robustness and} \\ Z_1 < \overline{Z}_1,\ Z_2 < \overline{Z}_2,\ \ldots, Z_m < \overline{Z}_m \end{array} \qquad [3.4]$$

Note that the use of the standard approach for H_∞ synthesis requires the stability of the weighting functions $W_o(s)$ and $W_i(s)$. This constraint is not necessary in the context of the loop-shaping approach. However, we tend to implement stable controllers (with near pure integrators), the final controller being determined by the recombination of the (stable) pre-compensator and weights; we tend to also impose the stability of weights in the loop-shaping approach (outside of pure integrators).

2 Note that with a fixed frequency template, the tuning of weighting functions is itself a more or less lengthy iterative process.

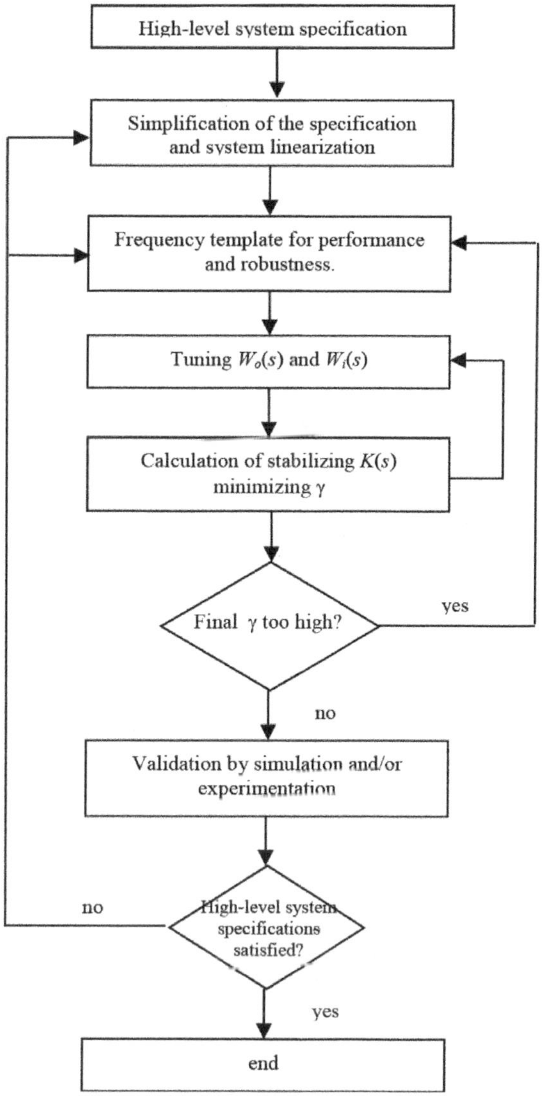

Figure 3.7. *Iterative tuning process for H∞ controllers*

In order to avoid the expensive phase of specification frequency declination, we will now consider the use of stochastic optimization to address the problem [3.4] head on as depicted in Figure 3.8.

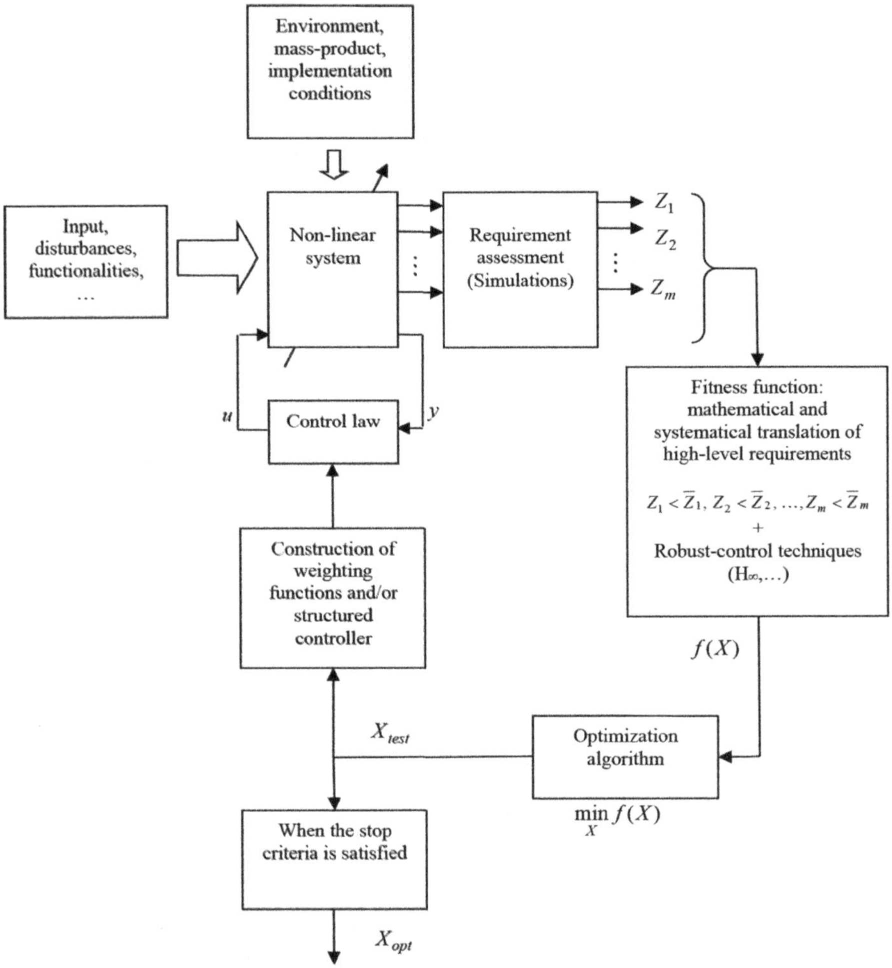

Figure 3.8. *"High-level" optimization*

We find several interesting works on this subject in the literature, for example, [CHI 96] concerning the low-order weighting functions tuned by an evolutionary algorithm and [SAN 10] showing that a complex specification with nonlinearity such as saturations can be effectively optimized with particle swarm optimization. These works clearly show that complex criteria can be considered even if their gradient is not exploitable (as it is too complex to formulate or generally undefined), with the only constraint being the possibility of evaluating the cost function in question. While of undeniable interest, the implementation of these works suffers from two major drawbacks that we will raise here:

– First, the cost function to be optimized is obtained by replacing the optimization problem under constraints [3.4] by an optimization problem without constraint. This can be achieved through the use of penalties. However, experience shows that the tuning of these penalties is a delicate task as it greatly influences the optimization process: clearly, the tuning process already requires significant expertise from the engineer. Supposing that the replacement of a weighting function tuning problem by a penalty tuning problem is of limited interest, particularly where there are many constraints, we will propose a systematic and generic tuning method for cost function penalties.

– The searched weights are frequently of a low order and structured in poles/zeros/damping. The advantage is the positivity of all the searched parameters, which reduces the search space. The disadvantage is that this *a priori* structuration of the searched weights does not guarantee the best possible solution to problem [3.4], with a fixed order of weights. Here, we propose a search for the weighting functions directly in the state space using the tridiagonal form while improving the speed of the optimization process by the transformation of the search space. These techniques were presented in the first section. Similarly, concerning MIMO weighting functions, we can consider structures that are centralized or not.

We will focus on the loop-shaping approach since it has proved itself to be better adapted for controller optimization than the standard approach for the following reasons:

– In the full-order case, the pre-compensator $K_p(s)$ is deduced directly from the resolution of two Riccati equations unlike in the standard approach based on the γ-iteration, which consists of a dichotomy on the resolution of Riccati equations. In terms of the calculation time, the loop-shaping approach proves to be far better adapted than stochastic optimization, which requires a significant number of cost function evaluations according to special evolution laws.

– As seen earlier, the number of degrees of freedom of the approach is limited by system dimensions. The number of decision variables is consequently much lower with this approach than with the standard approach, which proves to be decisive for the optimization process.

– One fundamental property of the loop-shaping approach is that the stability margins of the loop are directly linked to the value of γ [VIN 00]. Recall that:

$$\Delta G \geq \frac{1+\gamma^{-1}}{1-\gamma^{-1}},\ \Delta\Phi \geq 2\arcsin\left(\gamma^{-1}\right) \qquad [3.5]$$

where ΔG and $\Delta\Phi$ are the gain margins and phase margins, respectively (generalized in the case of MIMO). For $\gamma = 3$, we obtain $\Delta G \geq 6dB$, $\Delta\Phi \geq 39°$. So, minimizing γ by [3.4], we are sure to simultaneously maximize ΔG and $\Delta\Phi$; at the same time, we should also implement supplementary constraints in order to tune the stability margins with the standard approach as seen in [SAN 10].

Thus, in the context of loop-shaping, we observe the following reformulation:

$$\begin{gathered}\min_{W_o(s),W_i(s)} \gamma \\ \text{Subject to:} \\ K(s) \text{ ensures the required robustness and} \\ Z_1 < \overline{Z}_1, Z_2 < \overline{Z}_2, \ldots, Z_m < \overline{Z}_m \\ \\ \Rightarrow \\ \\ \max_{W_o(s),W_i(s)} \{\Delta G, \Delta\varphi\} \\ \text{Subject to:} \\ K(s) \text{ ensures the required robustness and} \\ Z_1 < \overline{Z}_1, Z_2 < \overline{Z}_2, \ldots, Z_m < \overline{Z}_m\end{gathered} \quad [3.6]$$

In the next section, we present an overview of the H_∞ loop-shaping synthesis approach.

3.2. Loop-shaping H_∞ synthesis

Here, we will briefly consider the loop-shaping approach for H_∞ synthesis. While it indeed originates in the robust stabilization problem of a system modeled by coprime factorization [MCF 90, VIN 00], we can present the approach without making reference to these concepts [MCF 92].

3.2.1. *Approach principle*

The principle of the loop-shaping approach is to proceed by shaping the open loop by implementing *frequency weights* such as those in the four-block schema in Figure 3.9.

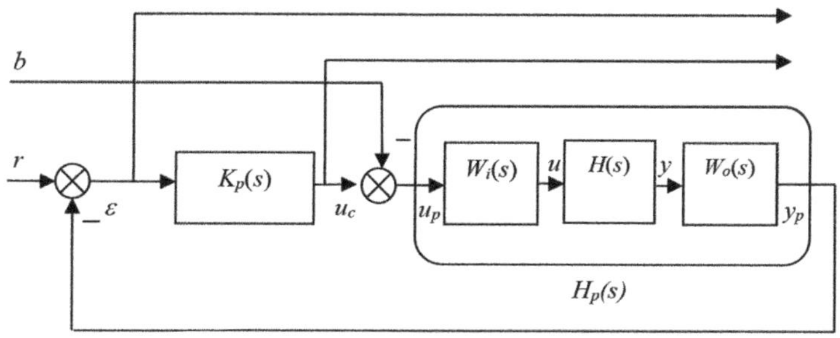

Figure 3.9. *Loop-shaping synthesis*

We write $H_p(s) = W_o(s)H(s)W_i(s)$. $H_p(s)$ is called the loop shape. The loop shape is made up of the system $H(s)$ pre-compensated by an input weight $W_i(s)$ and post-compensated by an output weight $W_o(s)$.

y_p is the measurement vector and u_p the control vector for the weighted system[3]. We should also note that b and r are external inputs analogous to state noise and observation noise of the Kalman filter of the LQG control [KWA 72].

In this schema, r is taken here as representative of a reference signal (the most-frequent specific case). It can also represent measurement noise, in which case, ε is the noisy measure.

It is known that:

$$\begin{pmatrix} \varepsilon(s) \\ u_c(s) \end{pmatrix} = \begin{pmatrix} S_p(s) & S_p(s)H_p(s) \\ K_p(s)S_p(s) & K_p(s)S_p(s)H_p(s) \end{pmatrix} \begin{pmatrix} r(s) \\ b(s) \end{pmatrix} \tag{3.7}$$

With $S_p(s)$, the direct sensitivity function at system's output is defined by:

$$S_p(s) = \left(I + H_p(s)K_p(s)\right)^{-1} \tag{3.8}$$

The aim is to search for a stabilizing $K_p(s)$ controller while ensuring that for a given value of $\gamma > 0$:

$$\left\| \begin{pmatrix} S_p(s) & S_p(s)H_p(s) \\ K_p(s)S_p(s) & K_p(s)S_p(s)H_p(s) \end{pmatrix} \right\|_\infty < \gamma \tag{3.9}$$

3 All signals that are related to the loop-shape are signed with a "p".

Assuming we have calculated such a $K_p(s)$ controller, in this case:

$$\begin{aligned}&\|S_p(s)\|_\infty < \gamma\\&\|S_p(s)H_p(s)\|_\infty < \gamma\\&\|K_p(s)S_p(s)\|_\infty < \gamma\\&\|K_p(s)S_p(s)H_p(s)\|_\infty < \gamma\end{aligned} \tag{3.10}$$

The open loop at the plant's output[4] is written as:

$$H_{BO_y}(s) = H_p(s)K_p(s) \tag{3.11}$$

– In the zones in which $\bar{\sigma}\left(H_{BO_y}(j\omega)\right) << 1$ (e.g. at high frequency):

$$\begin{aligned}&\bar{\sigma}\left(H_p(j\omega)K_p(j\omega)\right) << 1\\&\Rightarrow S_{p_y}(s) \approx I\\&\Rightarrow K_p(j\omega)S_{p_y}(j\omega) \approx K_p(j\omega)\\&\Rightarrow \bar{\sigma}\left(K_p(j\omega)\right) < \gamma\\&\Rightarrow \bar{\sigma}\left(H_p(j\omega)K_p(j\omega)\right) < \bar{\sigma}\left(K_p(j\omega)\right)\bar{\sigma}\left(H_p(j\omega)\right) < \gamma\bar{\sigma}\left(H_p(j\omega)\right)\end{aligned} \tag{3.12}$$

Finally,

$$\bar{\sigma}\left(H_{BO_y}(j\omega)\right) < \gamma\bar{\sigma}\left(H_p(j\omega)\right) \tag{3.13}$$

– In the zones in which $\underline{\sigma}\left(H_{BO_y}(j\omega)\right) >> 1$ (e.g. at low frequency):

$$\begin{aligned}&\underline{\sigma}\left(H_p(j\omega)K_p(j\omega)\right) >> 1\\&\Rightarrow \bar{\sigma}\left(H_p(j\omega)K_p(j\omega)\right) >> 1\\&\Rightarrow \bar{\sigma}\left(S_{p_y}(j\omega)H_p(j\omega)\right) \approx \bar{\sigma}\left(\left(H_p(j\omega)K_p(j\omega)\right)^{-1}H_p(j\omega)\right) \approx \bar{\sigma}\left(K_p(j\omega)^{-1}\right) < \gamma\end{aligned} \tag{3.14}$$

et :

$$\left.\begin{aligned}&\bar{\sigma}\left(K_p(j\omega)^{-1}\right) < \gamma\\&\bar{\sigma}\left(K_p(j\omega)^{-1}\right) = \frac{1}{\underline{\sigma}\left(K_p(j\omega)\right)}\end{aligned}\right\} \Rightarrow \underline{\sigma}\left(K_p(j\omega)\right) = \frac{1}{\bar{\sigma}\left(K_p(j\omega)\right)} > \frac{1}{\gamma}$$

4 On opening the loop at the output of $Hp(s)$

Finally,

$$\underline{\sigma}\left(H_p(j\,\omega)K_p(j\,\omega)\right) \geq \underline{\sigma}\left(H_p(j\,\omega)\right)\underline{\sigma}\left(K_p(j\,\omega)\right) \geq \frac{1}{\gamma}\underline{\sigma}\left(H_p(j\,\omega)\right) \qquad [3.15]$$

Or rather,

$$\underline{\sigma}\left(H_{BO_y}(j\,\omega)\right) \geq \frac{1}{\gamma}\underline{\sigma}\left(H_p(j\,\omega)\right) \qquad [3.16]$$

So, the loop shape forms an open-loop shaping target. Indeed, the open-loop singular values follow those of the loop shape described by $H_p(s)$ to the close γ ratio, and we can show [MCF 92] that this ratio moves asymmetrically toward the value $\sqrt{\gamma^2-1}$ if the weighting functions of the loop shape are such that $\lim_{\omega\to\infty}\overline{\sigma}\left(H_p(j\,\omega)\right)=0$ and $\lim_{\omega\to 0}\underline{\sigma}\left(H_p(j\,\omega)\right)=\infty$. Thus,

$$\begin{aligned}
&\frac{\underline{\sigma}\left(H_{BO_y}(j\,\omega)\right)}{\underline{\sigma}\left(H_p(j\,\omega)\right)} \to \frac{1}{\sqrt{\gamma^2-1}}, \quad \lim_{\omega\to 0}\overline{\sigma}\left(H_p(j\,\omega)\right)=\infty \\
&\frac{\overline{\sigma}\left(H_{BO_y}(j\,\omega)\right)}{\overline{\sigma}\left(H_p(j\,\omega)\right)} \to \sqrt{\gamma^2-1}, \quad \lim_{\omega\to\infty}\underline{\sigma}\left(H_p(j\,\omega)\right)=0
\end{aligned} \qquad [3.17]$$

That is,

$$\begin{aligned}
&\underline{\sigma}\left(K_p(j\,\omega)\right) \to \frac{1}{\sqrt{\gamma^2-1}}, \quad \lim_{\omega\to 0}\overline{\sigma}\left(H_p(j\,\omega)\right)=\infty \\
&\overline{\sigma}\left(K_p(j\,\omega)\right) \to \sqrt{\gamma^2-1}, \quad \lim_{\omega\to\infty}\underline{\sigma}\left(H_p(j\,\omega)\right)=0
\end{aligned} \qquad [3.18]$$

Using the following properties of the singular values:

$$\begin{aligned}
&\underline{\sigma}(A+B) \geq \underline{\sigma}(A)-\overline{\sigma}(B) \\
&\underline{\sigma}(A+B) \leq \underline{\sigma}(A)+\overline{\sigma}(B) \\
&\underline{\sigma}(AB) \geq \underline{\sigma}(A)\underline{\sigma}(B) \\
&\overline{\sigma}(AB) < \overline{\sigma}(A)\overline{\sigma}(B)
\end{aligned} \qquad [3.19]$$

We can easily deduce [VIN 00]:

$$\bar{\sigma}\left(S_p(j\,\omega)\right) \to \frac{1}{\dfrac{\underline{\sigma}\left(H_p(j\,\omega)\right)}{\sqrt{\gamma^2-1}}+1}, \quad \lim_{\omega\to 0}\bar{\sigma}\left(H_p(j\,\omega)\right)=\infty$$
$$\bar{\sigma}\left(T_p(j\,\omega)\right) \to \frac{1}{\dfrac{1}{\bar{\sigma}\left(H_p(j\,\omega)\right)\sqrt{\gamma^2-1}}+1}, \quad \lim_{\omega\to\infty}\underline{\sigma}\left(H_p(j\,\omega)\right)=0 \qquad [3.20]$$

So, γ will be higher as the loop shape is restrictive, with the open loop increasingly away from it. Accordingly, γ is a performance indicator of the loop-shape fit.

Finally, a controller $K(s)$ is obtained by combining the weights and the pre-compensator stabilizing the loop shape:

$$K(s)=W_i(s)K_p(s)W_o(s) \qquad [3.21]$$

The logic presented here and the previous results can be readily applied to the input open loop $H_{BO_u}(s)=K_p(s)H_p(s)$.

3.2.2. Generalized gain and phase margins

$\left\|\left(I_p+H_p(s)K_p(s)\right)^{-1}\right\|_\infty<\gamma$ guarantees a stability margin (in other words, in the mono variable case, the modulus margin is greater than γ^{-1}) while providing an indication of the generalized gain and phase margin values in the MIMO case.

If we note ΔG as the gain margin and $\Delta\Phi$ as the phase margin, then we have:

$$\Delta G \geq \frac{1+\gamma^{-1}}{1-\gamma^{-1}}$$
$$\Delta\Phi \geq 2\arcsin\left(\gamma^{-1}\right) \qquad [3.22]$$

So, taking a usual value of 3 for γ, we obtain $\Delta G\geq 6dB,\ \Delta\Phi\geq 39°$.

3.2.3. *Four-block interpretation of the method*

Allowing for inversible weights:

$$\begin{aligned} H_p(s) &= W_o(s)H(s)W_i(s) \\ K_p(s) &= W_i^{-1}(s)K(s)W_o^{-1}(s) \end{aligned} \qquad [3.23]$$

Assuming that they are more diagonal, we have [MCF 92]:

$$\begin{aligned} &\left\| \left(\begin{bmatrix} I \\ K_p(s) \end{bmatrix} S_{p_y}(s) \begin{bmatrix} I & H_p(s) \end{bmatrix} \right) \right\|_\infty \\ &= \left\| \left(\begin{bmatrix} I \\ W_i^{-1}(s)K(s)W_o^{-1}(s) \end{bmatrix} \left(I + W_o(s)H(s)K(s)W_o^{-1}(s)\right)^{-1} \begin{bmatrix} I & W_o(s)H(s)W_i(s) \end{bmatrix} \right) \right\|_\infty \\ &= \left\| \left(\begin{bmatrix} W_o(s) \\ W_i^{-1}(s)K(s) \end{bmatrix} W_o^{-1}(s)\left(I + W_o(s)H(s)K(s)W_o^{-1}(s)\right)^{-1} W_o(s) \begin{bmatrix} W_o(s)^{-1} & H(s)W_i(s) \end{bmatrix} \right) \right\|_\infty \\ &= \left\| \left(\begin{bmatrix} W_o(s) \\ W_i^{-1}(s)K(s) \end{bmatrix} (I + H(s)K(s))^{-1} \begin{bmatrix} W_o(s)^{-1} & H(s)W_i(s) \end{bmatrix} \right) \right\|_\infty \end{aligned} \qquad [3.24]$$

In this way, we obtain an equivalent four-block criteria for which the addition of loop-shape modeling weights is equivalent to the set weights on the exogenous inputs and monitored signals:

$$\left\| \left(\begin{bmatrix} W_o(s) \\ W_i^{-1}(s)K(s) \end{bmatrix} (I + H(s)K(s))^{-1} \begin{bmatrix} W_o(s)^{-1} & H(s)W_i(s) \end{bmatrix} \right) \right\|_\infty < \gamma \qquad [3.25]$$

So, we have an equivalence with the control schemas of Figure 3.10.

The H∞ norm of direct and complementary sensitivity functions is given in advance by the γ value, thus providing the loop stability margin before the controller calculation. In addition, note that this formulation requires that the weighting functions satisfy several hypotheses: stable and proper with inverse stable and proper, a hypothesis that does not need to be satisfied in the loop-shape formulation.

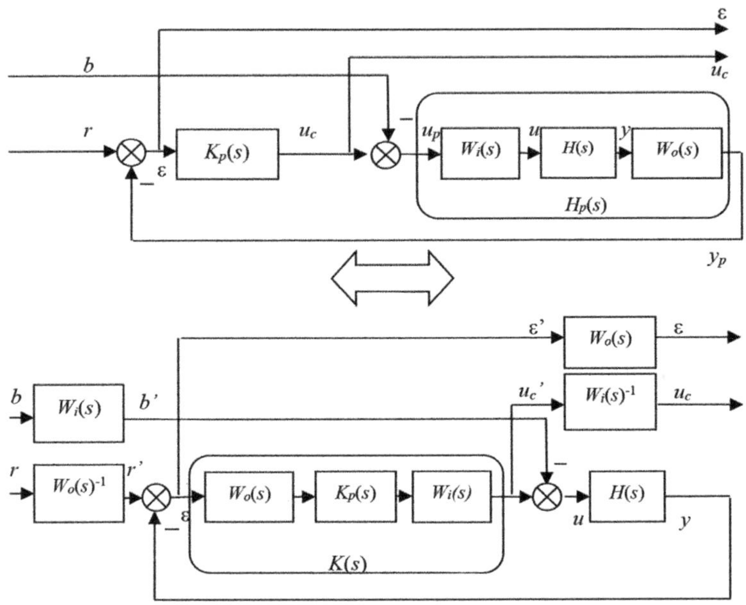

Figure 3.10. *Four-block interpretation of the method*

3.2.4. *Practical implementation*

3.2.4.1. *Specifications*

It is widely accepted that the loop-shaping specification to be used should have the form:

$$\begin{aligned} &\underline{\sigma}(H_{BO}(j\omega))_{BF} > |W_{BF}(j\omega)| \\ &\omega_{BP} > \omega_{BP\min} \\ &\Delta\Phi > \Delta\Phi_{\min} \\ &\Delta G > \Delta G_{\min} \\ &\overline{\sigma}(H_{BO}(j\omega))_{HF} < |W_{HF}(j\omega)| \end{aligned} \qquad [3.26]$$

where ω_{BP} is the expected bandwidth.

3.2.4.2. *Weighting function setting*

Weighting functions must be chosen so that the augmented system $H_p(s) = W_o(s)H(s)W_i(s)$ presents a satisfactory shape, matching the loop shape specified during the frequency specification phase. Weighting function selection is made according to the rules of conventional automatic frequency control[5]:

– In general, $W_o(s)$ is used for the "low-frequency" shaping components, that is, any possible integrators, phase leads, etc., so that the lower singular value of the low-frequency loop shape is such that $\underline{\sigma}\left(H_p(j\omega)\right) > |W_{BF}(j\omega)|$. Assuming that at a low frequency, $H(s) \approx H_{BF}(s)$, then the first reasonable choice could be where:

$$\underline{\sigma}\left(W_o(j\omega)H_{BF}(j\omega)\right) \geq |W_{BF}(j\omega)| \qquad [3.27]$$

– Typically, $W_i(s)$ is used for the roll-off components, that is, the high frequency filters, so that the higher singular value of the high-frequency loop shape is such that $\overline{\sigma}\left(H_p(j\omega)\right) < |W_{HF}(j\omega)|$. Assuming that at a high frequency, $H(s) \approx H_{HF}(s)$, then the reasonable first choice could be where:

$$\overline{\sigma}\left(H_{HF}(j\omega)W_i(j\omega)\right) \leq |W_{HF}(j\omega)| \qquad [3.28]$$

A very common exception to this is where H is low gain at high frequency (most physical systems are low pass). In this case, we can simply select:

$$\overline{\sigma}\left(W_i(j\omega)\right) = |W_{HF}(j\omega)| \qquad [3.29]$$

– The weighting functions are placed so that the loop-shape gain presents an acceptable shape without having to address the phase, managed by a significant part of the H∞ synthesis, accepting that the loop shape toward an integrator is shifted to a neighboring bandwidth.

– When the loop shape has a certain number of unstable p_i poles and/or a certain number of unstable zeros z_i, then the achievable bandwidth ω_{BP} is such that [SKO 01]:

$$\max_{\mathrm{Re}(p_i)>0}\left(|p_i|\right) < \omega_{BP} < \min_{\mathrm{Re}(z_i)>0}\left(|z_i|\right) \qquad [3.30]$$

– Weighting function tuning is performed with the results of classic automatic frequency control (PID-type tuning): despite the lack of flexibility relative to standard four-block synthesis, it bridges the gap between classic automatic frequency control and robust control.

5 The low-frequency domain is defined by $\omega < \omega BP$. The high-frequency domain is defined by $\omega > \omega BP$.

– No restriction is made as *to the stability* of the weighting functions which, in particular, can contain pure integrators.

Once the weighting functions are defined, there remains only the calculation of the pre-compensator according to the following formulae.

3.2.4.3. *Pre-compensator calculation*

The calculation is made by resolving the H_∞ problem presented earlier and with the scheme depicted in Figure 3.9. Thus, we write:

$$\left.\begin{array}{l}\dot{x}(t) = Ax(t) + Bu(t)\\ y(t) = Cx(t) + Du(t)\end{array}\right\} = H(s),\ \dim(A) = n$$

$$\left.\begin{array}{l}\dot{x}_i(t) = A_i x_i(t) + B_i u_p(t)\\ u(t) = C_i x_i(t) + D_i u_p(t)\end{array}\right\} = W_i(s),\ \ \dim(A_i) = n_i \qquad [3.31]$$

$$\left.\begin{array}{l}\dot{x}_o(t) = A_o x_o(t) + B_o y(t)\\ y_p(t) = C_o x_o(t) + D_o y(t)\end{array}\right\} = W_o(s),\ \dim(A_o) = n_o$$

The loop shape has the following state-space representation:

$$\left.\begin{array}{l}\dot{x}(t) = Ax(t) + B\left(C_i x_i(t) + D_i u_p(t)\right)\\ \dot{x}_i(t) = A_i x_i(t) + B_i u_p(t)\\ \dot{x}_o(t) = A_o x_o(t) + B_o\left(Cx(t) + D\left(C_i x_i(t) + D_i u_p(t)\right)\right)\\ \\ y_p(t) = C_o x_o(t) + D_o\left(Cx(t) + D\left(C_i x_i(t) + D_i u_p(t)\right)\right)\end{array}\right\} \qquad [3.32]$$

That is,

$$\frac{d}{dt}\begin{pmatrix} x \\ x_i \\ x_o \end{pmatrix} = \begin{pmatrix} A & BC_i & 0 \\ 0 & A_i & 0 \\ B_o C & B_o DC_i & A_o \end{pmatrix}\begin{pmatrix} x \\ x_i \\ x_o \end{pmatrix} + \begin{pmatrix} BD_i \\ B_i \\ B_o DD_i \end{pmatrix} u_p(t)$$

$$y_p(t) = \begin{pmatrix} D_o C & D_o DC_i & C_o \end{pmatrix}\begin{pmatrix} x \\ x_i \\ x_o \end{pmatrix} + D_o DD_i u_p(t) \qquad [3.33]$$

Note that:

$$\frac{dx_p}{dt} = A_p x_p + B_p u_p(t)$$
$$y_p(t) = C_p x_p + D_p u_p(t) \qquad [3.34]$$

And:

$$\dim\left(A_p\right) = n_p = n_i + n_o + n \qquad [3.35]$$

Assuming that the (A_p, B_p) pair can be stabilized and that the (C_p, A_p) pair is detectable. In this case, the pre-compensator is calculated directly by the resolution of the following two Riccati equations:

$$S = I + D_p{}^T D_p$$
$$R = I + D_p D_p{}^T$$

$$X\left(A_p - B_p S^{-1} D_p{}^T C_p\right) + \left(A_p - B_p S^{-1} D_p{}^T C_p\right)^T X$$
$$- X B_p S^{-1} B_p{}^T X + C_p{}^T R^{-1} C_p = 0$$

$$Y\left(A_p - B_p D_p{}^T R^{-1} C_p\right)^T + \left(A_p - B_p D_p{}^T R^{-1} C_p\right) Y$$
$$- Y C_p{}^T R^{-1} C_p Y + B_p S^{-1} B_p{}^T = 0 \qquad [3.36]$$

The γ_{opt} coefficient is calculated by the relationship:

$$\gamma_{opt} = \sqrt{1 + \lambda_{\sup}(YX)} \qquad [3.37]$$

For the coefficient $\gamma_0 = \varepsilon \gamma_{opt} > \gamma_{opt}$ (typically $\varepsilon = 1.01$), the pre-compensator is determined by:

$$K_p(s) = \left[\begin{array}{c|c} A_{K_p} & B_{K_p} \\ \hline C_{K_p} & D_{K_p} \end{array}\right] \qquad [3.38]$$

With:

$$\left.\begin{array}{l} A_{K_p} = A_p + B_p F + \gamma_o^2 ZYC_p^T\left(C_p + D_p F\right) \\ B_{K_p} = -\gamma_o^2 ZYC_p^T \\ C_{K_p} = B_p^T X \\ D_{K_p} = D_p^T \\ Z = \left(I + YX - \gamma_o^2 I\right)^{-1} \\ F = -S^{-1}\left(D_p^T C_p + B_p^T X\right) \end{array}\right\} \qquad [3.39]$$

The controller calculation does not involve the γ-iteration and is consequently direct.

3.2.4.4. *Open-loop and controller calculation*

Since the pre-compensator $K_p(s)$ was calculated earlier, the correction structure retained $K(s)$ is obtained by combining the weighting functions and the pre-compensator:

$$K(s) = W_i(s)K_p(s)W_0(s) \qquad [3.40]$$

The open loop (at input) is written as:

$$H_{BO_y}(s) = W_i(s)K_p(s)W_o(s)H(s) = K(s)H(s) \qquad [3.41]$$

3.2.4.5. *Comments concerning resolution*

– Since the weights and the system can be of a high order, we can normalize the frequency in order to reduce calculation problems that may occur through the resolution of the Riccati equations.

– Let $M(s)$ be a transfer matrix. The term $\partial^{\circ} M$ designates the order of M. The pre-compensator order is equal to that of the weighted system, that is:

$$\partial^{\circ} K_p = \partial^{\circ} H + \partial^{\circ} W_i + \partial^{\circ} W_o \Rightarrow \partial^{\circ} K = \partial^{\circ} H + 2\partial^{\circ} W_i + 2\partial^{\circ} W_o \qquad [3.42]$$

3.2.5. *Implementation of controllers*

There are various implementation methods. We will address frequency implementation and implementation in the form of state-feedback/observer.

3.2.5.1. *Frequency implementation*

This is a frequent form of implementation whereby the loop's stability only relies on the open loop, which, whatever the controller implementation, always has the same expression, with several implementation methods available. For the controller, we usually prefer to implement the controller using a serial product of the pre-compensator with the weights:

$$K(s) = W_i(s)K_p(s)W_o(s) \qquad [3.43]$$

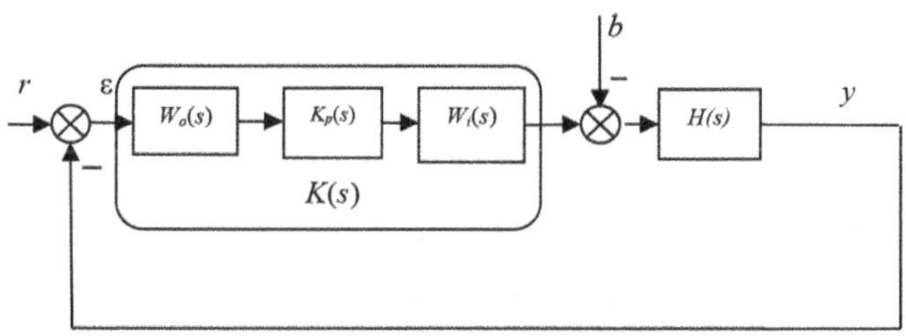

Figure 3.11. *Controller frequency implementation*

However, sometimes, the roll-off weights component $W_i(s)$ can be placed in the feedback according to requirements, as in the example in Figure 3.12.

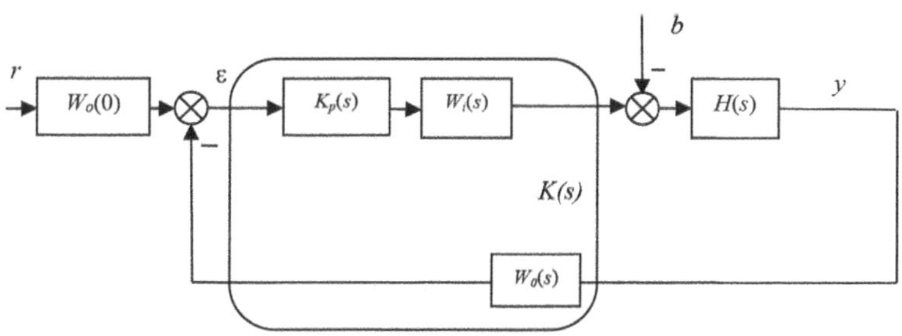

Figure 3.12. *Particular frequency implementation*

For example, this representation allows the bandwidth to be increased in a closed-loop response related to input r since at low frequency, the closed-loop response acts as the inverse of the output feedback, that is, $W_o(s)^{-1}$.

Finally, we should highlight that in [SKO 01], there is the possibility to simply implement an anti-windup structure on the $W_o(s)$ weighting function when it contains integrators and is at a non-minimal phase shift.

3.2.5.2. *State-feedback/observer implementation*

By its nature, loop-shaping H_∞ synthesis allows for immediate implementation in the form of state observer/feedback: this type of implementation is frequently seen in aeronautics where the observer is also employed for fault detection.

Here the loop shape is:

$$\left.\begin{aligned}\frac{dx_p}{dt} &= A_p x_p + B_p u_p \\ y_p &= C_p x_p + D_p u_p\end{aligned}\right\} = H_p(s) \qquad [3.44]$$

Assuming that we have a stabilizing state feedback:

$$u_p = K_c x_p \qquad [3.45]$$

When the state is not completely measurable, we can estimate it by the following observer:

$$\left.\begin{aligned}\frac{d\hat{x}_p}{dt} &= A_p \hat{x}_p + B_p u_p + K_f\left(\hat{y}_p - y_p\right) \\ \hat{y}_p &= C_p \hat{x}_p + D_p u_p\end{aligned}\right\} \qquad [3.46]$$

The use of the estimated state to elaborate the command leads to:

$$\left.\begin{aligned}\frac{d\hat{x}_p}{dt} &= A_p \hat{x}_p + B_p K_c \hat{x}_p + K_f\left(C_p \hat{x}_p + D_p K_c \hat{x}_p - y_p\right) \\ u_p &= K_c \hat{x}_p\end{aligned}\right\} \qquad [3.47]$$

That is,

$$\left.\begin{aligned}\frac{d\hat{x}_p}{dt} &= \left(A_p + B_p K_c + K_f\left(C_p + D_p K_c\right)\right)\hat{x}_p - K_f y_p \\ u_p &= K_c \hat{x}_p\end{aligned}\right\} \qquad [3.48]$$

In this way, the pre-compensator can be implemented in the form of state feedback/observer by using [VIN 00]:

$$\begin{aligned}K_f &= \gamma_o^2 ZYC_p^T \\ K_c &= F\end{aligned} \qquad [3.49]$$

By adding the direct transfer matrix of the system, we obtain the schema presented in Figure 3.13.

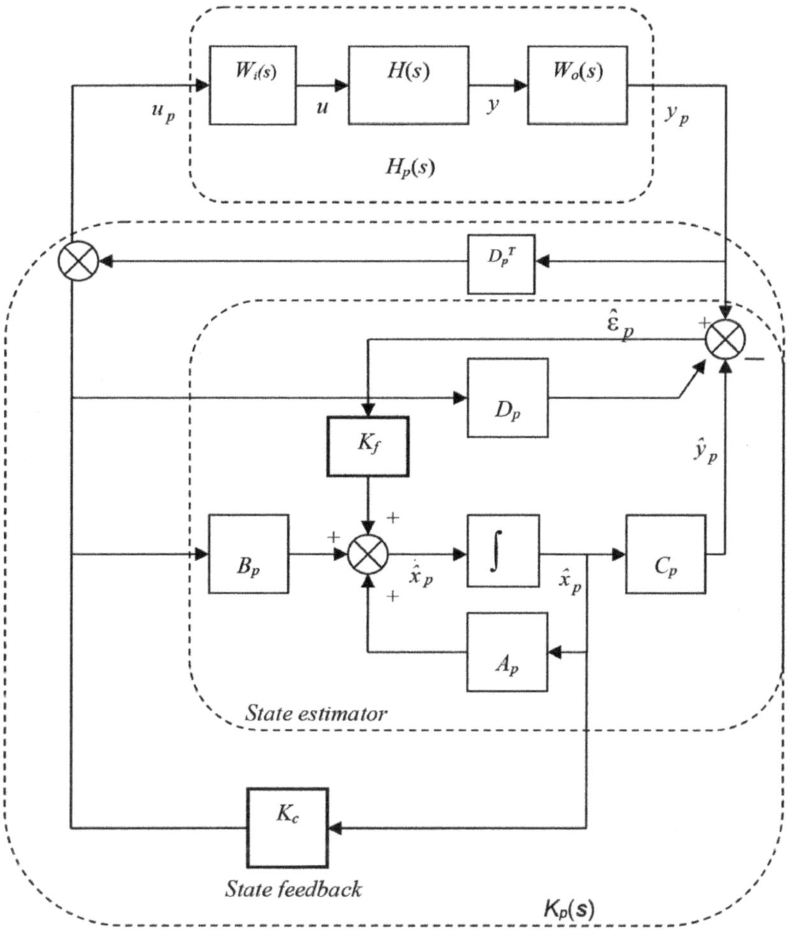

Figure 3.13. *State-feedback/observer implementation*

The advantage of this type of representation is that it gives pre-compensator states a physical meaning from the observer, thereby allowing for control over the system's state evolution. It also allows us to take into account certain nonlinearities (saturation of control signal for instance) directly in the observer.

3.3. A generic method for the declination of requirements

3.3.1. *General principles*

In industrial contexts, engineers must regularly address specifications that take the form of m requirements $Z_1 < \overline{Z}_1, Z_2 < \overline{Z}_2, \ldots, Z_m < \overline{Z}_m$. In order to avoid the frequency specification phase that inevitably leads to various approximations, we propose looking for the stable weighting functions that arise as solutions to the following optimization problem:

$$\begin{gathered} \min_{W_o(s), W_i(s)} \gamma \\ \text{Subject to:} \\ W_o(s), W_i(s) \in \mathrm{R}H_\infty \\ Z_1 < \overline{Z}_1, Z_2 < \overline{Z}_2, \ldots, Z_m < \overline{Z}_m \end{gathered} \qquad [3.50]$$

Assuming that we can evaluate each of these requirements (simulation for instance) for a given controller, this problem under constraint can be transformed into a non-constrained one by introducing penalty functions. Its general form is written as:

$$\min_{W_o(s), W_i(s) \in \mathrm{R}H_\infty} \gamma + f_c(H,K),\ f_c(H,K) = \sum_{j=1}^{m} \frac{e^{\alpha_j (Z_j - \overline{Z}_j')}}{\beta_j} \qquad [3.51]$$

The function of constraints $f_c(H, K)$ is evaluated when the system H is looped with the controller K.

In this way, we have three parameters $\alpha_j, \beta_j, \overline{Z}_j'$ to tune for each constraint. In order to satisfy the requirements on the Z_j, it is clear that we must use:

$$\begin{gathered} \overline{Z}_j' \leq \overline{Z}_j \\ \alpha_j >> 1 \end{gathered} \qquad [3.52]$$

However, there is no sense in transforming a weighting function tuning problem into a problem of tuning cost function parameters: a systematic tuning method is required.

Let us consider a problem with a single requirement:

$$\min_{W_o(s),W_i(s)\in RH_\infty} \gamma + e^{\alpha(Z-\overline{Z}')} \qquad [3.53]$$

In order to satisfy this requirement more easily, let us apply a security margin ε, whereby:

$$\overline{Z} = \overline{Z}' + \varepsilon \qquad [3.54]$$

At the end of the optimization process, γ is equal to γ_{opt}, Z is at most equal to $\overline{Z}$ and the penalty function must be of the same order of magnitude as γ_{opt}, that is:

$$e^{\alpha(Z-\overline{Z}')}\Big|_{Z=\overline{Z}} \approx \gamma_{opt} \Leftrightarrow e^{\alpha(\overline{Z}-\overline{Z}')} \approx \gamma_{opt} \Leftrightarrow e^{\alpha\varepsilon} \approx \gamma_{opt} \Leftrightarrow \alpha \approx \frac{\ln(\gamma_{opt})}{\varepsilon} \qquad [3.55]$$

In general, it should suffice to use:

$$\begin{array}{l} 0,1\overline{Z} \le \varepsilon \le 0,3\overline{Z},\ \overline{Z} > 0 \\ -0,3\overline{Z} \le \varepsilon \le -0,1\overline{Z},\ \overline{Z} < 0 \end{array} \qquad [3.56]$$

For example, with a 20% margin:

$$\varepsilon = 0,2\left|\overline{Z}\right| \qquad [3.57]$$

Furthermore, in the context of loop-shaping synthesis, a good tuning objective is characterized by (section 1.2.2):

$$\gamma_{obj} = 3 \qquad [3.58]$$

Let us return to the problem at m requirements [3.51]. At the end of optimization, f_c must be of the same order of magnitude as γ_{obj}. assuming that the m requirements are of the same importance, which would mean that:

$$\begin{array}{l} \alpha_j = \dfrac{\ln(\gamma_{obj})}{\varepsilon_j},\ \beta_j = m \\ \varepsilon_j = 0,2\left|\overline{Z}_j\right|,\ \gamma_{obj} = 3 \end{array} \qquad [3.59]$$

So:

$$f_c(H,K)=\frac{1}{m}\sum_{j=1}^{m} e^{\frac{\ln(\gamma_{obj})}{\varepsilon_j}(Z_j-\bar{Z}_j')} \qquad [3.60]$$

Note that:

$$\begin{aligned} f_c(H,K)&=\frac{1}{m}\sum_{j=1}^{m} e^{\frac{\ln(\gamma_{obj})}{\varepsilon_j}(Z_j-\bar{Z}_j')} \\ &=\frac{1}{m}\sum_{j=1}^{m} e^{\frac{\ln(\gamma_{obj})}{\varepsilon_j}(Z_j-\bar{Z}_j+\varepsilon_j)} \\ &=\frac{1}{m}\sum_{j=1}^{m} e^{\frac{\ln(\gamma_{obj})}{\varepsilon_j}(Z_j-\bar{Z}_j)} e^{\ln(\gamma_{obj})} \\ &=\frac{\gamma_{obj}}{m}\sum_{j=1}^{m} e^{\frac{\ln(\gamma_{obj})}{\varepsilon_j}(Z_j-\bar{Z}_j)} \end{aligned} \qquad [3.61]$$

Accordingly, this comprises a means of systematic declination for high-level specifications with no loss of generality. The penalty coefficients are formally established by the relationships given above. Finally, we can rewrite:

$$\begin{gathered} \min_{W_o(s),W_i(s)\in RH_\infty} \gamma \\ \text{Subject to:} \\ Z_1<\bar{Z}_1,\, Z_2<\bar{Z}_2,\,\ldots,Z_m<\bar{Z}_m \\ \Downarrow \\ \min_{W_o(s),W_i(s)\in RH_\infty} \gamma+f_c(H,K),\; f_c(H,K)=\frac{\gamma_{obj}}{m}\sum_{j=1}^{m} e^{\frac{\ln(\gamma_{obj})}{\varepsilon_j}(Z_j-\bar{Z}_j)} \end{gathered} \qquad [3.62]$$

3.3.2. *Special cases*

3.3.2.1. *Zero constraint*

The first special case concerns the optimization of a null requirement, that is:

$$Z_j<0 \Leftrightarrow \bar{Z}_j=0 \qquad [3.63]$$

In this case, we can address such a requirement by observing that:

$$Z_j < 0 \Leftrightarrow Z_j + \lambda < \lambda,\ \lambda << 1 \in \mathbf{R} \Leftrightarrow \Im_j < \overline{\Im}_j \qquad [3.64]$$

3.3.2.2. *Equality-type constraint*

The second special case concerns the optimization of equality-type requirements, that is:

$$Z_j = \overline{Z}_j \qquad [3.65]$$

In this case, we can envisage addressing such a requirement by observing that:

$$Z_j = \overline{Z}_j \Leftrightarrow \begin{cases} Z_j < \overline{Z}_j \\ Z_j > \overline{Z}_j \end{cases} \Leftrightarrow \begin{cases} Z_j < \overline{Z}_j \\ -Z_j < -\overline{Z}_j \end{cases} \Leftrightarrow \begin{cases} Z_j < \overline{Z}_j \\ \Im_j < \overline{\Im}_j \end{cases} \qquad [3.66]$$

which amounts to optimizing both sub-requirements simultaneously.

3.3.3. *Management of requirement priority level*

Part of the interest of the declination method given earlier is the possibility of easily managing the priority levels of different requirements. Indeed, we can write the penalty function in the form of a limited development:

$$e^{\alpha(z-\overline{z}')} = \sum_{k=0}^{+\infty} \frac{\left(\alpha\left(z-\overline{z}'\right)\right)^k}{k\,!} \qquad [3.67]$$

In this way, we can establish the priority order n of a constraint by proceeding with an order n limited development of the penalty function. Nonetheless:

– the resulting penalty must remain positive;

– the penalty must remain monotonous, which restricts the use of limited development to odd orders.

So, we can define the penalty of priority n as follows:

$$\begin{array}{c} \text{High priority } (\infty)\text{: } e^{\alpha(z-\overline{z}')} \\ \Downarrow \\ \text{Priority of } (n)\text{: } \max\left(0,\ 1+\sum_{k=1}^{2n+1} \frac{\left(\alpha\left(z-\overline{z}'\right)\right)^k}{k\,!}\right) \end{array} \qquad [3.68]$$

The advantage of this method is that it allows us to keep the tuning rule of α_j and β_j by initially considering all of the requirements as being of high priority.

In Figure 3.14, we show the first orders of development in terms of the corresponding exponential function. Only the odd orders need to be used (here, orders 1 and 3).

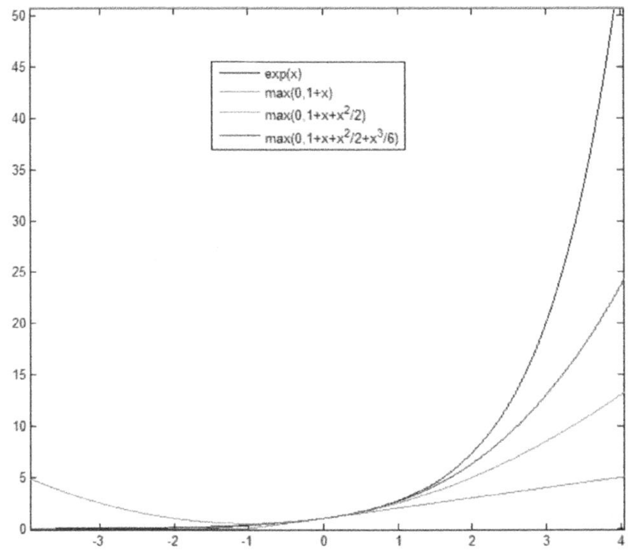

Figure 3.14. *Management of requirements priority. For a color version of this figure, see www.iste.co.uk/feyel/optimization.zip*

3.4. Optimal tuning of weighting functions

3.4.1. *Optimization on nominal plant*

3.4.1.1. *Definition of the cost function*

Here, we will take the same approach as that proposed in [FEY 13b], where a solution is sought for the following problem:

$$\begin{gathered} \min_{W_o(s),W_i(s)} \gamma \\ \text{Subject to:} \\ W_o(s),W_i(s) \in RH_\infty \\ Z_1 < \overline{Z}_1,\, Z_2 < \overline{Z}_2,\, \ldots, Z_m < \overline{Z}_m \end{gathered} \qquad [3.69]$$

We showed that this problem can be approached by the following:

$$\min_{W_o(s),W_i(s)\in RH_\infty} \gamma + f_c(H,K),\ f_c(H,K) = \frac{\gamma_{obj}}{m}\sum_{j=1}^{m} e^{\frac{\ln(\gamma_{obj})}{0.2\left|\bar{z}_j\right|}(z_j-\bar{z}_j)} \quad [3.70]$$

Recall that loop-shaping synthesis consists of calculating the loop controller on the basis of a set of weighting functions by means of the following equations:

$$\begin{aligned} &H_p(s) = W_o(s)H(s)W_i(s) \\ &K_p(s) = \arg\min \gamma,\ \left\| \begin{bmatrix} I \\ K_p(s) \end{bmatrix} \left(I + H_p(s)K_p(s)\right)^{-1} \begin{bmatrix} I & H_p(s) \end{bmatrix} \right\|_\infty < \gamma \\ &K(s) = W_i(s)K_p(s)W_o(s) \end{aligned} \quad [3.71]$$

We have decided not to structure the weighting functions in terms of poles/zeros/damping. Accordingly, we have decided to search for the weights (centralized or not) directly in state-space form:

$$W_i(s) \overset{ss}{=} \left[\begin{array}{c|c} A_i & B_i \\ \hline C_i & D_i \end{array}\right],\ W_o(s) \overset{ss}{=} \left[\begin{array}{c|c} A_o & B_o \\ \hline C_o & D_o \end{array}\right] \quad [3.72]$$

Writing:

$$\bar{\lambda}_o = \max_j\left(\mathrm{Re}\left(\lambda_j\left(A_o\right)\right)\right),\ \bar{\lambda}_i = \max_j\left(\mathrm{Re}\left(\lambda_j\left(A_i\right)\right)\right) \quad [3.73]$$

And for the sake of simplicity, noting that:

$$\Theta_i = \begin{pmatrix} A_i & B_i \\ C_i & D_i \end{pmatrix},\ \Theta_o = \begin{pmatrix} A_o & B_o \\ C_o & D_o \end{pmatrix} \quad [3.74]$$

Thus, we can calculate the H_∞-optimal controller, solution to problem [3.70], by considering the following optimization problem:

$$\min_{\Theta_i,\Theta_o} f(\Theta_i,\Theta_o)$$

If $\max\left(\overline{\lambda}_o, \overline{\lambda}_i\right) \geq 0$

$$f(\Theta_i, \Theta_o) = \max\left(\overline{\lambda}_o, \overline{\lambda}_i\right)$$

Else

$$\begin{aligned} &H_p(s) = W_o(s)H(s)W_i(s) \\ &K_p(s) = \arg\min \gamma, \left\| \begin{bmatrix} I \\ K_p(s) \end{bmatrix} \left(I + H_p(s)K_p(s)\right)^{-1} \begin{bmatrix} I & H_p(s) \end{bmatrix} \right\|_\infty < \gamma \\ &K(s) = W_i(s)K_p(s)W_o(s) \\ &f(\Theta_i, \Theta_o) = -\frac{1}{\gamma + f_c(H, K)} \end{aligned} \quad [3.75]$$

This optimization scheme is identical to that of H∞ synthesis. It is broken down into two phases: a first phase of weighting function stabilization and a second phase related to H_∞ optimization proper. Here again, the cost function is non-differentiable but quasi-continuous.

Clearly, according to the architecture of the weighting functions retained (centralized or not), the only degree of expertise required for engineers is reduced to the choice of the weighting function order.

3.4.1.2. *Implementation*

In order to optimize the rate of convergence of the optimization process, weighting functions are sought in a tridiagonal form and the search space is transformed so as to increase the algorithm's sensitivity (see section 2.2.3.3).

As for the fixed order controller synthesis, weighting functions can be decentralized or not. Their reconstruction follows exactly the same scheme. In both cases, the decision vector will collect the state representation coefficients for the two weighting functions in question:

$$X = \begin{pmatrix} X_o & X_i \end{pmatrix}^T \quad [3.76]$$

For example:

– If $W_o(s)$ is centralized, then we write:

$$\begin{aligned} X_o &= \left(x_{A_o}, x_{B_o}, x_{C_o}, x_{D_o}\right) \\ x_{A_o} &= \left(a_{1,1}, \ldots, a_{n,n}, a_{1,2}, \ldots, a_{n-1,n}, a_{2,1}, \ldots, a_{n,n-1}\right) \\ x_{B_o} &= \left(b_{1,1}, \ldots, b_{n,1}, b_{1,2}, \ldots, b_{n,2}, \ldots, b_{1,n_y}, \ldots, b_{n,n_y}\right) \\ x_{C_o} &= \left(c_{1,1}, \ldots, c_{1,n}, c_{2,1}, \ldots, c_{2,n}, \ldots, c_{n_y,1}, \ldots, c_{n_y,n}\right) \\ x_{D_o} &= \left(d_{1,1}, \ldots, d_{1,n_y}, d_{2,1}, \ldots, d_{2,n_y}, \ldots, d_{n_y,1}, \ldots, d_{n_y,n_y}\right) \end{aligned} \quad [3.77]$$

– If $W_o(s)$ is decentralized[6], then we write:

$$W_o(s) = \begin{bmatrix} W_{o1,1}(s) & \cdots & W_{o1,n_y}(s) \\ \vdots & \ddots & \vdots \\ W_{on_y,1}(s) & \cdots & W_{on_y,n_y}(s) \end{bmatrix} \quad [3.78]$$

The order of each transfer is defined by:

$$\partial^\circ W_o = \begin{bmatrix} n_{o1,1} & \cdots & n_{o1,n_y} \\ \vdots & \ddots & \vdots \\ n_{on_y,1} & \cdots & n_{on_y,n_y} \end{bmatrix} \quad [3.79]$$

The decision vector is written as:

$$\begin{aligned} X &= \left(X_{o_{1,1}}, \ldots, X_{o_{1,n_y}}, X_{o_{2,1}}, \ldots, X_{o_{2,n_y}}, \ldots, X_{o_{n_y,1}}, \ldots, X_{o_{n_y,n_y}}\right)^T \\ X_{o_{i,j}} &= \left(x_D\right)_{i,j} \;\; if \;\; n_{oi,j} = 0 \\ X_{o_{i,j}} &= \left(x_A, x_B, x_C, x_D\right)_{i,j} \;\; si \;\; n_{oi,j} \geq 1 \\ &\left.\begin{aligned} x_A &= \left(a_{1,1}, \ldots, a_{n,n}, a_{1,2}, \ldots, a_{n-1,n}, a_{2,1}, \ldots, a_{n,n-1}\right) \\ x_B &= \left(b_{1,1}, \ldots, b_{n,1}\right) \\ x_C &= \left(c_{1,1}, \ldots, c_{1,n}\right) \\ x_D &= d \end{aligned}\right\} n = n_{oi,j} \end{aligned} \quad [3.80]$$

6 In accordance with the section 2.2.3.4.

We will to attempt to resolve the following optimization problem:

$$\min_{X\in[\underline{x};\overline{x}]} f(X) \tag{3.81}$$

where the variable X is already defined and where function $f(X)$ consists of performing the following steps:

– build $W_o(s) \overset{ss}{=} \left[\begin{array}{c|c} A_o & B_o \\ \hline C_o & D_o \end{array}\right]$ and $W_i(s) \overset{ss}{=} \left[\begin{array}{c|c} A_i & B_i \\ \hline C_i & D_i \end{array}\right]$ from X;

– calculate the following spectral abscissa:

$$\overline{\lambda}_o = \max_j\left(\mathrm{Re}\left(\lambda_j\left(A_o\right)\right)\right),\ \overline{\lambda}_i = \max_j\left(\mathrm{Re}\left(\lambda_j\left(A_i\right)\right)\right),\ \overline{\lambda} = \max(\overline{\lambda}_o, \overline{\lambda}_i) \tag{3.82}$$

– if $\overline{\lambda} \geq 0$, then $f(X) = \overline{\lambda}$,

– else:

- build the loop shape $H_p(s) = W_o(s)H(s)W_i(s)$;

- calculate the pre-compensator $K_p(s)$ that minimizes γ so that:

$$\left\| \begin{bmatrix} I \\ K_p(s) \end{bmatrix} \left(I + H_p(s)K_p(s)\right)^{-1} \begin{bmatrix} I & H_p(s) \end{bmatrix} \right\|_\infty < \gamma \tag{3.83}$$

- calculate the final controller $K(s) = W_i(s)K_p(s)W_o(s)$;

- evaluate $f(X) = -\dfrac{1}{\gamma + f_c(H,K)}$.

3.4.2. *Multiple plant optimization*

In this section, we assume the availability of the following ball of plants centered on a nominal plant $H_0(s)$:

$$\wp = \{H_i(s),\ i = 0,\ldots,N\} \tag{3.84}$$

3.4.2.1. *Stability robustness optimization problem*

We need the controller obtained previously, working on the nominal plant, to also stabilize the set $\wp$. This stabilization objective is equivalent to the following:

$$\begin{gathered} K(s) \text{ stabilizes } \wp \\ \Leftrightarrow \\ H_{cl_j}(s) \in \mathbf{R}H_\infty,\ j = 1,\ldots,N \\ H_{cl_j}(s) = \begin{bmatrix} I \\ K(s) \end{bmatrix} (I + H_j(s)K(s))^{-1} \begin{bmatrix} I & H_j(s) \end{bmatrix} \end{gathered} \quad [3.85]$$

The stability robustness optimization problem consists of resolving the following problem:

$$\begin{gathered} \min_{W_o(s),W_i(s)} \gamma \\ \text{Subject to:} \\ W_o(s),\ W_i(s) \in \mathbf{R}H_\infty, \\ H_{cl_j}(s) \in \mathbf{R}H_\infty,\ j = 1,\ldots,N, \\ Z_1 < \overline{Z}_1,\ Z_2 < \overline{Z}_2, \ldots, Z_m < \overline{Z}_m \text{ for } H_0(s). \end{gathered} \quad [3.86]$$

Note that the controller already stabilizes the nominal plant.

Then, we calculate the H_∞-optimal controller, solution to problem [3.86], by considering the following optimization problem:

$$\min_{\Theta_i,\Theta_o} f(\Theta_i,\Theta_o)$$

If $\max(\overline{\lambda}_o, \overline{\lambda}_i) \geq 0$

$$f(\Theta_i,\Theta_o) = \max(\overline{\lambda}_o, \overline{\lambda}_i)$$

Else

$$\begin{aligned} & H_p(s) = W_o(s)H_0(s)W_i(s) \\ & K_p(s) = \arg\min \gamma, \left\| \begin{bmatrix} I \\ K_p(s) \end{bmatrix} (I + H_p(s)K_p(s))^{-1} \begin{bmatrix} I & H_p(s) \end{bmatrix} \right\|_\infty < \gamma \\ & K(s) = W_i(s)K_p(s)W_o(s) \\ & \left. \begin{aligned} & H_{cl_j}(s) = \begin{bmatrix} I \\ K(s) \end{bmatrix} (I + H_j(s)K(s))^{-1} \begin{bmatrix} I & H_j(s) \end{bmatrix} \\ & \overline{\lambda}_{cl\,j} = \max\left(\mathrm{Re}\left(\lambda\left(A_{cl\,j}\right)\right)\right) \end{aligned} \right\} j = 1,\ldots,N \end{aligned} \quad [3.87]$$

If $\max_{j=1,\ldots,N} \overline{\lambda}_{cl\,j} \geq 0$

$$f(\Theta_i,\Theta_o) \rightarrow \infty$$

Else

$$f(\Theta_i,\Theta_o) = -\frac{1}{\gamma + f_c(H_0,K)}$$

End If

End If

3.4.2.2. *Stability margin robustness optimization problem*

We need the controller obtained previously, working on the nominal plant and stabilizing $\wp$, to also maximize the "worst case" stability margins when it loops the $H_j(s)$, which amounts to minimizing the γ worst case.

The stability margin robustness optimization problem consists of resolving the following problem:

$$\begin{gathered} \min_{W_o(s),W_i(s)} \gamma,\ \gamma = \max_{i=0,\ldots,N} \gamma_i \\ \text{Subject to:} \\ W_o(s),\ W_i(s) \in \mathbf{R}H_\infty \\ H_{cl\,j}(s) \in \mathbf{R}H_\infty,\ j=1,\ldots,N \\ Z_1 < \overline{Z}_1,\ Z_2 < \overline{Z}_2,\ \ldots, Z_m < \overline{Z}_m \ \text{ for } H_o(s) \end{gathered} \qquad [3.88]$$

We then calculate the H_∞-optimal controller, solution to problem [3.88], by considering the following optimization problem:

$$\min_{\Theta_i,\Theta_o} f(\Theta_i,\Theta_o)$$

If $\max\left(\overline{\lambda}_o, \overline{\lambda}_i\right) \geq 0$

$$f(\Theta_i,\Theta_o) = \max\left(\overline{\lambda}_o, \overline{\lambda}_i\right)$$

Else

$$
\left.
\begin{aligned}
&H_p(s) = W_o(s)H_0(s)W_i(s)\\
&K_p(s) = \arg\min \gamma_0, \left\| \begin{bmatrix} I \\ K_p(s) \end{bmatrix} \left(I + H_p(s)K_p(s)\right)^{-1} \begin{bmatrix} I & H_p(s) \end{bmatrix} \right\|_\infty < \gamma_0\\
&K(s) = W_i(s)K_p(s)W_o(s)\\
&\left.\begin{aligned}
&H_{pj}(s) = W_o(s)H_j(s)W_i(s)\\
&H_{clj}(s) = \begin{bmatrix} I \\ K_p(s) \end{bmatrix} \left(I + H_{pj}(s)K_p(s)\right)^{-1} \begin{bmatrix} I & H_{pj}(s) \end{bmatrix}\\
&\bar{\lambda}_{clj} = \max\left(\mathrm{Re}\left(\lambda\left(A_{clj}\right)\right)\right)
\end{aligned}\right\} j = 1,\ldots,N
\end{aligned}
\right. \quad [3.89]
$$

If $\max\limits_{j=1,\ldots,N} \bar{\lambda}_{clj} \geq 0$

$$f(\Theta_i, \Theta_o) \to \infty$$

Else

$$\gamma_j = \left\| H_{clj}(s) \right\|_\infty, j = 1,\ldots,N$$

$$f(\Theta_i, \Theta_o) = -\frac{1}{\max\limits_{j=0,\ldots,N} \gamma_j + f_c(H_0, K)}$$

End If

End If

3.4.2.3. *Performance robustness optimization problem*

We need the controller obtained previously, working on the nominal plant, stabilizing $\wp$ and maximizing the "worst case" stability margins on $\wp$, to also be performing across all $\wp$. Accordingly, this is an issue of performance robustness.

An ambitious objective would be to look for the controller that is the solution to the following problem:

$$\min_{W_o(s),W_i(s)} \gamma,\ \gamma = \max_{i=0,\ldots,N} \gamma_i$$

Subject to:

$$\begin{aligned} & W_o(s),\ W_i(s) \in \mathrm{R}H_\infty, \\ & \left.\begin{aligned} & H_{cl}(s) \in \mathrm{R}H_\infty, \\ & Z_1 < \overline{Z}_1,\ Z_2 < \overline{Z}_2,\ \ldots, Z_m < \overline{Z}_m \end{aligned}\right\} \forall H_j(s) \in \wp \end{aligned} \tag{3.90}$$

There are several more or less conservative formulations with which we can address this problem.

The most direct one consists of establishing the H_∞-optimal controller, solution to problem [3.90], by considering the following optimization problem:

$$\min_{\Theta_i,\Theta_o} f(\Theta_i,\Theta_o)$$

If $\max\left(\overline{\lambda}_o,\ \overline{\lambda}_i\right) \geq 0$

$$f(\Theta_i,\Theta_o) = \max\left(\overline{\lambda}_o, \overline{\lambda}_i\right)$$

Else

$$\begin{aligned} & H_p(s) = W_o(s)H_0(s)W_i(s) \\ & K_p(s) = \arg\min \gamma_0,\ \left\| \begin{bmatrix} I \\ K_p(s) \end{bmatrix} \left(I + H_p(s)K_p(s)\right)^{-1} \begin{bmatrix} I & H_p(s) \end{bmatrix} \right\|_\infty < \gamma_0 \\ & K(s) = W_i(s)K_p(s)W_o(s) \\ & \left.\begin{aligned} & H_{pj}(s) = W_o(s)H_j(s)W_i(s) \\ & H_{clj}(s) = \begin{bmatrix} I \\ K_p(s) \end{bmatrix} \left(I + H_{pj}(s)K_p(s)\right)^{-1} \begin{bmatrix} I & H_{pj}(s) \end{bmatrix} \\ & \overline{\lambda}_{clj} = \max\left(\mathrm{Re}\left(\lambda\left(A_{clj}\right)\right)\right) \end{aligned}\right\} j = 1,\ldots,N \end{aligned} \tag{3.91}$$

If $\max\limits_{j=1,\ldots,N} \overline{\lambda}_{cl\,j} \geq 0$

$$f(\Theta_i, \Theta_o) \rightarrow \infty$$

Else

$$\gamma_j = \left\| H_{cl\,j}(s) \right\|_\infty, j = 1, \ldots, N$$

$$f(\Theta_i, \Theta_o) = -\frac{1}{\max\limits_{j=0,\ldots,N} \left(\gamma_j + f_c\left(H_j, K\right)\right)}$$

End If

End If

3.4.3. *Applicative example – inertial stabilization of line of sight*

3.4.3.1. *Observation viewfinder*

3.4.3.1.1. System description and requirements

In order to illustrate the synthesis method, we propose applying the theory developed in this part to the line of sight stabilization problem.

The aim of a two-axis gyrostabilized platform is to maintain as inertial (that is, insensitive to the motion of a host vehicle movement relative to the support) the observation device's orientation in elevation and azimuth. This inertiality is made possible by the use of gyrometers (respectively, gyroscopes) to measure the absolute rate (respectively, the absolute orientation) of the gimbals, together with electric motors as actuators (Figure 3.15).

For more details concerning gyrostabilized viewfinders, we refer to [MAS 08, HIL 08], which describe the different available architectures in a more exhaustive manner. Each axis can be considered to be a SIMO system $H_0(s)$ with input u (motor input voltage) and with outputs i and Ω_a, which are, respectively, the motor current and absolute rate of the load measured by the gyrometer.

$$\begin{aligned} Y(s) &= \begin{pmatrix} \Omega_a(s) \\ I(s) \end{pmatrix} = H_0(s)U(s) \\ H_0(s) &= \begin{pmatrix} H_m(s) \\ 1 \end{pmatrix} H_{el}(s) = \begin{pmatrix} H_{u\rightarrow\Omega}(s) \\ H_{u\rightarrow i}(s) \end{pmatrix} \end{aligned} \qquad [3.92]$$

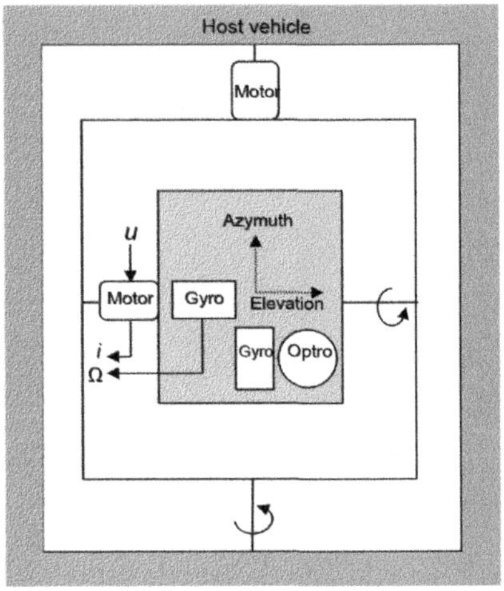

Figure 3.15. *Simplified block diagram of a gyrostabilized platform*

An example of frequency representation for $H_0(s)$ is given in Figure 3.16. The system in question is order 17.

For confidentiality reasons, the frequencies and signals involved are normalized in this study.

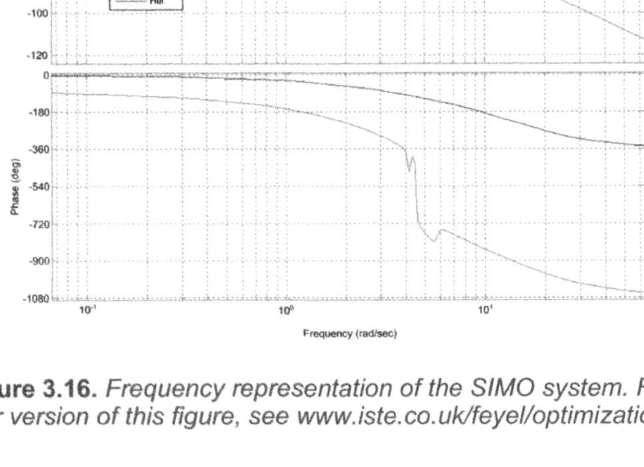

Figure 3.16. *Frequency representation of the SIMO system. For a color version of this figure, see www.iste.co.uk/feyel/optimization.zip*

Typically, the electromechanical platform comprises an optronic device that confers the viewfinder with certain fundamental capabilities: observation, detection, observation, etc. In order to satisfy these capabilities, the viewfinder must have sufficient range. This range is directly linked to the stabilization performance of the gyrometric servo-loop toward mechanical environments and more or less severe working conditions and must be compatible with the sizing of the optronic sensors; these sensors are characterized by an integration time (T_i) (time in which the image is electronically acquired and generated) and by the size of their pixels (IFOV). Accordingly, in order to obtain a high-quality observation, we must limit the absolute angle variations of the line of sight to a value compatible with IFOV during the integration time T_i in response to all disturbances.

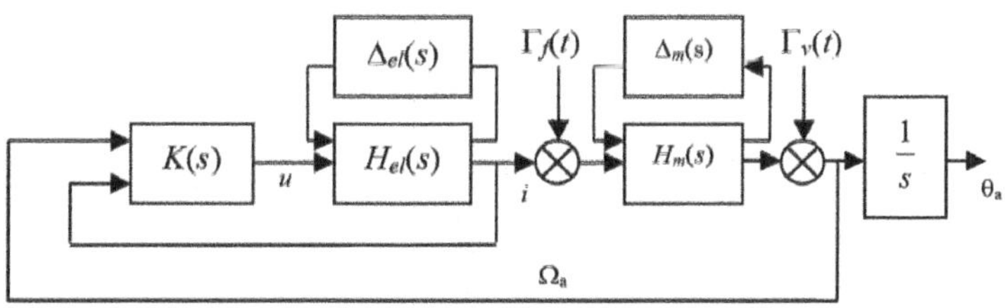

Figure 3.17. *Observation viewfinder control scheme*

As shown in the control scheme in Figure 3.17, the issue of inertial stabilization consists primarily of rejecting two types of external disturbances by means of a gyrometric servo-loop:

– The first of these disturbances is the friction torque created during a relative motion between the viewfinder host vehicle and the gimbals. Most of the time, this torque is modeled by an input plant disturbance step Γ_{f0} for which the sign depends on the sign of the relative velocity (Coulomb model):

$$\Gamma_f(t) = \Gamma_{f0}\, sign(\Omega_{relative}(t)),\ \Gamma_{f0} < 0 \tag{3.93}$$

This disturbance appears when the attitude of the support and/or the precession reference of the viewfinder changes. In order to avoid picture motion during observation (and thereby ensure visual comfort), we must make the angular displacement of the line of sight in response to the friction (when it increases from $-\Gamma_{f0}$ to $+\Gamma_{f0}$, which is a relative step of $2\Gamma_{f0}$) compatible with IFOV. At the same time, the steady-state error created must be zero. More specifically, the temporal dynamic rejection of the friction must be compatible with the maximal dynamic of the support's oscillating regime, which is only several Hz or a decade at most. The steady-state error in response to friction must consequently be zero (or at least very low) at the end of time T_f:

$$\left.\begin{array}{l} |\theta_a(t)| \le \bar{\theta}_f \\ \max\limits_{t>T_f} |\theta_a(t)| < \varepsilon_f \end{array}\right\} \text{ in response to step } \Gamma_f(t) = 2\Gamma_{f0} \qquad [3.94]$$

– The second disturbance comes from the mechanical deformations of the viewfinder's structure, whose flexibility transmits part of the mechanical environment of the viewfinder to the line of sight, which occurs in three orthogonal directions: the longitudinal direction (X axis), the transversal direction (Y axis) and the vertical direction (Z axis). For each excitation direction, we can obtain a temporal acquisition of the disturbance induced on the (non-looped) line of sight (Figure 3.18), here called Γ_{vx}, Γ_{vy} and Γ_{vz}, respectively.

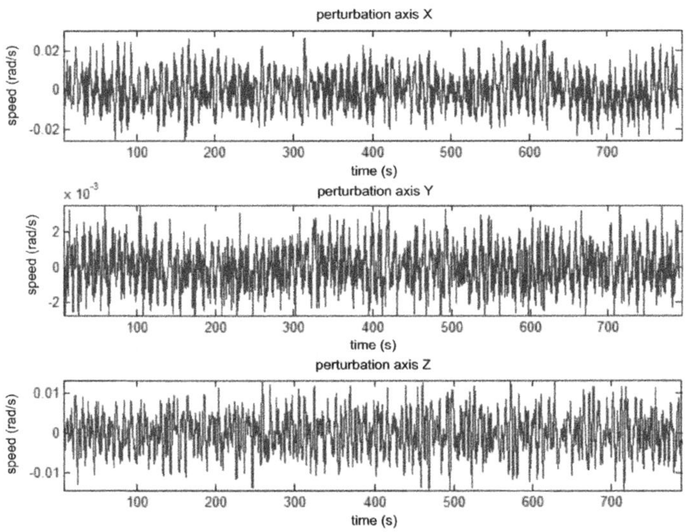

Figure 3.18. *Temporal acquisition of Γ_{vx}, Γ_{vy} and Γ_{vz}*

In order to ensure the observation capabilities of the viewfinder, we must reduce the blur generated by this "high-frequency" disturbance to a value compatible with IFOV during the integration time T_i. Thus, the specification relative to this disturbance can be of the form:

$$\sigma_{\theta vx}(t) = \sqrt{\int_t^{t+T_i} \theta_a(\tau)^2 d\tau} \le \bar{\theta}_v \text{ in response to excitation } \Gamma_{vx}(t)$$

$$\sigma_{\theta vy}(t) = \sqrt{\int_t^{t+T_i} \theta_a(\tau)^2 d\tau} \leq \bar{\theta}_v \text{ in response to excitation } \Gamma_{vy}(t) \quad [3.95]$$

$$\sigma_{\theta vz}(t) = \sqrt{\int_t^{t+T_i} \theta_a(\tau)^2 d\tau} \leq \bar{\theta}_v \text{ in response to excitation } \Gamma_{vz}(t)$$

Note that often, we are more conservative when taking $T_i \rightarrow \infty$.

Another significant requirement concerns the servo-loop power consumption. Indeed, the motors have a maximal input voltage that needs to be respected in order to avoid damage to them. Thus, we must ensure the disturbance rejection while limiting the maximum value of motor current, that is, by adding requirement [3.96].

$$|i(t)| < i_{\max} \text{ in response to disturbances} \quad [3.96]$$

The specification, noted here as χ, resembles a set of system requirements:

$$|\theta_a(t)| \leq \bar{\theta}_f,\ \forall t > 0 \text{ in response to step } \Gamma_f(t)$$

$$\max_{t > T_f} |\theta_a(t)| \leq \varepsilon_f \text{ in response to step } \Gamma_f(t)$$

$$|i(t)| \leq i_{\max},\ \forall t > 0 \text{ in response to step } \Gamma_f(t)$$

$$\sigma_{\theta vx} \leq \theta_v \text{ in response to excitation } \Gamma_{vx}(t)$$

$$|i(t)| \leq i_{\max},\ \forall t > 0 \text{ in response to excitation } \Gamma_{vx}(t)$$

$$\sigma_{\theta vy} \leq \bar{\theta}_v \text{ in response to excitation } \Gamma_{vy}(t) \quad [3.97]$$

$$|i(t)| \leq i_{\max},\ \forall t > 0 \text{ in response to excitation } \Gamma_{vy}(t)$$

$$\sigma_{\theta vz} \leq \bar{\theta}_v \text{ in response to excitation } \Gamma_{vz}(t)$$

$$|i(t)| \leq i_{\max},\ \forall t > 0 \text{ in response to excitation } \Gamma_{vz}(t)$$

Write:

– $\theta_f(t)$ as the position response to excitation $\Gamma_f(t)$;

– $i_f(t)$ as the current response to excitation $\Gamma_f(t)$;

– $\theta_{vx}(t)$ as the position response to excitation $\Gamma_{vx}(t)$;

– $\theta_{vy}(t)$ as the position response to excitation $\Gamma_{vy}(t)$;

– $\theta_{vz}(t)$ as the position response to excitation $\Gamma_{vz}(t)$;

– $i_{vx}(t)$ as the current response to excitation $\Gamma_{vx}(t)$;

– $i_{vy}(t)$ as the current response to excitation $\Gamma_{vy}(t)$;

– $i_{vz}(t)$ as the current response to excitation $\Gamma_{vz}(t)$.

So, the specification can easily be reformulated as follows:

$$\left.\begin{array}{c}\max\limits_{t>0}\left(\left|\theta_f(t)\right|\right)\leq\bar{\theta}_f,\ \max\limits_{t>T_f}\left(\left|\theta_f(t)\right|\right)\leq\varepsilon_f,\max\limits_{t>0}\left(\left|i_f(t)\right|\right)\leq i_{\max}\\ \sigma_{\theta vx}\leq\bar{\theta}_v,\ \max\limits_{t>0}\left(\left|i_{vx}(t)\right|\right)\leq i_{\max}\\ \sigma_{\theta vy}\leq\bar{\theta}_v,\ \max\limits_{t>0}\left(\left|i_{vy}(t)\right|\right)\leq i_{\max}\\ \sigma_{\theta vz}\leq\bar{\theta}_v,\ \max\limits_{t>0}\left(\left|i_{vz}(t)\right|\right)\leq i_{\max}\end{array}\right\}:=\chi \qquad [3.98]$$

For confidentiality reasons, the numerical specifications are hidden. We will measure the requirements' achievability relative to the specification.

3.4.3.1.2. Weighting function optimization on the nominal plant

We propose to use the loop-shaping approach to search for the H_∞ controller, solution to the following optimization problem (Figure 3.19):

$$\begin{array}{c}\min\limits_{W_o(s),W_i(s)\in RH_\infty}\gamma\\ \text{Subject to }\chi\end{array} \qquad [3.99]$$

As we have seen, an advantage of the loop-shaping approach is in the direct link between γ and the servo-loop stability margins. Thus, via problem [3.99], we are sure of having suitable stability margins when constraints are satisfactory.

Accordingly, the following is a direct application of problem [3.75]:

$$f_c\left(H_0,K\right)=\frac{\gamma_{obj}}{9}\left[\begin{array}{l} e^{\frac{\ln(\gamma_{obj})}{0.2\bar{\theta}_f}\left(\max\limits_{t>0}\left(\left|\theta_f(t)\right|\right)-\bar{\theta}_f\right)}+e^{\frac{\ln(\gamma_{obj})}{0.2\varepsilon_f}\left(\max\limits_{t>T_f}\left(\left|\theta_f(t)\right|\right)-\varepsilon_f\right)} \\ +e^{\frac{\ln(\gamma_{obj})}{0.2i_{\max}}\left(\max\limits_{t>0}\left(\left|i_f(t)\right|\right)-i_{\max}\right)} \\ \\ +e^{\frac{\ln(\gamma_{obj})}{0.2\bar{\theta}_v}\left(\sigma_{\theta vx}-\bar{\theta}_v\right)}+e^{\frac{\ln(\gamma_{obj})}{0.2i_{\max}}\left(\max\limits_{t>0}\left(\left|i_{vx}(t)\right|\right)-i_{\max}\right)} \\ +e^{\frac{\ln(\gamma_{obj})}{0.2\bar{\theta}_v}\left(\sigma_{\theta vy}-\bar{\theta}_v\right)}+e^{\frac{\ln(\gamma_{obj})}{0.2i_{\max}}\left(\max\limits_{t>0}\left(\left|i_{vy}(t)\right|\right)-i_{\max}\right)} \\ +e^{\frac{\ln(\gamma_{obj})}{0.2\bar{\theta}_v}\left(\sigma_{\theta vz}-\bar{\theta}_v\right)}+e^{\frac{\ln(\gamma_{obj})}{0.2i_{\max}}\left(\max\limits_{t>0}\left(\left|i_{vz}(t)\right|\right)-i_{\max}\right)} \end{array}\right] \qquad [3.100]$$

$$\gamma_{obj}=3$$

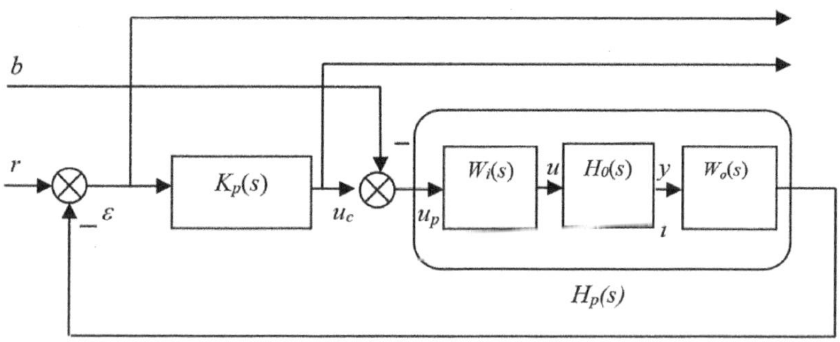

$$K_p(s)=\arg\min\gamma,\ \left\|\begin{bmatrix} I \\ K_p(s)\end{bmatrix}\left(I+H_p(s)K_p(s)\right)^{-1}\begin{bmatrix} I & H_p(s)\end{bmatrix}\right\|_{\infty}<\gamma$$

Figure 3.19. *Loop-shaping synthesis*

At the plant's output, we select a decentralized MIMO weighting function:

$$W_o(s)=\begin{bmatrix} W_{\Omega}(s) & 0 \\ 0 & W_I(s)\end{bmatrix} \qquad [3.101]$$
$$W_i(s)=W_u(s)$$

We select the following orders:

$$\partial^{\circ}W_{\Omega} = 6,\ \partial^{\circ}W_{u} = 6,\ \partial^{\circ}W_{I} = 1 \qquad [3.102]$$

which leads to 62 parameters to be optimized.

We solve the problem using several algorithms and then perform the statistical analysis.

The cost function f is defined by [3.100].

Since the optimization algorithms tested are stochastic, a statistical analysis of the results is required for each test.

Thus, the performance measurement rule is as follows.

Each test is subjected to 10 successive runs: accordingly, we note the minimum value of f obtained (f_{min}), the mean value of f (f_{mean}), its standard deviation (f_{std}) and, finally, the average calculation time (t_{cpu}) taken to perform these 10 runs.

$$\left.\begin{aligned} f_{min} &= \min_{i=1,\dots,n\ runs} f_i \\ f_{mean} &= \frac{1}{n}\sum_{i=1}^{n\ runs} f_i \\ f_{std} &= \sqrt{\frac{1}{n}\sum_{i=1}^{n\ runs}\left(f_i - f_{mean}\right)^2} \end{aligned}\right\} \qquad [3.103]$$

This allows us to establish a "mean" rank for each algorithm based on the ranks obtained for values [3.103] as was done previously for structured H_∞ synthesis in part one. The results are presented in Table 1.1.

We consider that the algorithm with the lowest mean rank is ultimately the best for our problem of weighting function tuning.

Here, the measurement of calculation time is interesting because of the following reasons:

– Some algorithms have a local search phase associated with a global search phase (ABC, CS); thus, we can evaluate the cost function through one iteration more than for more direct approaches (PSO, QPSO, DE).

– The cost function [3.75] occurs in two stages: stabilization and performance. These two stages do not require the same run time (the second takes longer).

So, for an equal number of iterations, the computation time used by each algorithm for each test can vary significantly.

The entire test was performed on a machine equipped with an Intel Core i7-3740QM (2.7 GHz) processor and 8 GB RAM.

The tests were performed with parallel computation. In this last case, [MAT 10] was used with 8 cores (pool). Each run was subject to 10,000 iterations.

$IT\ /\ run = 10000$	f_{min}	f_{mean}	f_{std}	$t_{cpu/run}$ (mn)
StaticCEC_QPSO	−0.27	−0.21	0.030	130.27
TVDecCEC_QPSO	−0.23	−0.18	0.038	89.70
Rnd_DE	−0.24	−0.15	0.081	62.76
Rnd2Bst_DE	−0.29	**−0.26**	**0.020**	91.63
Rnd2Bst_PDE	**−0.32**	−0.24	0.045	136.9
Std_PSO	−0.22	−0.14	0.077	93.84
TVRndIW_PSO	−0.22	−0.15	0.065	70.46
ABC	−0.16	−0.14	0.029	132.33
CS	−0.14	−0.06	0.046	21.21
CS/DE	−0.27	−0.23	0.049	290.63
FF	−0.18	−0.035	0.058	15.79

Table 3.1. *Algorithm comparison for the weighting function optimization problem*

We see that the evolutionary computation, and *Rnd2Bst_PDE* in particular, is the best algorithm in terms of results for weighting function tuning. *CS* leads to a much lower calculation time but to noticeably poorer results.

The best *runs* lead to the result presented in Figure 3.20.

In Figure 3.20, we see from left to right the Black-Nichols chart of the (input) open loop and various frequency closed-loop responses together with the temporal responses of the optimized signals, that is, residual of angular stabilization in vibration environment, motor current in vibration environment, residual of angular stabilization in response to friction and corresponding current.

The results concerning the evaluation of temporal requirements are shown in Table 3.2. Note that for confidentiality purposes, the results are presented relative to the maximal temporal values specified (confidential), that is, relative to 1.

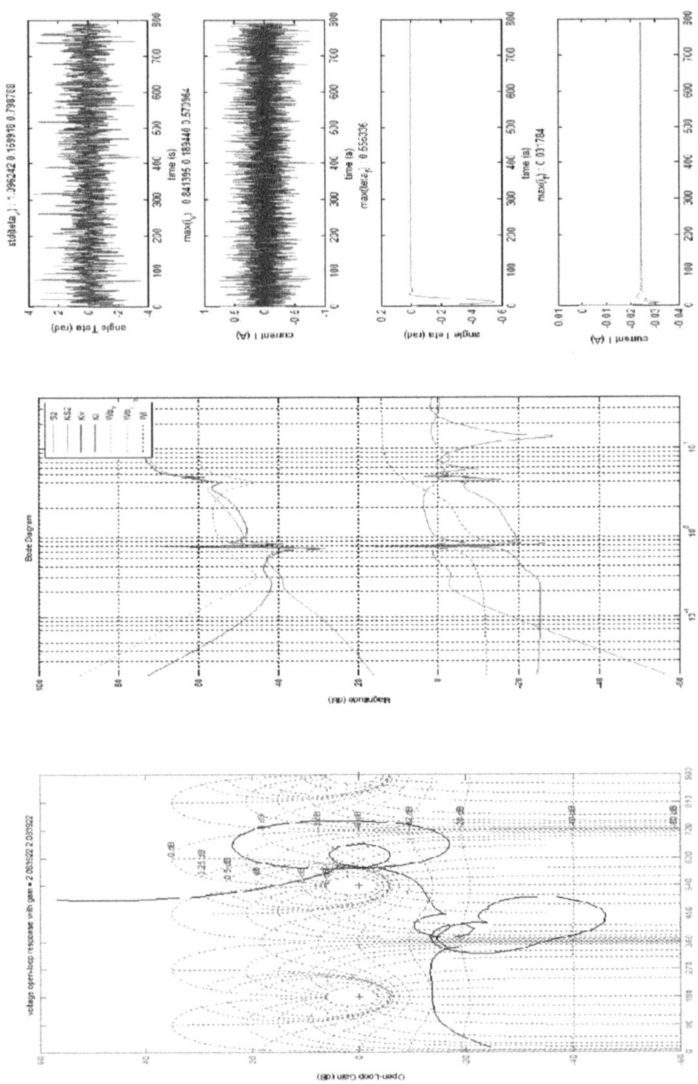

Figure 3.20. *Optimal tuning of weighting functions. For a color version of this figure, see www.iste.co.uk/feyel/optimization.zip*

Requirements	*Final evaluation*
γ	2.08
$\max_{t>0}\left(\lvert\theta_f(t)\rvert\right) \le \bar{\theta}_f$	0.55
$\max_{t>T_f}\left(\lvert\theta_f(t)\rvert\right) \le \varepsilon_f$	0
$\max_{t>0}\left(\lvert i_f(t)\rvert\right) \le i_{\max}$	0.03
$\sigma_{\theta vx} \le \bar{\theta}_v$	1.09
$\max_{t>0}\left(\lvert i_{vx}(t)\rvert\right) \le i_{\max}$	0.84
$\sigma_{\theta_{vy}} \le \bar{\theta}_v$	0.16
$\max_{t>0}\left(\lvert i_{vy}(t)\rvert\right) \le i_{\max}$	0.19
$\sigma_{\theta vz} \le \bar{\theta}_v$	0.80
$\max_{t>0}\left(\lvert i_{vz}(t)\rvert\right) \le i_{\max}$	0.60

Table 3.2. *Evaluation of the requirements derived from weighting function optimization*

We observe that all of the requirements are met with the exception of the fourth, which only exceeds 9% of the specified requirement.

3.4.3.1.3. Multiple plant optimization

We can add a requirement of stability or even performance robustness to the set of requirements χ over a ball of N plants centered on the nominal plant $H_0(s)$:

$$\wp = \{H_i(s),\ i = 0,\dots,N\} \qquad [3.104]$$

Here, we assume that the N plants are not necessarily measured but can be obtained from the nominal response shown in Figure 3.16.

– Concerning the "mechanical" block $H_m(s)$:

There are several methods for dispersing this system. Here, we propose making random variations inside a bounded interval around the nominal value of the proper frequency and the pole damping of the system. Indeed, this is simple and physical and proves to be well adapted to systems with form [3.105].

$$\frac{\Omega_a(s)}{C_m(s)}=\frac{1}{Js}+\sum_{k=1}^{n}\frac{\alpha_k s}{s^2+2\xi_k\omega_{0k}s+\omega_{0k}^{\ 2}} \tag{3.105}$$

An easy way to ensure this is by expressing the state representation of $H(s)$ in its modal base. In this base, the state matrix of system H(s) is of the form:

$$A=\begin{bmatrix}\lambda_1 & 0 & 0 & \\ 0 & \sigma_2 & \omega_2 & \\ 0 & -\omega_2 & \sigma_2 & \\ & & & \ddots\end{bmatrix} \tag{3.106}$$

$$\sigma_i+j\omega_i=-\xi_i\omega_{0i}\pm j\omega_{0i}\sqrt{1-\xi_i^2}$$

Thus, supposing that the proper frequencies and damping vary by ± δ_ω % and ± δ_ξ %, respectively:

$$\begin{aligned}\sigma_i+j\omega_i&=-\xi_i\left(1+\delta_\xi\right)\omega_{0i}\left(1+\delta_\omega\right)\pm j\omega_{0i}\left(1+\delta_\omega\right)\sqrt{1-\xi_i^2\left(1+\delta_\xi\right)^2}\\&=-\xi_i\omega_{0i}\left(1+\delta_\xi+\delta_\omega+\delta_\xi\delta_\omega\right)\pm j\omega_{0i}\left(1+\delta_\omega\right)\sqrt{1-\xi_i^2\left(1+\delta_\xi\right)^2}\\&\approx-\xi_i\omega_{0i}\left(1+\delta_\xi+\delta_\omega\right)\pm j\omega_{0i}\left(1+\delta_\omega\right)\sqrt{1-\xi_i^2}\\&=\sigma_i\left(1+\delta_\xi+\delta_\omega\right)\pm j\omega_i\left(1+\delta_\omega\right)\end{aligned} \tag{3.107}$$

which equates to making the real part of the corresponding pole vary by $\pm\left(\delta_\xi+\delta_\omega\right)$ % and its imagined part by $\pm\delta_\omega$ %.

– Concerning the "electric" block H_{el}(s):

The main contributor is the motor's electric resistance. Its variation of $\pm\delta_R$ % is induced by the thermal environment (around ± 25% when the temperature varies from −40 °C to +70 °C). Considering that this block's static gain varies by R^{-1}, at the first order, we can consider this to induce a gain variation of H_{el}(s) from $\pm\delta_R$ %. Accordingly, we can generate plants by causing random variation in the static gain of $\pm\delta_R$ % around its nominal value.

For example, let us assume here that $\delta_\omega = \pm 15$ %, $\delta_\xi = \pm 15$% and $\delta_R = \pm 25$%. We can define the ball of 25 plants whose frequency responses are provided in Figure 3.21 and Figure 3.22.

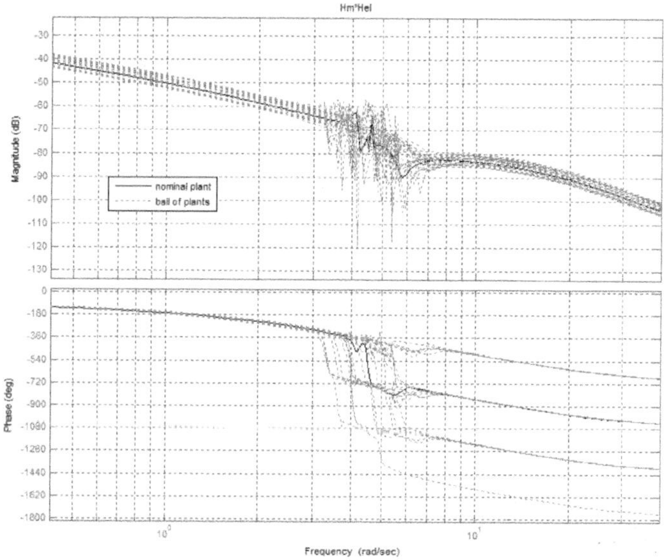

Figure 3.21. *Dispersed mechanic block. For a color version of this figure, see www.iste.co.uk/feyel/optimization.zip*

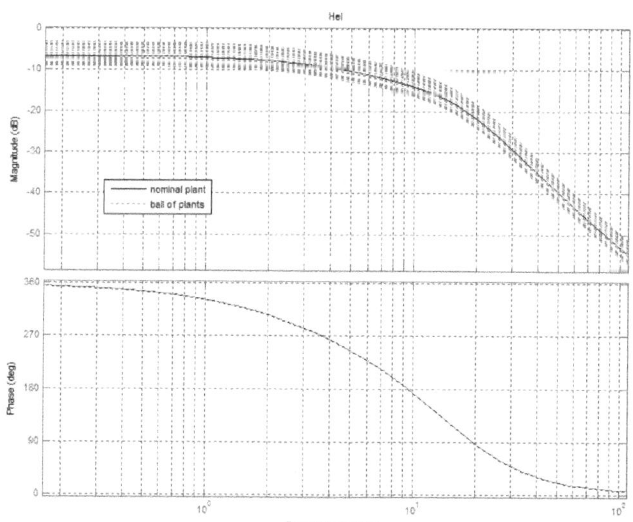

Figure 3.22. *Dispersed electric block. For a color version of this figure, see www.iste.co.uk/feyel/optimization.zip*

In Figure 3.23, we observe that the controller obtained earlier does not stabilize the ball of plants. The servo-loop is at its stability limit for the ball of plants. In fact, the worst case value of γ obtained for the whole of the ball of plants is 19.77. Consequently, we must now attempt to make the controller more robust by means of the multiple plant techniques developed earlier.

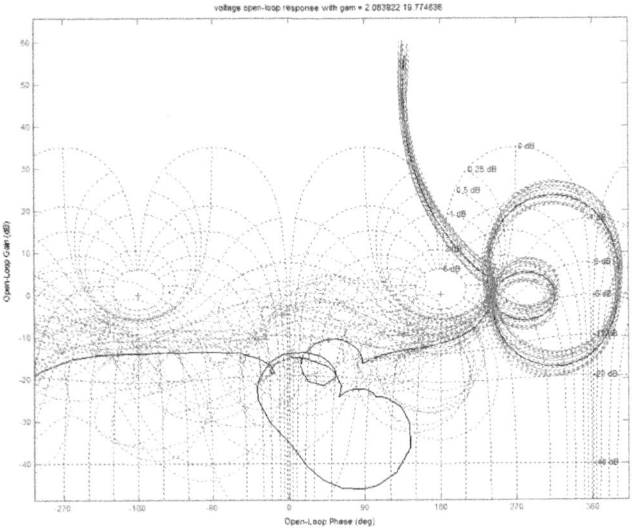

Figure 3.23. *Frequency response of the open loop for the ball of plants. For a color version of this figure, see www.iste.co.uk/feyel/optimization.zip*

We can find (with *Rnd2Bst_PDE*) a controller solution to the following problems:

– the problem of stability robustness:

$$\begin{gathered}\min_{W_o(s),W_i(s)\in \mathbf{R}H_\infty} \gamma \\ \text{Subject to } \chi \\ \text{and } K(s) \text{ stabilizes all } \wp\end{gathered} \tag{3.108}$$

We obtain the results shown in Table 3.3.

IT / run = 10000	f_{min}	f_{mean}	f_{std}	$t_{cpu/run}$ (mn)
Rnd2Bst_PDE	–0.26	–0.21	–0.04	159

Table 3.3. *Optimization of weighting functions – stability robustness*

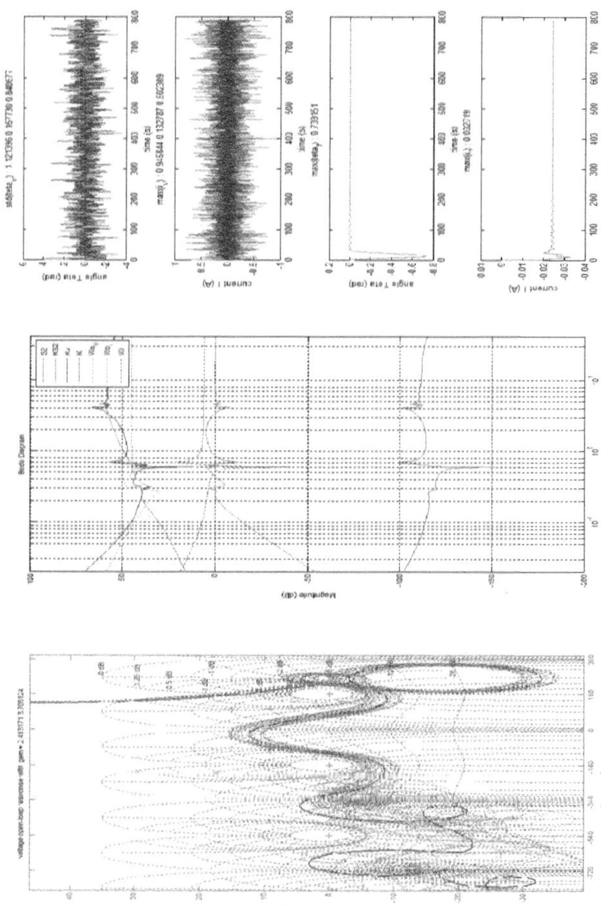

Figure 3.24. *The problem of stability robustness. For a color version of this figure, see www.iste.co.uk/feyel/optimization.zip*

We observe that the computation time of the approach on the nominal plant is less than that of the stability robustness problem; this is due to the stabilization of N plants, which consists of evaluating N eigenvalues (rather than only one) in the cost function.

The best of 10 runs leads to the results in Figure 3.24.

The final evaluation of the requirements is presented in Table 3.4.

Requirements	*Final evaluations on the nominal plant*
$\gamma_{nom}; \gamma_{pire}$	2.41; 3.70
$\max_{t>0}\left(\left\|\theta_f(t)\right\|\right) \leq \bar{\theta}_f$	0.74
$\max_{t>T_f}\left(\left\|\theta_f(t)\right\|\right) \leq \varepsilon_f$	0
$\max_{t>0}\left(\left\|i_f(t)\right\|\right) \leq i_{\max}$	0.03
$\sigma_{\theta vx} \leq \bar{\theta}_v$	1.12
$\max_{t>0}\left(\left\|i_{vx}(t)\right\|\right) \leq i_{\max}$	0.94
$\sigma_{\theta vy} \leq \bar{\theta}_v$	0.17
$\max_{t>0}\left(\left\|i_{vy}(t)\right\|\right) \leq i_{\max}$	0.13
$\sigma_{\theta vz} \leq \bar{\theta}_v$	0.84
$\max_{t>0}\left(\left\|i_{vz}(t)\right\|\right) \leq i_{\max}$	0.50

Table 3.4. *Evaluation of requirements – stability robustness*

Here again, we see that all requirements are satisfied (albeit somewhat degraded, the increase in stability robustness being accompanied by a decrease in performances that remains quite acceptable), with the exception of the fourth. The worst case value of γ is reduced to 3.70.

– problem of stability margin robustness:

$$\begin{gathered}\min_{W_o(s),W_i(s)\in\mathbf{R}H_\infty} \gamma,\ \gamma = \max_{j=1,\dots,N} \gamma_j \\ \text{Subject to } \chi \\ \text{and } K(s) \text{ stabilizes all } \wp\end{gathered} \quad [3.109]$$

We obtain the results shown in Table 3.5.

IT / run = 10000	f_{min}	f_{mean}	f_{std}	$t_{cpu/run}$ (mn)
Rnd2Bst_PDE	–0.22	–0.18	0.03	691.11

Table 3.5. *Optimization of weighting function – stability margin robustness*

We see that the computation time for the problem of stability margin robustness is much greater than for the nominal approach; this is due to the number of evaluations of the closed-loop H_∞ norm being now N times greater.

The best 10 runs give the results in Figure 3.25.

We see here again that all requirements with the exception of the fourth are satisfied (this, however, is not degraded relative to the problem of stability robustness). The worst case value of γ is reduced to 3.27.

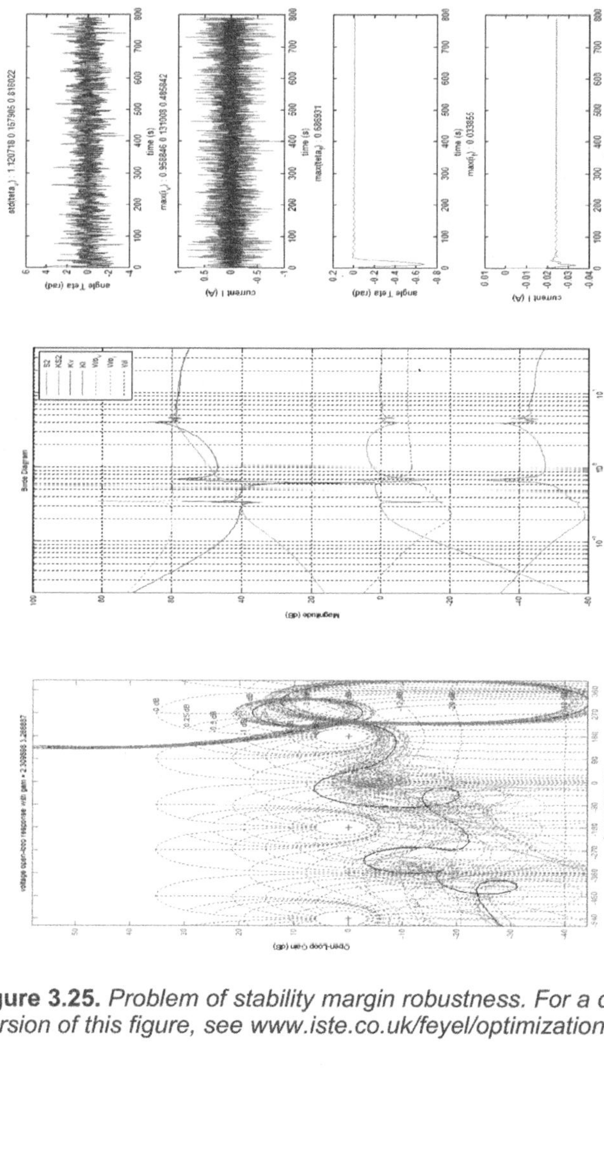

Figure 3.25. *Problem of stability margin robustness. For a color version of this figure, see www.iste.co.uk/feyel/optimization.zip*

The final evaluation of the requirements is presented in Table 3.6.

Requirements	*Final evaluation on nominal system*
γ_{nom}; γ_{pire}	2.30; 3.27
$\max_{t>0}\left(\lvert\theta_f(t)\rvert\right) \leq \bar{\theta}_f$	0.69
$\max_{t>T_f}\left(\lvert\theta_f(t)\rvert\right) \leq \varepsilon_f$	0
$\max_{t>0}\left(\lvert i_f(t)\rvert\right) \leq i_{\max}$	0.03
$\sigma_{\theta vx} \leq \bar{\theta}_v$	1.12
$\max_{t>0}\left(\lvert i_{vx}(t)\rvert\right) \leq i_{\max}$	0.96
$\sigma_{\theta vy} \leq \bar{\theta}_v$	0.16
$\max_{t>0}\left(\lvert i_{vy}(t)\rvert\right) \leq i_{\max}$	0.13
$\sigma_{\theta vz} \leq \bar{\theta}_v$	0.81
$\max_{t>0}\left(\lvert i_{vz}(t)\rvert\right) \leq i_{\max}$	0.48

Table 3.6. *Evaluation of requirements – stability margin robustness*

3.4.3.2. *The viewfinder for shooting*

3.4.3.2.1. System description and requirements

We will now address the viewfinder for shooting, whose frequency representation of the transfer function between the control voltage and the gyrometric rate is given in Figure 3.26. This system is of order 20.

For confidentiality reasons, the frequencies and signals involved are normalized in this study.

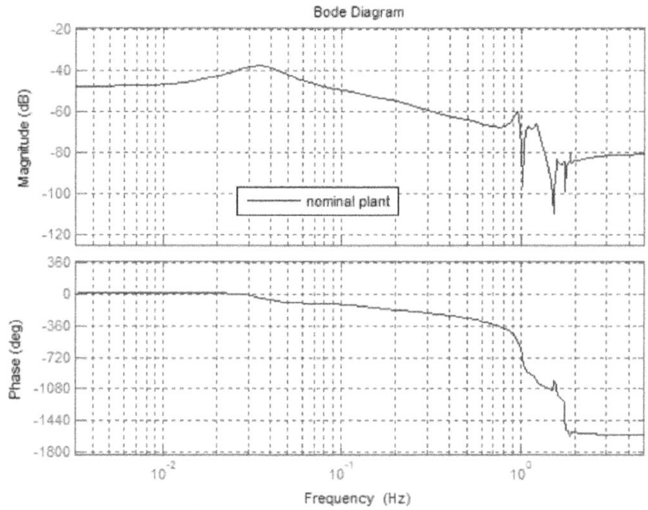

Figure 3.26. *Frequency response of the viewfinder for shooting*

Here, we will consider the control scheme in Figure 3.27.

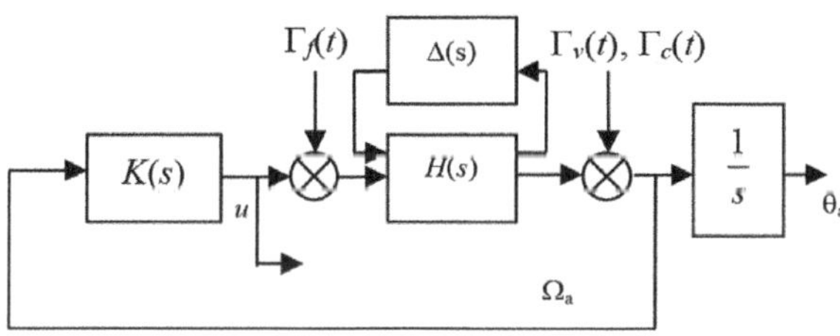

Figure 3.27. *Viewfinder for shooting control scheme*

There are residual stabilization requirements concerning vibratory excitations along directions x and y, for which we can dispose of a temporal acquisition $\Gamma_{vx}(t)$ and $\Gamma_{vy}(t)$ induced on absolute rate (Figure 3.28).

We also have a requirement of friction rejection. Note that we can make friction homogeneous as a voltage if we know the resistance and torque constant of the motorization, allowing us to bring friction $\Gamma_f(t)$ to the plant's input. Also, rather than limiting motor current, we will attempt to limit the control voltage u on the controller's output to a value u_{max}.

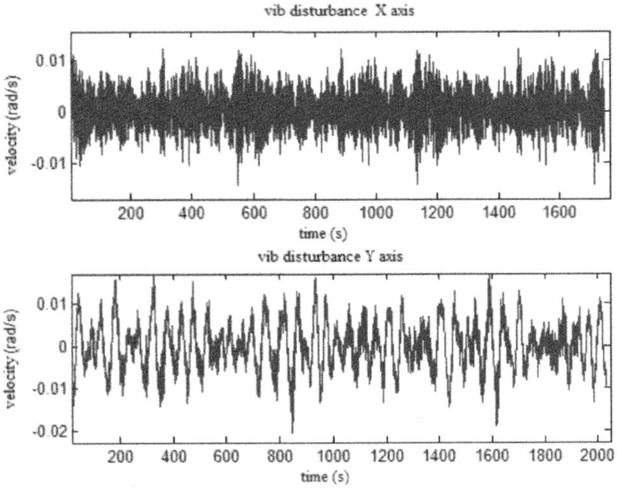

Figure 3.28. *Temporal acquisition of Γ_{vx} and Γ_{vy}*

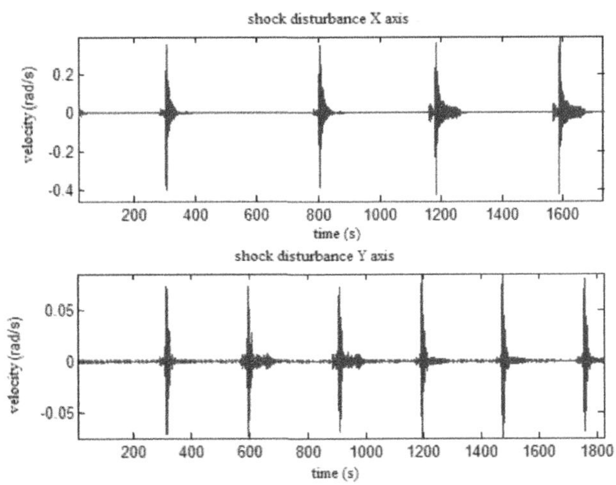

Figure 3.29. *Acquisition of shock type disturbances*

In addition to these "nominal" operating requirements, on occasion, we also have "degraded" requirements according to viewfinder peculiarities. A common example of this is the viewfinder for shooting, where each shot can lead to a "shock" acting on the loop of the line of sight. In this case, in order to avoid the depointing induced by the controller's conditioning, we attempt to reduce the control voltage in response to these shocks. As with nominal operation, we can have temporal acquisitions in the response of the line of sight (non-looped) toward these shocks, occurring over the two axes, which we will refer to here as Γ_{cx}(t) and Γ_{cy}(t). These disturbances on model output are of the type pseudo-dampened to the frequency of the first frequency mode of the viewfinder's structure as seen in Figure 3.29.

Since this is a degraded operation, these additional requirements are not of the same priority as the nominal requirements, which allows us to illustrate the priority management mechanism section 1.3.3. Accordingly, the system specification, noted here as χ', is as follows:

High priority:

$|\theta_a(t)| \leq \overline{\theta}_f$ in response to step $\Gamma_f(t)$

$\max\limits_{t>T_f} |\theta_a(t)| \leq \varepsilon_f$ in response to step $\Gamma_f(t)$

$|u(t)| \leq u_{\max}$ in response to step $\Gamma_f(t)$

$\sigma_{\theta x} \leq \overline{\theta}_v$ in response to excitation $\Gamma_{vx}(t)$ [3.110]

$|u(t)| \leq u_{\max}$ in response to excitation $\Gamma_{vx}(t)$

$\sigma_{\theta y} \leq \overline{\theta}_v$ in response to excitation $\Gamma_{vy}(t)$

$|u(t)| \leq u_{\max}$ in response to excitation $\Gamma_{vy}(t)$

Low priority:

$|u(t)| \leq u_{\max}$ in response to excitation $\Gamma_{cx}(t)$

$|u(t)| \leq u_{\max}$ in response to excitation $\Gamma_{cy}(t)$

Write:

– $\theta_f(t)$ as the response in excitation position $\Gamma_f(t)$;

– $u_f(t)$ as the response in excitation voltage $\Gamma_f(t)$;

– $\theta_{vx}(t)$ as the response in excitation position $\Gamma_{vx}(t)$;

– $\theta_{vy}(t)$ as the response in excitation position $\Gamma_{vy}(t)$;

– $u_{vx}(t)$ as the response in excitation voltage $\Gamma_{vx}(t)$;

– $u_{vy}(t)$ as the response in excitation voltage $\Gamma_{vy}(t)$;

– $u_{cx}(t)$ as the response in excitation voltage $\Gamma_{cx}(t)$;

– $u_{cy}(t)$ as the response in excitation voltage $\Gamma_{cy}(t)$.

So, specification χ' can be easily reformulated as follows:

High priority:

$$\begin{gathered}
\max_{t>0}\left(\left|\theta_f(t)\right|\right)\le\overline{\theta}_f,\ \max_{t>T_f}\left(\left|\theta_f(t)\right|\right)\le\varepsilon_f, \\
\max_{t>0}\left(\left|u_f(t)\right|\right)\le u_{\max} \\
\sigma_{\theta vx}\le\overline{\theta}_v,\ \max_{t>0}\left(\left|u_{vx}(t)\right|\right)\le u_{\max} \\
\sigma_{\theta vy}\le\overline{\theta}_v,\ \max_{t>0}\left(\left|u_{vy}(t)\right|\right)\le u_{\max}
\end{gathered} \qquad [3.111]$$

Low priority:

$$\begin{aligned}
&\max_{t>0}\left(\left|u_{cx}(t)\right|\right)\le u_{\max} \\
&\max_{t>0}\left(\left|u_{cy}(t)\right|\right)\le u_{\max}
\end{aligned}$$

For confidentiality reasons, the numerical specifications are hidden here. We measure the requirements' achievability relative to the specification.

3.4.3.2.2. Optimization of weighting functions on the nominal plant

First, we propose using loop shaping to search for the H_∞ SISO controller solution to the following optimization problem:

$$\begin{gathered}
\min_{W_o(s),W_i(s)\in\mathbf{R}H_\infty}\gamma \\
\text{Subject to }\chi'
\end{gathered} \qquad [3.112]$$

This is again a direct application of problem [3.75] with:

$$f_c(H_0,K)=\frac{\gamma_{obj}}{9}\left[\begin{array}{l} e^{\frac{\ln(\gamma_{obj})}{0.2\bar{\theta}_f}\left(\max\limits_{t>0}(|\theta_f(t)|)-\bar{\theta}_f\right)}+e^{\frac{\ln(\gamma_{obj})}{0.2\varepsilon_f}\left(\max\limits_{t>T_f}(|\theta_f(t)|)-\varepsilon_f\right)} \\ +e^{\frac{\ln(\gamma_{obj})}{0.2u_{\max}}\left(\max\limits_{t>0}(|u_f(t)|)-u_{\max}\right)} \\ \\ +e^{\frac{\ln(\gamma_{obj})}{0.2\bar{\theta}_v}\left(\sigma_{\theta vx}-\bar{\theta}_v\right)}+e^{\frac{\ln(\gamma_{obj})}{0.2u_{\max}}\left(\max\limits_{t>0}(|u_{vx}(t)|)-u_{\max}\right)} \\ +e^{\frac{\ln(\gamma_{obj})}{0.2\bar{\theta}_v}\left(\sigma_{\theta vy}-\bar{\theta}_v\right)}+e^{\frac{\ln(\gamma_{obj})}{0.2u_{\max}}\left(\max\limits_{t>0}(|u_{vy}(t)|)-u_{\max}\right)} \\ \\ +\max\left(0,\ 1+\frac{\ln(\gamma_{obj})}{0.2u_{\max}}\left(\max\limits_{t>0}(|u_{vx}(t)|)-u_{\max}\right)\right) \\ +\max\left(0,\ 1+\frac{\ln(\gamma_{obj})}{0.2u_{\max}}\left(\max\limits_{t>0}(|u_{vy}(t)|)-u_{\max}\right)\right) \end{array}\right] \quad [3.113]$$

$$\gamma_{obj}=3$$

Since the requirements relative to the shocks are of low priority, we choose to develop the penalty functions concerning them to order 1 (see section 1.3.3).

Since a SISO controller is used here, the output and input plant weights $W_\Omega(s)$ and W_u(s), respectively, are also SISO. We choose the following orders:

$$\partial^\circ W_\Omega=7,\ \partial^\circ W_u=1 \quad [3.114]$$

This leads to 38 parameters to optimize.

We use the *Rnd2Bst_PDE* algorithm as the previous example shows that this algorithm is more effective for the problem of weighting function optimization.

The entire test was performed on a machine equipped with an Intel Core i7-3740QM (2.7 GHz) and 8 GB RAM.

The tests were performed with parallel computation. In this last case, [MAT 10] was used with 8 cores (pool). Each run was subject to 2,000 iterations.

We obtain the results presented in Table 3.7.

IT / run = 2000	f_{min}	f_{mean}	f_{std}	$t_{cpu/run}$ (mn)
Rnd2Bst_PDE	−0.21	−0.20	0.019	29.99

Table 3.7. *Optimization of viewfinder for shooting weighting functions*

The best 10 runs give the results in Figure 3.30.

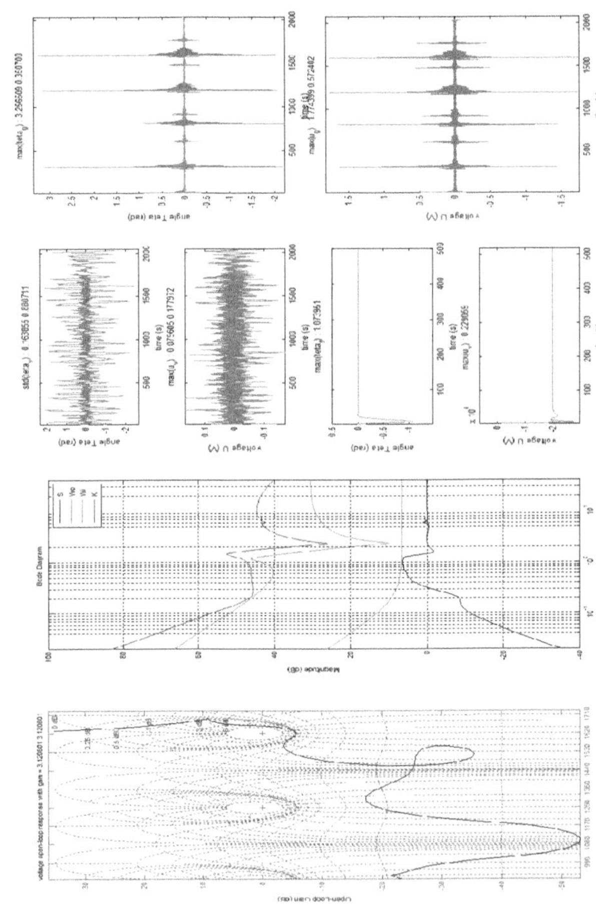

Figure 3.30. *Weighting function optimization. For a color version of this figure, see www.iste.co.uk/feyel/optimization.zip*

The results concerning the final evaluation of the temporal requirements are presented in Table 3.8. Note that for confidentiality reasons, we present the results relative to the maximum temporal values specified (confidential), that is, relative to 1.

Requirements	*Final evaluation*
γ	3.12
$\max\limits_{t>0}\left(\lvert\theta_f(t)\rvert\right)\leq\bar{\theta}_f$	1.07
$\max\limits_{t>T_f}\left(\lvert\theta_f(t)\rvert\right)\leq\varepsilon_f$	0
$\max\limits_{t>0}\left(\lvert u_f(t)\rvert\right)\leq u_{\max}$	0.23
$\sigma_{\theta vx}\leq\bar{\theta}_v$	0.16
$\max\limits_{t>0}\left(\lvert u_{vx}(t)\rvert\right)\leq u_{\max}$	0.08
$\sigma_{\theta vy}\leq\bar{\theta}_v$	0.88
$\max\limits_{t>0}\left(\lvert u_{vy}(t)\rvert\right)\leq u_{\max}$	0.18
$\max\limits_{t>0}\left(\lvert u_{cx}(t)\rvert\right)\leq u_{\max}$	1.77
$\max\limits_{t>0}\left(\lvert u_{cy}(t)\rvert\right)\leq u_{\max}$	0.57

Table 3.8. *Evaluation of requirements following weighting function optimization*

We see that the high-priority requirements are almost all satisfied. One low-priority requirement is not satisfied but remains suitable nonetheless: the algorithm attempted to optimize the low-priority requirements to the maximum once the high-priority requirements were met (which explains the controller's rejection behavior toward 2 rad/s, the proper frequency mode of the viewfinder that is solicited for shocks). In the absence of requirements concerning response toward shocks, this rejection would not be calculated.

3.4.3.2.3. Multiple plant optimization

To the requirement set χ', we can add the need for stability or performance robustness over a ball of N plants centered on the nominal model $H_0(s)$:

$$\wp = \{H_i(s), i = 0,...,N\} \tag{3.115}$$

$N = 25$ models are obtained using a technique that is strictly identical to that of section 3.4.3.1 with $\delta_\omega = \pm15\%$, $\delta_\xi = \pm15\%$ and $\delta_R = \pm10\%$. The frequency response of the ball of plants is traced in Figure 3.31.

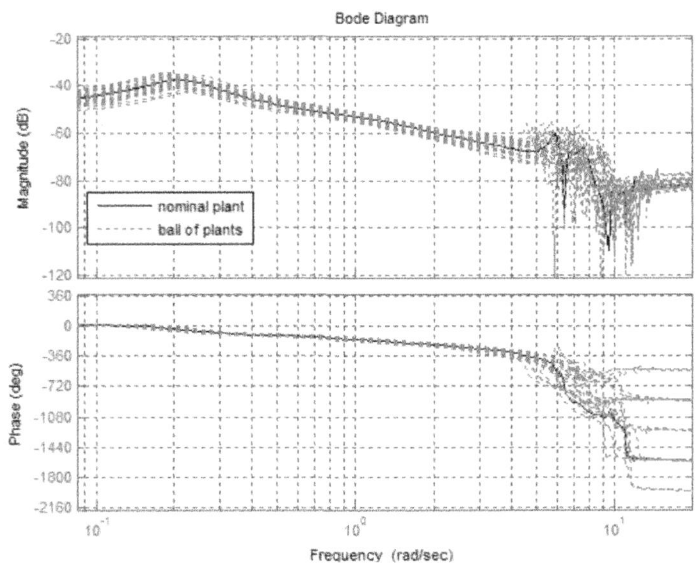

Figure 3.31. *Ball of plants. For a color version of this figure, see www.iste.co.uk/feyel/optimization.zip*

We see in Figure 3.32 that the controller obtained earlier is actually not robust with the ball of plants despite all of the ball being stabilized. Indeed, the worst case value of γ obtained for the ball of plants is 4.59. Accordingly, we must search to make the stability margins robust over the entire ball of plants.

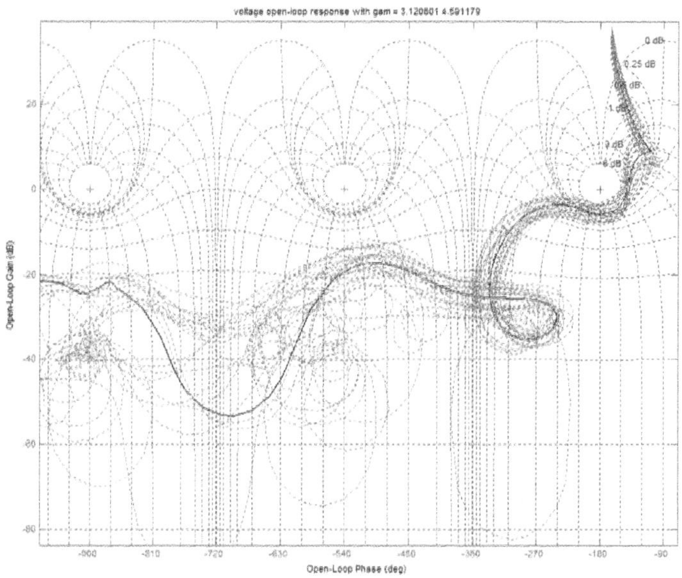

Figure 3.32. *Examination of weighting functions robustness. For a color version of this figure, see www.iste.co.uk/feyel/optimization.zip*

Thus, we can find (using *Rnd2Bst_PDE*) a controller solution for the following problem:

– Problem of stability margin robustness:

$$\min_{W_o(s),W_i(s)\in \mathbf{R}H_\infty} \gamma,\ \gamma = \max_{j=1,\ldots,N} \gamma_j$$
$$\text{Subject to } \chi' \qquad [3.116]$$
$$\text{and } K(s) \text{ stabilizes all } \wp$$

We obtain the results shown in Table 3.9.

IT / run = 2000	f_{min}	f_{mean}	f_{std}	$t_{cpu/run}$ (mn)
Rnd2Bst_PDE	−0.16	−0,15	0.011	120.14

Table 3.9. *Optimization of weighting functions – stability margin robustness*

We see that the computation time for the stability margin robustness problem is much greater than for the approach using nominal plant: this is due to a number of evaluations of the closed-loop H_{∞} norm being now N times greater.

The best 10 runs give the results in Figure 3.33.

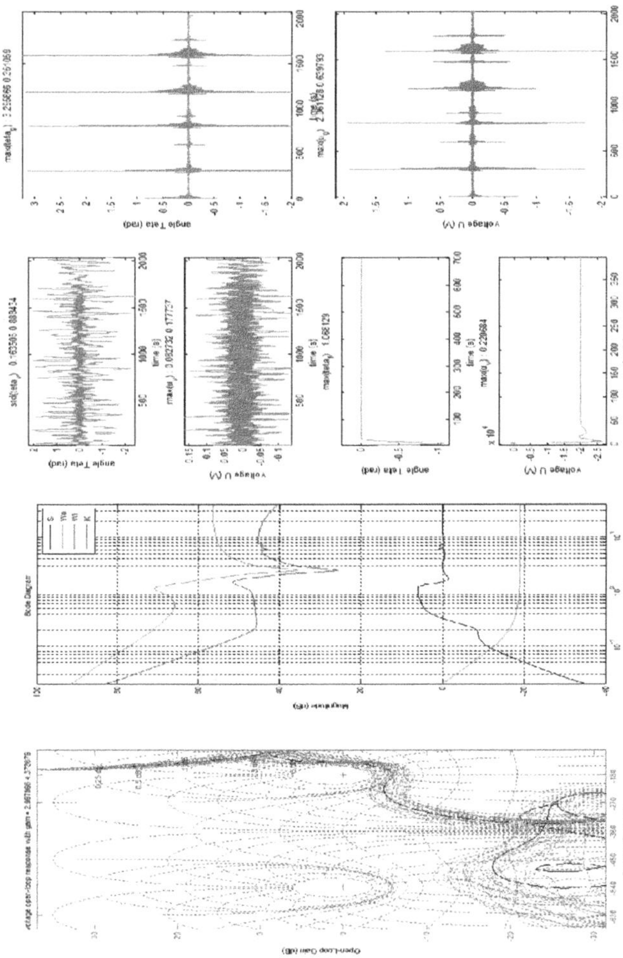

Figure 3.33. *Problem of stability margin robustness. For a color version of this figure, see www.iste.co.uk/feyel/optimization.zip*

The final evaluation of requirements is presented in Table 3.10.

Requirements	*Final evaluations on nominal system*
$\gamma_{nom}; \gamma_{pire}$	2.99; 4.37
$\max\limits_{t>0}\left(\lvert\theta_f(t)\rvert\right) \leq \bar{\theta}_f$	1.06
$\max\limits_{t>T_f}\left(\lvert\theta_f(t)\rvert\right) \leq \varepsilon_f$	0
$\max\limits_{t>0}\left(\lvert u_f(t)\rvert\right) \leq u_{\max}$	0.22
$\sigma_{\theta vx} \leq \bar{\theta}_v$	0.16
$\max\limits_{t>0}\left(\lvert u_{vx}(t)\rvert\right) \leq u_{\max}$	0.082
$\sigma_{\theta vy} \leq \bar{\theta}_v$	0.89
$\max\limits_{t>0}\left(\lvert u_{vy}(t)\rvert\right) \leq u_{\max}$	0.18
$\max\limits_{t>0}\left(\lvert u_{cx}(t)\rvert\right) \leq u_{\max}$	2.06
$\max\limits_{t>0}\left(\lvert u_{cy}(t)\rvert\right) \leq u_{\max}$	0.63

Table 3.10. *Evaluation of requirements – stability margin robustness*

3.4.3.3. *Going even further…*

The high-level requirements of an optronic inertial device are related to its global Modulation Transfer Function (MTF), as MTF provides a synthetic presentation of the contrast and resolution of an optical or optronic system [MAY 99], as illustrated in Figure 3.34. Thus, this allows for the quantification of image quality (blur) and, accordingly, the *range* of the device in question.

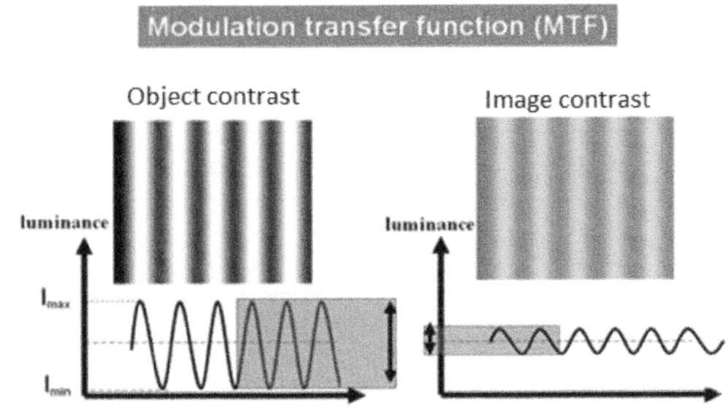

Figure 3.34. *MTF and contrast. For a color version of this figure, see www.iste.co.uk/feyel/optimization.zip*

In case of a complex optronic system, the global MTF of the system is equal to the product of the MTFs of the component's sub-systems [MAY 99, HOL 06].

$$MTF_{system}(\upsilon) = \Pi\left(MTF_{elements}\right) \qquad [3.117]$$

Indeed, one of the main contributors in MTF degradation is the device's residual stabilization. Consequently, in the case of a viewfinder, we typically decline the specification relative to the viewfinder range to a specification on the standard deviation of the residual stabilization that should not exceed the size of a demi-pixel in order to limit the blur generated. Indeed, by supposing that the residual stabilization is a Gaussian white noise, the MTF induced (that degrades the global MTF of the viewfinder and also its range) is written as follows:

$$MTF_{stab}(\upsilon) = e^{-2\pi^2\sigma_{\theta_v}{}^2\upsilon^2} \qquad [3.118]$$

where σ_{θ_v} is the standard deviation of the random movement (in rad) and υ the spatial frequency (in cycle/rad).

This specification can prove to be still more conservative as the desired ranges are greater as it supposes that the residual stabilization is white and Gaussian, which is never the case due to the presence of diverse spectral components according to the mechanical environment of the viewfinder (helicopter) and the modal landscape of the viewfinder's mechanical structure; accordingly, it is increasingly difficult in experimental terms to approximate the MTF of the residual of stabilization by a Gaussian for increasingly small residuals. Hence, a concern arises about the specification's conservatism relating to the stabilization stage of the standard deviation residual stabilization.

We now assume the availability of a specified MTF at the stabilization stage noted $ftm_{obj}(v)$ without any hypotheses on the nature of the residual of stabilization. From this point, nothing prevents the use of MTF deviation criteria between the stabilization stage looped by the controller and the specified MTF. Indeed, the specification is satisfied if the MTF of the residual of stabilization remains greater than the one specified. In this case, we must replace the requirements concerning the standard deviation of the residual of stabilization by the following requirements relative to MTFs[7]:

$$\left.\begin{array}{l}
\max\limits_{t>0}\left(\left|\theta_f(t)\right|\right)\le\bar{\theta}_f,\ \max\limits_{t>T_f}\left(\left|\theta_f(t)\right|\right)\le\varepsilon_f,\max\limits_{t>0}\left(\left|i_f(t)\right|\right)\le i_{\max}\\
\max\limits_{v}\left(\left|\dfrac{ftm_{obj}(v)}{ftm_{\theta_{vx}}(v)}\right|\right)<1,\ \max\limits_{t>0}\left(\left|i_{vx}(t)\right|\right)\le i_{\max}\\
\max\limits_{v}\left(\left|\dfrac{ftm_{obj}(v)}{ftm_{\theta_{vy}}(v)}\right|\right)<1,\ \max\limits_{t>0}\left(\left|i_{vy}(t)\right|\right)\le i_{\max}\\
\max\limits_{v}\left(\left|\dfrac{ftm_{obj}(v)}{ftm_{\theta_{vz}}(v)}\right|\right)<1,\ \max\limits_{t>0}\left(\left|i_{vz}(t)\right|\right)\le i_{\max}
\end{array}\right\}:=\chi \qquad [3.119]$$

Accordingly, we can resume the controller calculation with:

$$f_c(H_0,K)=\frac{\gamma_{obj}}{9}\left[\begin{array}{l}
e^{\frac{\ln(\gamma_{obj})}{0.2\bar{\theta}_f}\left(\max\limits_{t>0}\left(\left|\theta_f(t)\right|\right)-\bar{\theta}_f\right)}+e^{\frac{\ln(\gamma_{obj})}{0.2\varepsilon_f}\left(\max\limits_{t>T_f}\left(\left|\theta_f(t)\right|\right)-\varepsilon_f\right)}\\
+e^{\frac{\ln(\gamma_{obj})}{0.2i_{\max}}\left(\max\limits_{t>0}\left(\left|i_f(t)\right|\right)-i_{\max}\right)}\\
+e^{\frac{\ln(\gamma_{obj})}{0.2}\left(\max\limits_{v}\left(\left|\frac{ftm_{obj}(v)}{ftm_{\theta_{vx}}(v)}\right|\right)-1\right)}+e^{\frac{\ln(\gamma_{obj})}{0.2i_{\max}}\left(\max\limits_{t>0}\left(\left|i_{vx}(t)\right|\right)-i_{\max}\right)}\\
+e^{\frac{\ln(\gamma_{obj})}{0.2}\left(\max\limits_{v}\left(\left|\frac{ftm_{obj}(v)}{ftm_{\theta_{vy}}(v)}\right|\right)-1\right)}+e^{\frac{\ln(\gamma_{obj})}{0.2i_{\max}}\left(\max\limits_{t>0}\left(\left|i_{vy}(t)\right|\right)-i_{\max}\right)}\\
+e^{\frac{\ln(\gamma_{obj})}{0.2}\left(\max\limits_{v}\left(\left|\frac{ftm_{obj}(v)}{ftm_{\theta_{vz}}(v)}\right|\right)-1\right)}+e^{\frac{\ln(\gamma_{obj})}{0.2i_{\max}}\left(\max\limits_{t>0}\left(\left|i_{vz}(t)\right|\right)-i_{\max}\right)}
\end{array}\right] \qquad [3.120]$$

$\gamma_{obj}=3$

7 The division $\left|\dfrac{ftm_{obj}(v)}{ftm_{\theta_v}(v)}\right|$ is performed term to term.

The means of calculating MTF from a temporal signal are included in Appendix D.

The inconvenience of this approach is not its feasibility since a solution will be found but rather in terms of the computation time to find this solution.

3.5. Optimal tuning of the fixed-structure and fixed-order final controller

3.5.1. *Introduction*

One drawback that is inherent to modern controller synthesis methods is that the order of the obtained controller is high (it is directly linked to the order of weights and of the system). Accordingly, engineers must subsequently simplify the obtained controller; even if the full-order controller is satisfactory, questions may be posed about the performance of the reduced controller obtained against problem [3.4]. So, engineers have to constrain the controller's order *a priori* and thus resolve the problem as follows:

$$\begin{array}{c} \min\limits_{W_o(s),W_i(s)} \gamma \\ \text{Subject to:} \\ K(s) \text{ with order } n_k \text{, stabilizing and ensuring the} \\ \text{required robustness} \\ Z_1 < \overline{Z}_1,\ Z_2 < \overline{Z}_2, \ldots, Z_m < \overline{Z}_m . \end{array} \qquad [3.121]$$

It should be noted that the H_∞ problem [3.71] is equivalent to problem [3.122] [MCF 92]:

$$\begin{array}{l} K(s) = \arg\min \gamma, \\ \left\| \begin{bmatrix} W_o(s) \\ W_i(s)^{-1}K(s) \end{bmatrix} \left(I + H(s)K(s)\right)^{-1} \begin{bmatrix} W_o(s)^{-1} & H(s)W_i(s) \end{bmatrix} \right\|_\infty < \gamma \end{array} \qquad [3.122]$$

The use of this "standard equivalent" formulation at full order imposes the use of the γ-iteration [DOY 89] for the controller calculation, which is much slower than the simple resolution of the loop shape's robust stabilization problem [MCF 90].

Furthermore, it requires the stability of the weights and of their inverse.

We note that:

$$W_i(s) \overset{ss}{=} \left[\begin{array}{c|c} A_i & B_i \\ \hline C_i & D_i \end{array}\right] \Rightarrow W_i(s)^{-1} \overset{ss}{=} \left[\begin{array}{c|c} A_i - B_i D_i^{-1} C_i & B_i D_i^{-1} \\ \hline D_i^{-1} C_i & D_i^{-1} \end{array}\right]$$
$$W_o(s) \overset{ss}{=} \left[\begin{array}{c|c} A_o & B_o \\ \hline C_o & D_o \end{array}\right] \Rightarrow W_o(s)^{-1} \overset{ss}{=} \left[\begin{array}{c|c} A_o - B_o D_o^{-1} C_o & B_o D_o^{-1} \\ \hline D_o^{-1} C_o & D_o^{-1} \end{array}\right] \quad [3.123]$$

Writing:

$$\bar{\lambda}_o = \max_j \left(\mathrm{Re}\left(\lambda_j(A_o)\right)\right),\ \underline{\lambda}_o = \max_j \left(\mathrm{Re}\left(\lambda_j\left(A_o - B_o D_o^{-1} C_o\right)\right)\right)$$
$$\bar{\lambda}_i = \max_j \left(\mathrm{Re}\left(\lambda_j(A_i)\right)\right),\ \underline{\lambda}_i = \max_j \left(\mathrm{Re}\left(\lambda_j\left(A_i - B_i D_i^{-1} C_i\right)\right)\right) \quad [3.124]$$

Accordingly, the resolution framework of problem [3.75] can be carried over to problem [3.121], albeit at the cost of a longer calculation time. For this, it should suffice to determine the final controller $K(s)$ on each iteration using the techniques of structured synthesis presented in part one, considering the following optimization problem:

$$\min_{\Theta_i,\Theta_o} f(\Theta_i,\Theta_o)$$

If $\max\left(\bar{\lambda}_o,\ \underline{\lambda}_o, \bar{\lambda}_i,\ \underline{\lambda}_i\right) \geq 0$

$$f(\Theta_i,\Theta_o) = \max\left(\bar{\lambda}_o,\ \underline{\lambda}_o, \bar{\lambda}_i,\ \underline{\lambda}_i\right) \geq 0$$

Else

$$K(s) = \arg\min \gamma,$$
$$\left\| \begin{bmatrix} W_o(s) \\ W_i(s)^{-1} K(s) \end{bmatrix} (I + H(s)K(s))^{-1} \begin{bmatrix} W_o(s)^{-1} & H(s)W_i(s) \end{bmatrix} \right\|_\infty < \gamma \quad [3.125]$$
$$f(\Theta_i,\Theta_o) = -\frac{1}{\gamma + f_c(H,K)}$$

End if

However, this fixed-order synthesis approach presents some significant drawbacks:

– The use of structured synthesis techniques on each iteration, such as non-smooth optimization, presented in part one for the $K(s)$ calculation, makes the optimization process much longer since each time it necessitates several runs for the $K(s)$ calculation.

– Using such algorithms for the $K(s)$ calculation makes the function $f(\Theta_i,\Theta_o)$ random, thereby reducing the effectiveness of the global problem's optimization. Indeed, when always starting from different initial conditions for the $K(s)$ calculation, we never obtain the same result. An identical initial condition must be imposed for each iteration; however, this would reduce the effectiveness of the $K(s)$ calculation.

In Chapter 2, we described how to synthesize a fixed-order controller by stochastic optimization, an approach whose flexibility we demonstrated when accounting for complex constraints. Thus, the constraints in problem [3.121] can be addressed in the same manner, which allows us to obtain an effective method for the calculation of fixed-structure and order controllers using a specification of any form. Below, we will see that the use of frequency weights for the (final) fixed-order controller computation becomes redundant in the loop-shaping context, which of course eases the optimization process.

Two methods are available for managing the order constraint:

– Order optimization acts on the precompensator K_p(s). This approach is possible but leads to the final controller order being at least equal to that of the weights:

$$\partial°K = \partial°K_p + \partial°W_i + \partial°W_o \qquad [3.126]$$

So, if we are to manage the order of the final controller, we can act on the order of the pre-compensator and on that of the weighting functions, which can prove to be complex.

– The optimization of the order acts directly on the final controller $K(s)$: this is the approach we will use here.

3.5.2. *Toward eliminating weighting functions*

3.5.2.1. *Optimization with nominal plant*

Here, we resume the approach that we proposed in [FEY 13a], where we search for the weighting functions and final controller simultaneously by resolving the following problem:

$$\begin{gathered} \min_{W_o(s),W_i(s),K(s)} \gamma \\ \text{Subject to:} \\ W_o(s),\, W_i(s) \in \mathbf{R}H_\infty, \\ W_o(s)^{-1},\, W_i(s)^{-1} \in \mathbf{R}H_\infty, \\ T(s) \in \mathbf{R}H_\infty \\ Z_1 < \overline{Z}_1,\, Z_2 < \overline{Z}_2,\, \ldots, Z_m < \overline{Z}_m \end{gathered} \quad [3.127]$$

with:

$$\begin{aligned} T(s) &\overset{ss}{=} \left[\begin{array}{c|c} A_{cl} & B_{cl} \\ \hline C_{cl} & D_{cl} \end{array}\right] \\ &= \begin{bmatrix} W_o(s) \\ W_i(s)^{-1}K(s) \end{bmatrix} \left(I + H(s)K(s)\right)^{-1} \begin{bmatrix} W_o(s)^{-1} & H(s)W_i(s) \end{bmatrix} \end{aligned} \quad [3.128]$$

and:

$$\gamma = \|T(s)\|_\infty \quad [3.129]$$

For simplicity, we use:

$$\Theta_k = \begin{pmatrix} A_k & B_k \\ C_k & D_k \end{pmatrix},\ \Theta_i = \begin{pmatrix} A_i & B_i \\ C_i & D_i \end{pmatrix},\ \Theta_o = \begin{pmatrix} A_o & B_o \\ C_o & D_o \end{pmatrix} \quad [3.130]$$

Writing:

$$\overline{\lambda}_{cl} = \max_j \left(\operatorname{Re}\left(\lambda_j\left(A_{cl}\right)\right)\right) \quad [3.131]$$

we resolve problem [3.127] by considering the following optimization problem:

$$\begin{aligned} &\min_{\Theta_i,\Theta_o,\Theta_k} f(\Theta_i,\Theta_o,\Theta_k) \\ &\text{If } \max\left(\overline{\lambda}_o,\, \underline{\lambda}_o, \overline{\lambda}_i,\, \underline{\lambda}_i, \overline{\lambda}_{cl}\right) \geq 0 \\ &\quad f(\Theta_i,\Theta_o) = \max\left(\overline{\lambda}_o,\, \underline{\lambda}_o, \overline{\lambda}_i,\, \underline{\lambda}_i, \overline{\lambda}_{cl}\right) \\ &\text{Else} \\ &\quad \gamma = \|T(s)\|_\infty \\ &\quad f(\Theta_i,\Theta_o,\Theta_k) = -\frac{1}{\gamma + f_c(H,K)} \\ &\text{End if} \end{aligned} \quad [3.132]$$

This structure is identical to that of the structured H_∞ problems with constraints presented in section 2.2.3.9, whose effectiveness we have shown already. It is the flexibility in the approach proposed in part one that makes its implementation relatively straightforward.

In practice, we soon note that the final controller has all the frequency information required to guarantee the initial specification [3.1] and that the use of weights becomes redundant as *they tend to become constant.* Accordingly, it is worthwhile to simplify problem [3.132] to a simultaneous optimization of the controller and static weights (called scaling so), that is:

$$\begin{gathered}\min_{D_o, D_i, K(s)} \gamma \\ \text{Subject to:} \\ T(s) \in \mathbf{R}H_\infty \\ Z_1 < \overline{Z}_1,\ Z_2 < \overline{Z}_2,\ \ldots, Z_m < \overline{Z}_m\end{gathered} \qquad [3.133]$$

with:

$$\begin{aligned}T(s) &\overset{ss}{=} \left[\begin{array}{c|c} A_{cl} & B_{cl} \\ \hline C_{cl} & D_{cl}\end{array}\right] \\ &= \begin{bmatrix} D_o \\ D_i^{-1}K(s)\end{bmatrix} (I + H(s)K(s))^{-1} \begin{bmatrix} D_o^{-1} & H(s)D_i\end{bmatrix}\end{aligned} \qquad [3.134]$$

and

$$\gamma = \|T(s)\|_\infty \qquad [3.135]$$

Accordingly, we resolve problem [3.133] by considering the following optimization problem:

$$\begin{aligned}&\min_{D_i, D_o, \Theta_k} f(D_i, D_o, \Theta_k) \\ &\text{If } \overline{\lambda}_{cl} \geq 0 \\ &\quad f(D_i, D_o, \Theta_k) = \overline{\lambda}_{cl} \\ &\text{Else} \\ &\quad \gamma = \|T(s)\|_\infty \\ &\quad f(D_i, D_o, \Theta_k) = -\frac{1}{\gamma + f_c(H, K)} \\ &\text{End if}\end{aligned} \qquad [3.136]$$

This simplification is significant because:

– the number of unknowns is greatly reduced since the "dynamic aspect" of the weighting functions no longer needs finding, so there follows an increase in the speed of the optimization process;

– the weighting function stabilization step and that of its inverse no longer exists, which further shortens optimization.

The use of static scalings on the other hand is still required in order to allow for the evaluation of γ, which keeps all of its properties due to the structure of this particular H_∞ problem, which is the loop-shaping H_∞ synthesis (that is, a four-block problem with the weights and their inverse two by two); the function $f_c(H, K)$ keeps all of its meaning here. But without the use of scalings (e.g. by imposing them as unitary), we no longer have any notion as to the value of γ_{opt}, the penalty tuning and, finally, for the behavior of the optimization process.

3.5.2.2. *Multiple plant optimization*

In this section, it is assumed that we have the following ball of plants centered on a nominal model $H_0(s)$:

$$\wp = \{H_i(s),\ i = 0, \ldots, N\} \qquad [3.137]$$

3.5.2.2.1. Stability robustness optimization problem

The stability robustness optimization problem consists of resolving the following problem:

$$\begin{gathered} \min_{D_o, D_i, K(s)} \gamma \\ \text{Subject to:} \\ T_j(s) \in \mathbf{R}H_\infty, j = 0, \ldots, N \\ Z_1 < \overline{Z}_1, Z_2 < \overline{Z}_2, \ldots, Z_m < \overline{Z}_m \end{gathered} \qquad [3.138]$$

with:

$$\begin{aligned} T_j(s) &\overset{ss}{=} \left[\begin{array}{c|c} A_{cl\,j} & B_{cl\,j} \\ \hline C_{cl\,j} & D_{cl\,j} \end{array}\right] \\ &= \begin{bmatrix} D_o \\ D_i^{-1} K(s) \end{bmatrix} \left(I + H_j(s) K(s)\right)^{-1} \begin{bmatrix} D_o^{-1} & H_j(s) D_i \end{bmatrix} \end{aligned} \qquad [3.139]$$

and:

$$\gamma = \|T_0(s)\|_\infty \qquad [3.140]$$

Write:

$$\overline{\lambda}_{cl} = \max_{j=0,\dots,N}\left[\max_i\left(\mathrm{Re}\left(\lambda_i\left(A_{cl\,j}\right)\right)\right)\right] \qquad [3.141]$$

So, we can resolve problem [3.138] by considering the following optimization problem:

$$\begin{array}{l} \min\limits_{D_i,D_o,\Theta_k} f(D_i,D_o,\Theta_k) \\ \text{If } \overline{\lambda}_{cl} \geq 0 \\ \quad f(D_i,D_o,\Theta_k) = \overline{\lambda}_{cl} \\ \text{Else} \\ \quad \gamma = \|T_0(s)\|_\infty \\ \quad f(D_i,D_o,\Theta_k) = -\dfrac{1}{\gamma + f_c(H_0,K)} \\ \text{End if} \end{array} \qquad [3.142]$$

3.5.2.2.2. Stability margin robustness optimization problem

The stability margin robustness optimization problem consists of resolving the following problem:

$$\begin{array}{c} \min\limits_{D_o,D_i,K(s)} \gamma,\ \gamma = \max\limits_{i=0,\dots,N} \gamma_i \\ \text{Subject to:} \\ T_j(s) \in \mathbf{R}H_\infty, j = 0,\dots,N \\ Z_1 < \overline{Z}_1,\ Z_2 < \overline{Z}_2, \dots, Z_m < \overline{Z}_m \ \text{ for } H_0(s) \end{array} \qquad [3.143]$$

with:

$$\begin{aligned} T_j(s) &\overset{ss}{=} \left[\begin{array}{c|c} A_{cl\,j} & B_{cl\,j} \\ \hline C_{cl\,j} & D_{cl\,j} \end{array}\right] \\ &= \begin{bmatrix} D_o \\ D_i^{-1}K(s) \end{bmatrix}\left(I + H_j(s)K(s)\right)^{-1}\begin{bmatrix} D_o^{-1} & H_j(s)D_i \end{bmatrix} \end{aligned} \qquad [3.144]$$

and:

$$\gamma_j = \|T_j(s)\|_\infty \quad [3.145]$$

Write:

$$\bar{\lambda}_{cl} = \max_{j=0,\ldots,N}\left[\max_i\left(\mathrm{Re}\left(\lambda_i\left(A_{cl\,j}\right)\right)\right)\right] \quad [3.146]$$

So, we can resolve problem [3.138] by considering the following optimization problem:

$$\begin{array}{l} \min\limits_{D_i,D_o,\Theta_k} f(D_i,D_o,\Theta_k) \\ \text{If } \bar{\lambda}_{cl} \geq 0 \\ \quad f(D_i,D_o,\Theta_k) = \bar{\lambda}_{cl} \\ \text{Else} \\ \quad \gamma = \max\limits_{j=0,\ldots,N} \gamma_j,\ \gamma_j = \|T_i(s)\|_\infty \\ \quad f(D_i,D_o,\Theta_k) = -\dfrac{1}{\gamma + f_c(H_0,K)} \\ \text{End if} \end{array} \quad [3.147]$$

3.5.2.2.3. Performance robustness optimization problem

The performance robustness optimization problem consists of resolving the following problem:

$$\begin{array}{c} \min\limits_{D_o,D_i,K(s)} \gamma,\ \gamma = \max\limits_{i=0,\ldots,N} \gamma_i \\ \text{Subject to:} \\ \left.\begin{array}{r} T(s) \in \mathbf{R}H_\infty, \\ Z_1 < \bar{Z}_1,\ Z_2 < \bar{Z}_2,\ \ldots,\ Z_m < \bar{Z}_m \end{array}\right\} \forall H_j(s) \in \wp \end{array} \quad [3.148]$$

with:

$$\begin{aligned} T_j(s) &\overset{ss}{=} \left[\begin{array}{c|c} A_{cl\,j} & B_{cl\,j} \\ \hline C_{cl\,j} & D_{cl\,j} \end{array}\right] \\ &= \begin{bmatrix} D_o \\ D_i^{-1}K(s) \end{bmatrix} \left(I + H_j(s)K(s)\right)^{-1} \begin{bmatrix} D_o^{-1} & H_j(s)D_i \end{bmatrix} \end{aligned} \qquad [3.149]$$

Write:

$$\bar{\lambda}_{cl} = \max_{j=1,\ldots,N} \left[\max_i \left(\mathrm{Re}\left(\lambda_i \left(A_{cl\,j} \right) \right) \right) \right] \qquad [3.150]$$

So, we can resolve problem [3.138] by considering the following optimization problem:

$$\begin{aligned} &\min_{D_i,D_o,\Theta_k} f(D_i,D_o,\Theta_k) \\ &\text{If } \bar{\lambda}_{cl} \geq 0 \\ &\quad f(D_i,D_o,\Theta_k) = \bar{\lambda}_{cl} \\ &\text{Else} \\ &\quad \gamma_j = \left\| T_j(s) \right\|_\infty, \ j = 0,\ldots,N \\ &\quad f(D_i,D_o,\Theta_k) = -\frac{1}{\max\limits_{j=0,\ldots,N} \left[\gamma_j + f_c\left(H_j, K\right) \right]} \\ &\text{End if} \end{aligned} \qquad [3.151]$$

3.5.2.3. *Applicative example (continued)*

In this section, we resume the examples given in section 3.4.3. The various viewfinders all have unchanged specifications, as is the case for the problems in question. So, the functions $f(H,K)$ as defined remain unchanged. Similarly, the synthesis models considered (nominal and ball of plants) remain identical.

The difference is found in the fact that we optimize the structured controller directly without proceeding via frequency weighting functions. This is a direct application of algorithm [3.136] and of the multiple plant extensions.

3.5.2.3.1. Observation viewfinder

3.5.2.3.1.1. Structured optimization on the nominal plant

Here, we search for the controller directly in its structured cascade form as in Figure 3.35.

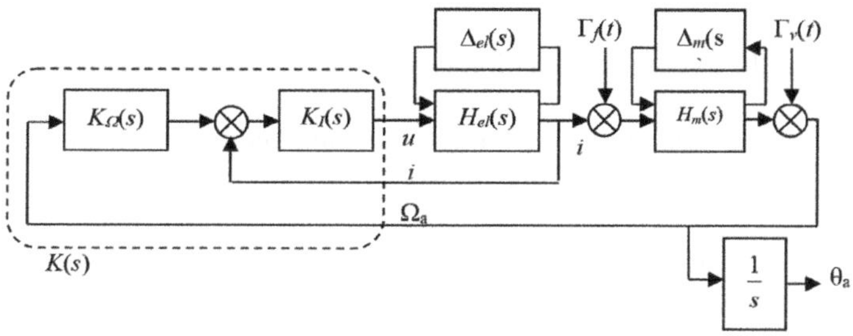

Figure 3.35. *Viewfinder control scheme in cascade form*

We choose to optimize the controllers with the following orders:

$$\partial K_I = 4,\ \partial K_\Omega = 12 \tag{3.152}$$

Note that:

$$K(s) = K_I(s)\begin{pmatrix}1 & K_\Omega(s)\end{pmatrix} \tag{3.153}$$

Thus, we write:

$$K_I(s) := \begin{pmatrix} A_I & B_I \\ C_I & D_I \end{pmatrix} = \Theta_I,\ K_\Omega(s) := \begin{pmatrix} A_\Omega & B_\Omega \\ C_\Omega & D_\Omega \end{pmatrix} = \Theta_\Omega \tag{3.154}$$

So:

$$K(s) := \begin{pmatrix} A_K & C_K \\ B_K & D_K \end{pmatrix} = \Theta_K = \left(\begin{array}{cc|cc} A_I & B_I C_\Omega & B_I D_\Omega & B_I \\ 0 & A_\Omega & B_\Omega & 0 \\ \hline C_I & D_I C_\Omega & D_I D_\Omega & D_I \end{array}\right) \tag{3.155}$$

Initially, we look for a resolution of the following problem:

$$\min_{D_o, D_i, (\Theta_I, \Theta_\Omega)} \gamma$$
$$\text{Subject to } \chi \text{ and } T(s) \in \mathbf{R}H_\infty \tag{3.156}$$

The requirements χ are defined in [3.98].

As with the approach by frequency weighting function optimization, we proceed to the resolution of the problem with several algorithms and to the statistical analysis of results in order to highlight the most efficient for this structured problem. The performance measurement rule is identical to [3.103].

We measure the average calculation time for information purposes.

The entire test was performed on a machine equipped with an Intel Core i7-3740QM (2.7 GHz) processor and 8 GB RAM.

The tests were performed with parallel computation. In this last instance, [MAT 10] was used with 8 cores (pool).

Each test was subject to 10 runs and each run was subject to 20,000 iterations.

IT / run = 20000	f_{min}	f_{mean}	f_{std}	$t_{cpu/run}$ (mn)
StaticCEC_QPSO	−0.22	−0.07	0.10	78.09
TVDecCEC_QPSO	−0.19	−0.11	0.04	46.79
Rnd_DE	−0.21	−0.10	0.07	61.88
Rnd2Bst_DE	−0.23	−0.18	0.04	125.07
Rnd2Bst_PDE	**−0.23**	**−0.19**	**0.03**	155.56
Std_PSO	−0.18	−0.06	0.07	143.05
TVRndIW_PSO	−0.18	−0.07	0.07	98.56
ABC	−0.15	−0.06	0.05	310.26
CS	−0.051	−0.01	0.02	22.62
CS/DE	−0.23	−0.13	0.07	565.37
FF				

Table 3.11. *Comparison of algorithms for structured controller optimization*

We observe that the performances vary greatly from one algorithm to the next and *FF* cannot optimize the cost function. The evolutionary calculation and particularly *Rnd2Bst_PDE* leads to the best results for this type of structured problem. The hybrid *CS/DE* leads to a good result albeit at the cost of a prohibitive calculation time. Next come algorithms *QPSO* and *PSO*.

The best runs lead to the result in Figure 3.36.

The results pertaining to the evaluation of temporal requirements can be collected in Table 3.12.

Note that for confidentiality reasons, we present results relative to the maximum temporal values specified (confidential), that is, relative to 1.

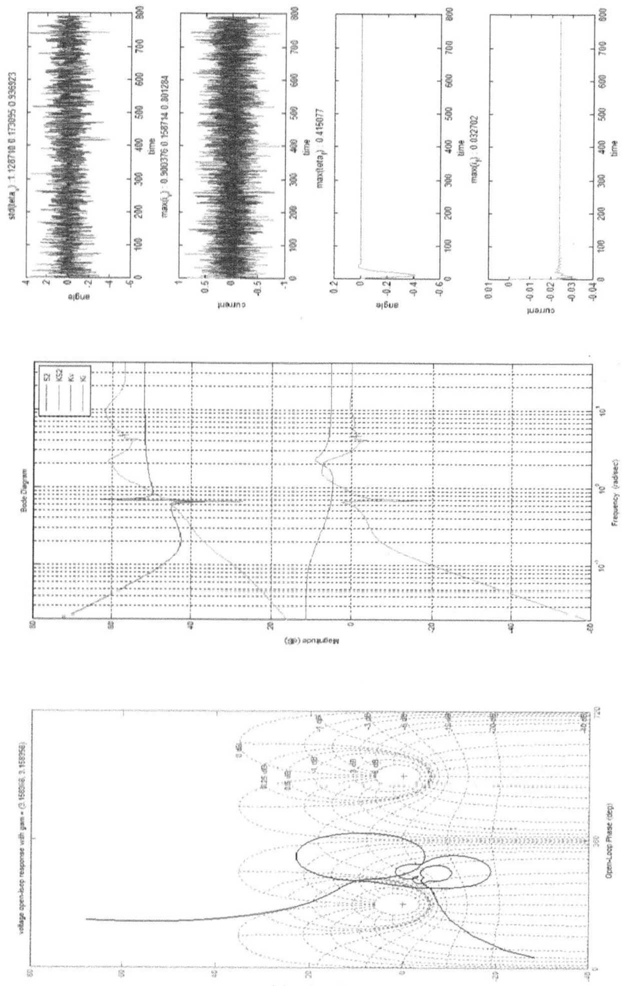

Figure 3.36. *Optimal tuning of the structured controller. For a color version of this figure, see www.iste.co.uk/feyel/optimization.zip*

Requirements	*Final evaluations*
γ	3.16
$\max_{t>0}\left(\lvert\theta_f(t)\rvert\right) \le \bar{\theta}_f$	0.41
$\max_{t>T_f}\left(\lvert\theta_f(t)\rvert\right) \le \varepsilon_f$	0
$\max_{t>0}\left(\lvert i_f(t)\rvert\right) \le i_{\max}$	0.03
$\sigma_{\theta_{vx}} \le \bar{\theta}_v$	1.12
$\max_{t>0}\left(\lvert i_{vx}(t)\rvert\right) \le i_{\max}$	0.90
$\sigma_{\theta_{vy}} \le \bar{\theta}_v$	0.17
$\max_{t>0}\left(\lvert i_{vy}(t)\rvert\right) \le i_{\max}$	0.16
$\sigma_{\theta_{vz}} \le \bar{\theta}_v$	0.94
$\max_{t>0}\left(\lvert i_{vz}(t)\rvert\right) \le i_{\max}$	0.80

Table 3.12. *Evaluation of requirements following the structured controller optimization*

We see that most of the requirements are satisfied, with only the fourth 11% above the specification, which is still acceptable.

The performances obtained are very similar to those obtained through weighting function optimization (full-order controller).

3.5.2.3.1.2. Multiple plant controller

The ball of plants $\wp$ used here is defined in section 3.4.3.1.3.

In Figure 3.37, we see that the controller obtained earlier is not robust with the ball of plants. Indeed, the lowest value of γ obtained for the ball of plants is 5.47. Accordingly, we should look to make the controller robust by means of the multiple plant techniques developed earlier.

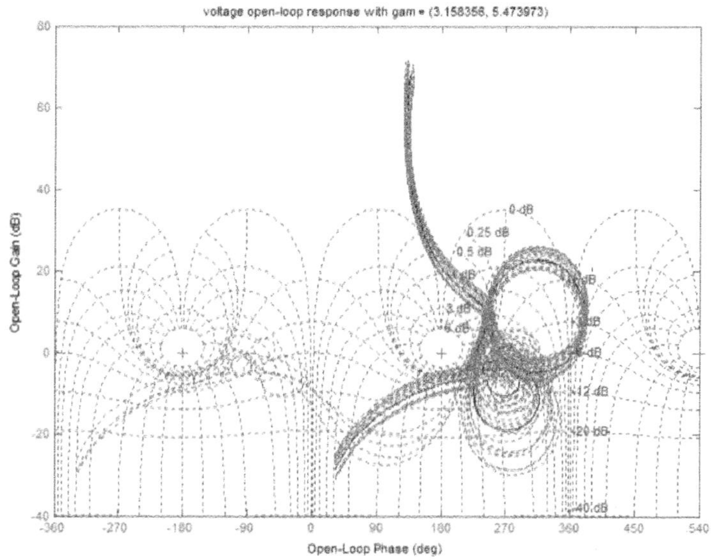

Figure 3.37. *Robustness controller examination. For a color version of this figure, see www.iste.co.uk/feyel/optimization.zip*

We can find (with *Rnd2Bst_PDE*) a controller solution to the following problems:

– Stability margin robustness problem:

$$\begin{gathered}\min_{D_o, D_i, K(s)} \gamma,\ \gamma = \max_{i=0,\dots,N} \gamma_i \\ \text{Subject to } 3\ \chi \\ \text{and } K(s) \text{ stabilizes all } \wp\end{gathered} \quad [3.157]$$

We obtain the results shown in Table 3.13.

IT / run = 20000	f_{min}	f_{mean}	f_{std}	$t_{cpu/run}$ (mn)
Rnd2Bst_PDE	−0.19	−0.12	0.05	731.53

Table 3.13. *Optimization of the controller – stability margin robustness*

We see that the calculation time for the stability margin robustness problem is much greater that the nominal model approach: this is due to a number of evaluations of the closed loop the H_∞ norm N times greater.

The best 10 runs give the results in Figure 3.38.

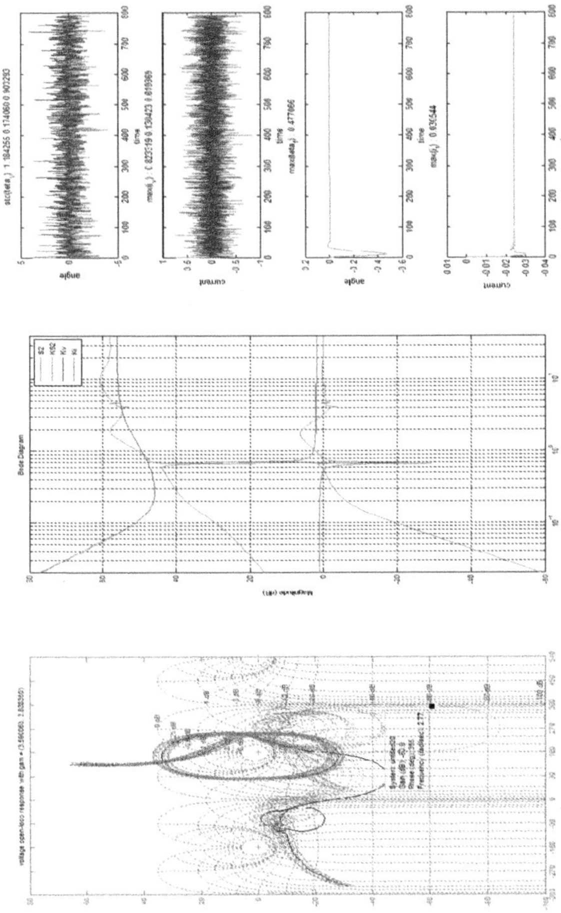

Figure 3.38. *Problem of stability margin robustness. For a color version of this figure, see www.iste.co.uk/feyel/optimization.zip*

The final evaluation of the requirements is presented in Table 3.14.

Requirements	*Final evaluation on nominal system*
$\gamma_{nom}; \gamma_{pire}$	3.50; 3.83
$\max\limits_{t>0}\left(\lvert\theta_f(t)\rvert\right) \le \bar{\theta}_f$	0.48
$\max\limits_{t>T_f}\left(\lvert\theta_f(t)\rvert\right) \le \varepsilon_f$	0
$\max\limits_{t>0}\left(\lvert i_f(t)\rvert\right) \le i_{\max}$	0.03
$\sigma_{\theta vx} \le \bar{\theta}_v$	1.18
$\max\limits_{t>0}\left(\lvert i_{vx}(t)\rvert\right) \le i_{\max}$	0.82
$\sigma_{\theta vy} \le \bar{\theta}_v$	0.17
$\max\limits_{t>0}\left(\lvert i_{vy}(t)\rvert\right) \le i_{\max}$	0.13
$\sigma_{\theta vz} \le \bar{\theta}_v$	0.90
$\max\limits_{t>0}\left(\lvert i_{vz}(t)\rvert\right) \le i_{\max}$	0.62

Table 3.14. *Evaluation of requirements – stability margin robustness*

We see that most of the requirements are satisfied and that the lowest value of γ is now 3.83.

3.5.2.3.2. The viewfinder for shooting

3.5.2.3.2.1. Structured optimization with nominal model

Let us resume the example of the viewfinder for shooting, now directly searching for the controller of Figure 3.39.

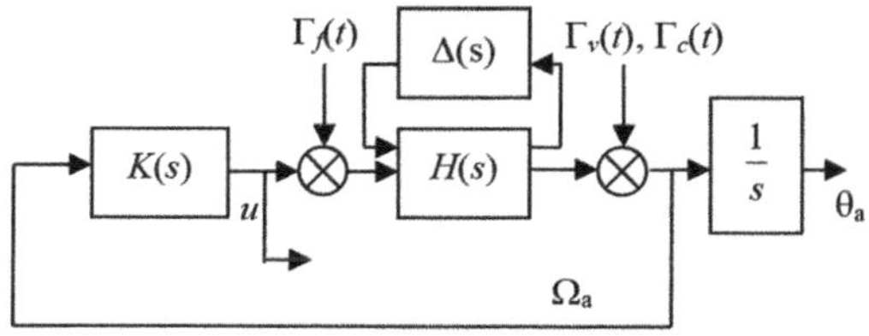

Figure 3.39. *Viewfinder for shooting control scheme*

We choose to optimize the controller with order 12.

Initially, we will look to resolve the following problem:

$$\begin{aligned} &\min_{D_o, D_i, K(s)} \gamma \\ &\text{Subject to } \chi\text{' and} \\ &T(s) \in \mathbf{R}H_\infty \end{aligned} \qquad [3.158]$$

The requirements χ' are defined in [3.111].

Since its effectiveness has already been established, we choose to approach the problem with the algorithm *Rnd2Bst_PDE*.

The entire test was performed on a machine equipped with an Intel Core i7-3740QM (2.7 GHz) processor and 8 GB RAM.

The tests were performed with parallel computation. In this last case, [MAT 10] was used with 8 cores (pool).

Each run was subject to 10,000 iterations.

We obtained the results shown in Table 3.15, which results in Figure 3.40.

$IT\ /\ run = 10000$	f_{min}	f_{mean}	f_{std}	$t_{cpu/run}$ (mn)
Rnd2Bst_PDE	−0.20	−0.18	0.012	96.59

Table 3.15. *Optimization of viewfinder for shooting controller*

The results pertaining to the final evaluation of the temporal requirements are presented in Table 3.16. Note that for confidentiality reasons, we present the results relative to the maximum temporal values specified (confidential), that is, relative to 1.

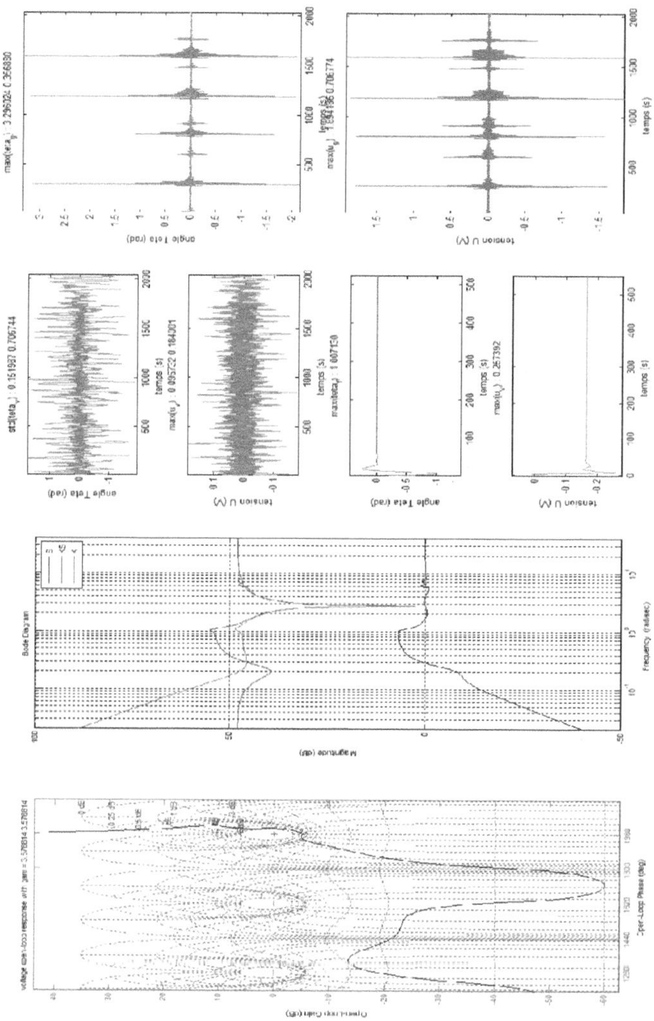

Figure 3.40. *Controller optimization. For a color version of this figure, see www.iste.co.uk/feyel/optimization.zip*

We observe that the high-priority requirements are almost all satisfied. One low-priority requirement is not satisfied but remains acceptable nonetheless: the algorithm sought to optimize the low-priority requirements to the maximum once the high-priority requirements were satisfied (which explains the controller's rejection behavior at 2 rad/s, the proper frequency mode of the viewfinder that applies during shocks). In the absence of requirements concerning response to shocks, this rejection would not have been calculated.

Requirements	*Final evaluations*
γ	3.58
$\max_{t>0}\left(\lvert\theta_f(t)\rvert\right) \leq \bar{\theta}_f$	1.00
$\max_{t>T_f}\left(\lvert\theta_f(t)\rvert\right) \leq \varepsilon_f$	0
$\max_{t>0}\left(\lvert u_f(t)\rvert\right) \leq u_{max}$	0.26
$\sigma_{\theta vx} \leq \bar{\theta}_v$	0.15
$\max_{t>0}\left(\lvert u_{vx}(t)\rvert\right) \leq u_{max}$	0.10
$\sigma_{\theta vy} \leq \bar{\theta}_v$	0.71
$\max_{t>0}\left(\lvert u_{vy}(t)\rvert\right) \leq u_{max}$	0.18
$\max_{t>0}\left(\lvert u_{cx}(t)\rvert\right) \leq u_{max}$	1.89
$\max_{t>0}\left(\lvert u_{cy}(t)\rvert\right) \leq u_{max}$	0.70

Table 3.16. *Evaluation of requirements following controller optimization*

3.5.2.3.2.2. Multiple plant optimization

The ball of plants $\wp$ used here is defined in section 3.4.3.2.3.

In Figure 3.41, we see that the controller obtained earlier is not robust with the ball of plants. Indeed, the lowest value of γ obtained for the ball of plants is 8.10. Accordingly, we should look to make the controller robust by means of the multiple plant techniques developed earlier.

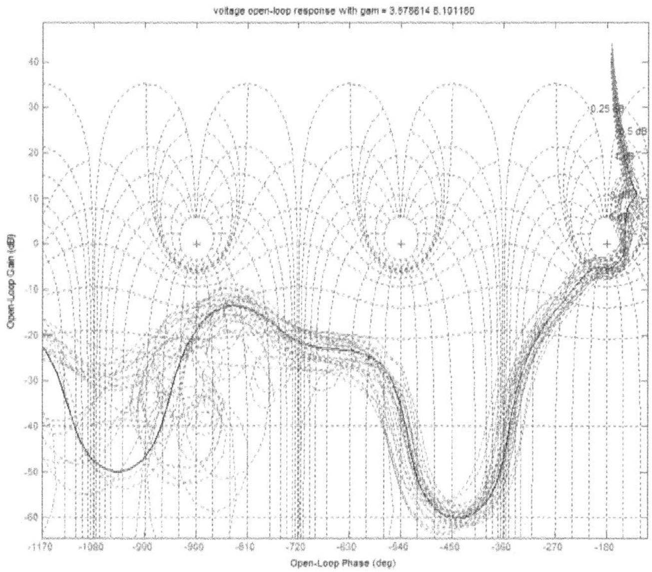

Figure 3.41. *Examination of controller robustness. For a color version of this figure, see www.iste.co.uk/feyel/optimization.zip*

We can find (with *Rnd2Bst_PDE*) a controller solution to the following problems:

– Problem of robustness stability margins:

$$\min_{D_o, D_i, K(s)} \gamma,\ \gamma = \max_{j=0,\ldots,N} \gamma_j \qquad [3.159]$$

Subject to χ' and $K(s)$ stabilizes all $\wp$

We obtain the results shown in Table 3.17.

IT / run = 10000	f_{min}	f_{mean}	f_{std}	$t_{cpu/run}$ (mn)
Rnd2Bst_PDE	−0.15	−0.087	0.037	256.04

Table 3.17. *Controller optimization – stability margin robustness*

We see that the calculation time of the stability margin robustness problem is much greater than the approach with the nominal model: this is due to a number of evaluations of the closed-loop H_∞ norm being N times greater.

The best 10 runs give the results in Figure 3.42.

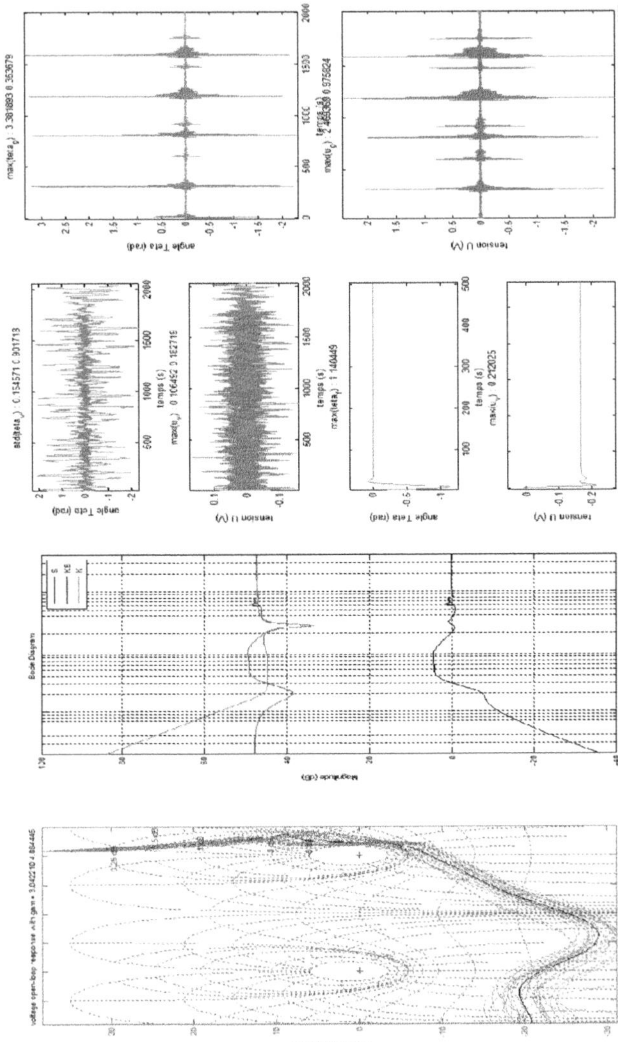

Figure 3.42. *Problem of stability margin robustness. For a color version of this figure, see www.iste.co.uk/feyel/optimization.zip*

The final evaluation of the requirements is presented in Table 3.18.

Requirements	*Final evaluation in nominal system*
$\gamma_{nom}; \gamma_{pire}$	3.04; 4.08
$\max_{t>0}\left(\lvert\theta_f(t)\rvert\right) \leq \bar{\theta}_f$	1.14
$\max_{t>T_f}\left(\lvert\theta_f(t)\rvert\right) \leq \varepsilon_f$	0
$\max_{t>0}\left(\lvert u_f(t)\rvert\right) \leq u_{\max}$	0.21
$\sigma_{\theta vx} \leq \bar{\theta}_v$	0.16
$\max_{t>0}\left(\lvert u_{vx}(t)\rvert\right) \leq u_{\max}$	0.11
$\sigma_{\theta vy} \leq \bar{\theta}_v$	0.90
$\max_{t>0}\left(\lvert u_{vy}(t)\rvert\right) \leq u_{\max}$	0.18
$\max_{t>0}\left(\lvert u_{cx}(t)\rvert\right) \leq u_{\max}$	2.47
$\max_{t>0}\left(\lvert u_{cy}(t)\rvert\right) \leq u_{\max}$	0.97

Table 3.18. *Evaluation of requirements – stability margin robustness*

We observe that the worst case value of γ is now 4.08. The value of γ for the nominal plant is also improved. In addition, the low-priority requirements are less satisfied. We can suppose that they would be satisfied by increasing the order of the controller or by increasing the number of iterations. High-priority requirements are almost all satisfied.

3.5.3. *Extensions to the approach*

3.5.3.1. *Extension to 2-degree-of-freedom controllers*

3.5.3.1.1. One-step approach

Reference tracking

Let us consider the scheme in Figure 3.43. Here, we see the loop-shaping H_∞ synthesis four-block equivalent scheme to which an anticipation component $K_{p1}(s)$

was added allowing the shaping of a loop transfer (here S_pH_p) vis-à-vis a given reference model $T_m(s)$ when the transfer connecting ε_m to r_m is constrained. The feedback controller is noted as $K_{p2}(s)$.

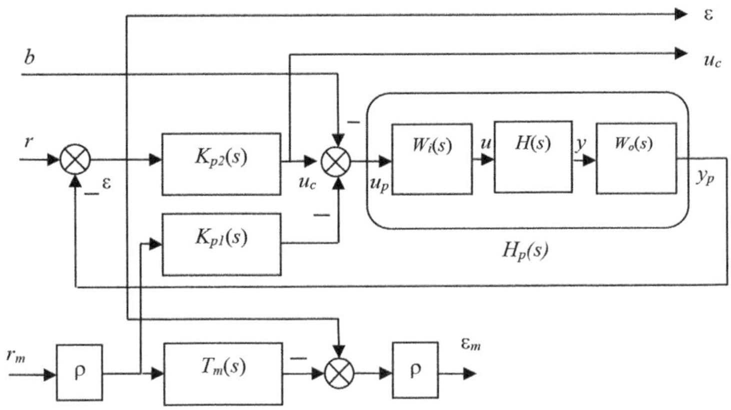

Figure 3.43. *Two-degree-of-freedom controller in the loop-shaping framework*

We can easily show that:

$$\begin{aligned}
&\varepsilon(s) = S_p(s)r(s) + S_p(s)H_p(s)b(s) + S_p(s)H_p(s)K_{p1}(s)\rho r_m(s)\\
&u_c(s) = K_{p2}(s)S_p(s)r(s) + K_{p2}(s)S_p(s)H_p(s)b(s) +\\
&\qquad K_{p2}(s)S_p(s)H_p(s)K_{p1}(s)\rho r_m(s)\\
&\varepsilon_m(s) = \rho S_p(s)r(s) + \rho S_p(s)H_p(s)b(s) + \rho\left(S_p(s)H_p(s)K_{p1}(s) - T_m(s)\right)\rho r_m(s)\\
&S_p(s) = \left(I + H_p(s)K_{p2}(s)\right)^{-1}
\end{aligned} \qquad [3.160]$$

This is, in matricial form:

$$\begin{pmatrix} \varepsilon(s) \\ u_c(s) \\ \hline \varepsilon_m(s) \end{pmatrix} =$$

$$\left(\begin{array}{cc|c} S_p(s) & S_p(s)H_p(s) & S_p(s)H_p(s)K_{p1}(s)\rho \\ K_{p2}(s)S_p(s) & K_{p2}(s)S_p(s)H_p(s) & K_{p2}(s)S_p(s)H_p(s)K_{p1}(s)\rho \\ \hline \rho S_p(s) & \rho S_p(s)H_p(s) & \rho\left(S_p(s)H_p(s)K_{p1}(s) - T_m(s)\right)\rho \end{array}\right) \begin{pmatrix} r(s) \\ b(s) \\ \hline r_m(s) \end{pmatrix} \qquad [3.161]$$

We can now find the controllers $K_{p2}(s)$ and $K_{p1}(s)$ that stabilize the previous transfer and minimize γ such that:

$$\left\| \begin{array}{cc|c} S_p(s) & S_p(s)H_p(s) & S_p(s)H_p(s)K_{p1}(s)\rho \\ K_{p2}(s)S_p(s) & K_{p2}(s)S_p(s)H_p(s) & K_{p2}(s)S_p(s)H_p(s)K_{p1}(s)\rho \\ \hline \rho S_p(s) & \rho S_p(s)H_p(s) & \rho\left(S_p(s)H_p(s)K_{p1}(s)-T_m(s)\right)\rho \end{array} \right\|_\infty < \gamma \qquad [3.162]$$

When $\rho = 0$, we find precisely the formulation of loop-shaping H_∞ synthesis to determine $K_{p2}(s)$. The ρ parameter is consequently a tuning parameter that allows for the decoupling of the reference tracking problem by $K_{p1}(s)$ from that of loop shape stabilization by $K_{p2}(s)$. Accordingly, the transfer $S_p(s)H_p(s)K_{p1}(s)$ is shaped by the reference model $T_m(s)$ in proportion to the increasing value of ρ. The other transfers of the criterion depend on ρ allowing to limit the impact on the determination of loop controller $K_{p2}(s)$ (that is the solution to the standard H_∞ loop-shaping problem).

We find a similar principle using the formalism of coprime factorization in [HOY 91, LIM 93, SKO 01].

When the pre-compensators K_{p1} and K_{p2} have been determined, the final controllers are reconstructed by relationships [3.163].

$$\begin{aligned} K_1(s) &= W_i(s)K_{p1}(s)W_o(s) \\ K_2(s) &= W_i(s)K_{p2}(s)W_o(s) \end{aligned} \qquad [3.163]$$

Let us justify [3.163] by considering Figure 3.44.

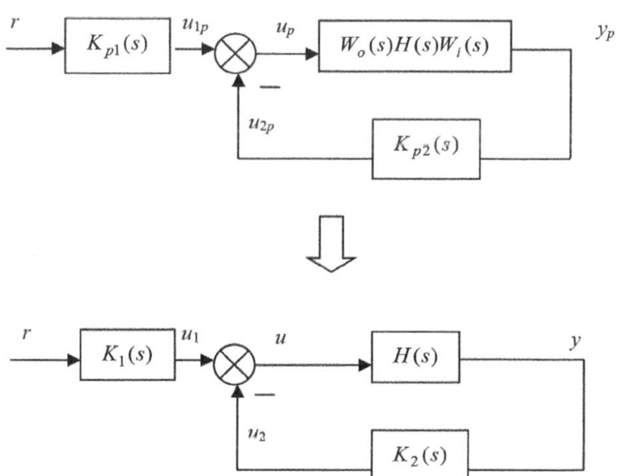

Figure 3.44. *Final reconstruction of the two-degree-of-freedom controller*

We can easily show that:

$$\begin{aligned} y(s) &= \left(I + H(s)K_2(s)\right)^{-1} H(s)K_1(s)r(s) \\ &= S(s)H(s)K_1(s)r(s) \\ &= T_{r\to y}(s)r(s) \end{aligned} \qquad [3.164]$$

On the other hand:

$$\begin{aligned} y_p(s) &= \left(I + H_p(s)K_{p2}(s)\right)^{-1} H_p(s)K_{p1}(s)r(s) \\ &= \left(I + W_o(s)H(s)W_i(s)K_{p2}(s)\right)^{-1} W_o(s)H(s)W_i(s)K_{p1}(s)r(s) \end{aligned} \qquad [3.165]$$

Assuming that the weights are inversible:

$$K_{p2}(s) = W_i(s)^{-1} K_2(s)W_o(s)^{-1} \qquad [3.166]$$

We must then assume that they are also diagonal:

$$\begin{aligned} y_p(s) &= \left(I + W_o(s)H(s)K_2(s)W_o(s)^{-1}\right)^{-1} W_o(s)H(s)W_i(s)K_{p1}(s)r(s) \\ &= W_o(s)\left(I + H(s)K_2(s)\right)^{-1} H(s)W_i(s)K_{p1}(s)r(s) \\ &= W_o(s)S(s)H(s)W_i(s)K_{p1}(s)r(s) \\ &= T_{r\to y_p}(s)r(s) \end{aligned} \qquad [3.167]$$

So:

$$\begin{aligned} \sigma\left(T_{r\to y_p}(s)\right) &= \sigma\left(W_o(s)S(s)H(s)W_i(s)K_{p1}(s)\right) \\ &= \sigma\left(S(s)H(s)W_i(s)K_{p1}(s)W_o(s)\right) \end{aligned} \qquad [3.168]$$

Thus, writing:

$$K_1(s) = W_i(s)K_{p1}(s)W_o(s) \qquad [3.169]$$

So:

$$\begin{aligned}\sigma\left(T_{r\to y_p}(s)\right) &= \sigma\left(S(s)H(s)K_1(s)\right)\\ &= \sigma\left(T_{r\to y}(s)\right)\end{aligned} \qquad [3.170]$$

This justifies the reconstruction [3.163].

Note that when we do not wish to impose a reference model on all of the measurements but on only several components, we introduce a channel selection matrix during the construction of ε_m.

Returning to our problem of m requirements, we can now find the weighting functions and the ρ parameter solutions of problem [3.171], a simple extension of the approach with 1 degree of freedom:

$$\begin{gathered}\min_{\rho, W_o(s), W_i(s)} \gamma\\ \text{Subject to:}\\ W_o(s),\ W_i(s) \in \mathbf{R}H_\infty\\ Z_1 < \overline{Z}_1,\ Z_2 < \overline{Z}_2, \ldots, Z_m < \overline{Z}_m\end{gathered} \qquad [3.171]$$

with γ defined in [3.162]. The formulation of this problem is identical to that of 1 degree of freedom. We calculate the controllers K_{p1} and K_{p2} H_∞-optimal, solution to problem [3.171], by considering the following optimization problem:

$$\begin{aligned}
&\min_{\Theta_i, \Theta_o, \rho} f(\Theta_i, \Theta_o, \rho)\\
&\text{If } \max\left(\overline{\lambda}_o,\ \underline{\lambda}_o, \overline{\lambda}_i,\ \underline{\lambda}_i\right) \geq 0\\
&\quad f(\Theta_i, \Theta_o, \rho) = \max\left(\overline{\lambda}_o,\ \underline{\lambda}_o, \overline{\lambda}_i,\ \underline{\lambda}_i\right)\\
&\text{Else}\\
&\quad H_p(s) = W_o(s)H(s)W_i(s)\\
&\quad K(s) = \left[K_{p1}(s)\ K_{p2}(s)\right] = \arg\min \gamma,\\
&\quad \left\| \begin{array}{cc|c} S_p(s) & S_p(s)H_p(s) & S_p(s)H_p(s)K_{p1}(s)\rho\\ K_{p2}(s)S_p(s) & K_{p2}(s)S_p(s)H_p(s) & K_{p2}(s)S_p(s)H_p(s)K_{p1}(s)\rho\\ \hline \rho S_p(s) & \rho S_p(s)H_p(s) & \rho\left(S_p(s)H_p(s)K_{p1}(s) - T_m(s)\right)\rho \end{array} \right\|_\infty < \gamma\\
&\quad K_1(s) = W_i(s)K_{p1}(s)W_o(s),\ K_2(s) = W_i(s)K_{p2}(s)W_o(s)\\
&\quad f(\Theta_i, \Theta_o, \rho) = -\frac{1}{\gamma + f_c(H, K)}\\
&\text{End if}
\end{aligned} \qquad [3.172]$$

As we did for the approach with 1 degree of freedom, nothing stops us from looking for the weighting functions and the final controllers simultaneously. Assuming that the loop shape's weights are invertible:

$$\left.\begin{array}{l} H_p(s)=W_o(s)H(s)W_i(s) \\ K_{p2}(s)=W_i(s)^{-1}K_2(s)W_o(s)^{-1} \end{array}\right\} \Rightarrow S_p(s)=\left(I+W_o(s)H(s)K_2(s)W_o(s)^{-1}\right)^{-1} \qquad [3.173]$$

$$K_{p1}(s)=W_i(s)^{-1}K_1(s)W_o(s)^{-1}$$

Note that:

$$\left\|\begin{array}{cc|c} S_p(s) & S_p(s)H_p(s) & S_p(s)H_p(s)K_{p1}(s)\rho \\ K_{p2}(s)S_p(s) & K_{p2}(s)S_p(s)H_p(s) & K_{p2}(s)S_p(s)H_p(s)K_{p1}(s)\rho \\ \hline \rho S_p(s) & \rho S_p(s)H_p(s) & \rho\left(S_p(s)H_p(s)K_{p1}(s)-T_m(s)\right)\rho \end{array}\right\|_\infty \qquad [3.174]$$

$$=\left\|\begin{pmatrix} I \\ K_{p2}(s) \\ \hline \rho I \end{pmatrix} S_p(s)\left(I \quad H_p(s) \mid H_p(s)K_{p1}(s)\rho\right)-\rho\begin{pmatrix} 0 & 0 & 0 \\ 0 & 0 & 0 \\ \hline 0 & 0 & T_m(s) \end{pmatrix}\rho\right\|_\infty$$

If the weights are also diagonal:

$$\begin{pmatrix} I \\ K_{p2}(s) \\ \hline \rho I \end{pmatrix} S_p(s)\left(I \quad H_p(s) \mid H_p(s)K_{p1}(s)\rho\right)$$

$$=\begin{pmatrix} I \\ W_i(s)^{-1}K_2(s)W_o(s)^{-1} \\ \hline \rho I \end{pmatrix} S_p(s)\left(I \quad W_o(s)H(s)W_i(s) \mid W_o(s)H(s)K_1(s)W_o(s)^{-1}\rho\right)$$

$$=\begin{pmatrix} W_o(s) \\ W_i(s)^{-1}K_2(s) \\ \hline \rho W_o(s) \end{pmatrix} W_o(s)^{-1}S_p(s)W_o(s)\left(W_o(s)^{-1} \quad H(s)W_i(s) \mid H(s)K_1(s)W_o(s)^{-1}\rho\right) \qquad [3.175]$$

$$=\begin{pmatrix} W_o(s) \\ W_i(s)^{-1}K_2(s) \\ \hline \rho W_o(s) \end{pmatrix} \left(I+H(s)K_2(s)\right)^{-1}\left(W_o(s)^{-1} \quad H(s)W_i(s) \mid H(s)K_1(s)W_o(s)^{-1}\rho\right)$$

$$=\begin{pmatrix} W_o(s) \\ W_i(s)^{-1}K_2(s) \\ \hline \rho W_o(s) \end{pmatrix} S(s)\left(W_o(s)^{-1} \quad H(s)W_i(s) \mid H(s)K_1(s)W_o(s)^{-1}\rho\right)$$

Thus:

$$\left(\begin{array}{cc|c} S_p(s) & S_p(s)H_p(s) & S_p(s)H_p(s)K_{p1}(s)\rho \\ K_{p2}(s)S_p(s) & K_{p2}(s)S_p(s)H_p(s) & K_{p2}(s)S_p(s)H_p(s)K_{p1}(s)\rho \\ \hline \rho S_p(s) & \rho S_p(s)H_p(s) & \rho\left(S_p(s)H_p(s)K_{p1}(s)-T_m(s)\right)\rho \end{array}\right)$$

$$= \left(\begin{array}{l} \left(\begin{array}{c} W_o(s) \\ W_i(s)^{-1}K_2(s) \\ \hline \rho W_o(s) \end{array}\right) S(s)\left(W_o(s)^{-1} \quad H(s)W_i(s) \mid H(s)K_1(s)W_o(s)^{-1}\rho\right) \\ -\rho\left(\begin{array}{cc|c} 0 & 0 & 0 \\ 0 & 0 & 0 \\ \hline 0 & 0 & T_m(s) \end{array}\right)\rho \end{array}\right) \qquad [3.176]$$

$$= T(s)$$

Note that the H_∞ criterion [3.176] corresponds to scheme Figure 3.45.

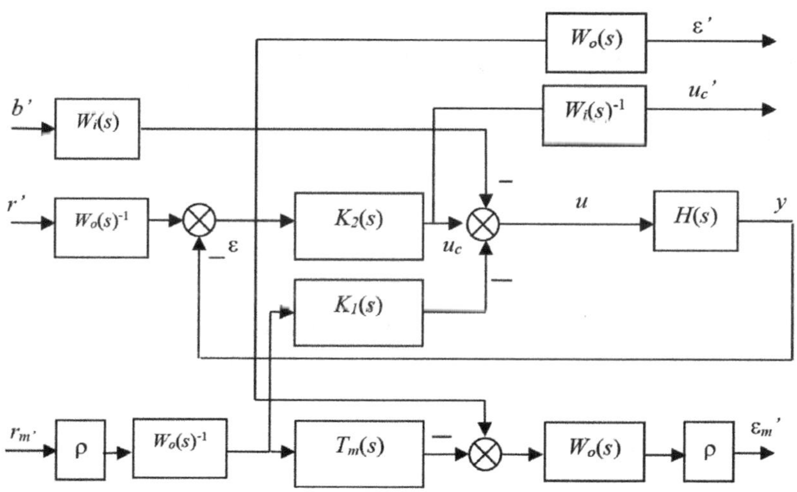

Figure 3.45. *Two-degree-of-freedom controller – externalized weights*

We can also directly look for the final structured controllers, weights and ρ parameter, solution to problem [3.177]:

$$\begin{gathered}\min_{\rho, W_o(s), W_i(s), K_1(s),\, K_2(s)} \gamma \\ \text{Subject to:} \\ W_o(s),\, W_i(s) \in \mathbf{R}H_\infty \\ T(s) \in \mathbf{R}H_\infty, \\ Z_1 < \overline{Z}_1,\, Z_2 < \overline{Z}_2, \ldots, Z_m < \overline{Z}_m .\end{gathered} \qquad [3.177]$$

with $\gamma = \|T(s)\|_\infty$.

As we had observed for the approach with 1 degree of freedom, the simultaneous search for weighting functions and final controllers leads to static weighting functions so that we can replace problem [3.177] with problem [3.178] with the implementation of static scalings in the place of frequency weights:

$$\begin{gathered}\min_{\rho, D_o, D_i, K_1(s),\, K_2(s)} \gamma \\ \text{Subject to:} \\ T(s) \in \mathbf{R}H_\infty, \\ Z_1 < \overline{Z}_1,\, Z_2 < \overline{Z}_2, \ldots, Z_m < \overline{Z}_m\end{gathered} \qquad [3.178]$$

With:

$$\begin{aligned} T(s) &= \begin{pmatrix} A_{cl} & B_{cl} \\ C_{cl} & D_{cl} \end{pmatrix} \\ &= \left(\begin{pmatrix} D_o \\ D_i^{-1} K_2(s) \\ \hline \rho D_o \end{pmatrix} S(s) \left(D_o^{-1} \quad H(s) D_i \;\middle|\; H(s) K_1(s) D_o^{-1} \rho \right) - \rho \left(\begin{array}{cc|c} 0 & 0 & 0 \\ 0 & 0 & 0 \\ \hline 0 & 0 & T_m(s) \end{array} \right) \rho \right) \end{aligned} \qquad [3.179]$$

$$S = (I + HK_2)^{-1}$$

Writing:

$$\overline{\lambda}_{cl} = \max_j \left(\mathrm{Re}\left(\lambda_j \left(A_{cl} \right) \right) \right) \qquad [3.180]$$

we resolve problem [3.178] by considering the following optimization problem:

$$\min_{\rho, D_i, D_o, \Theta_{k1}, \Theta_{k2}} f(\rho, D_i, D_o, \Theta_{k1}, \Theta_{k2})$$

If $\bar{\lambda}_{cl} \geq 0$

$$f(\rho, D_i, D_o, \Theta_{k1}, \Theta_{k2}) = \bar{\lambda}_{cl}$$

$$\gamma = \|T(s)\|_\infty \qquad [3.181]$$

$$f(\rho, D_i, D_o, \Theta_{k1}, \Theta_{k2}) = -\frac{1}{\gamma + f_c(H, K)}$$

End if

In this way, we can greatly simplify the problem's resolution.

Disturbance rejection

We can apply the principle above for the simultaneous calculation of a (structured) feedback controller and a (structured) feedforward one for a disturbance rejection problem on plant's output as shown in the scheme in Figure 3.46.

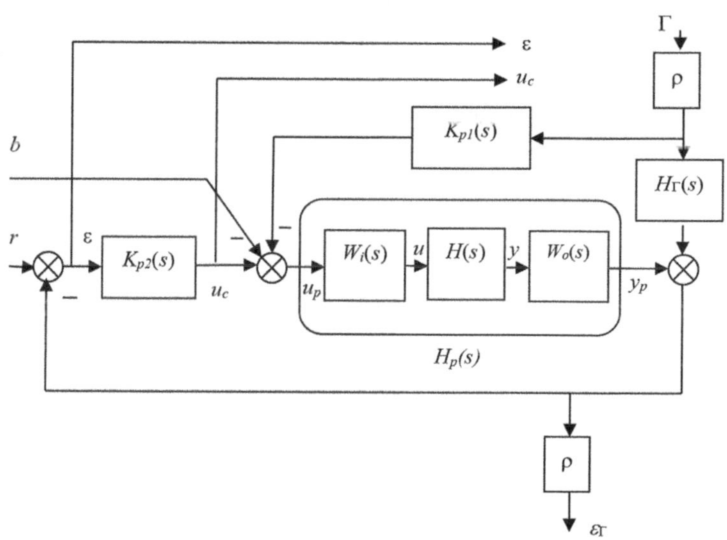

Figure 3.46. *Two-degree-of-freedom correction – disturbance rejection*

We show that:

$$
\begin{aligned}
&\varepsilon(s) = S_p(s)r(s) + S_p(s)H_p(s)b(s) + \\
&\qquad S_p(s)\big(-H_\gamma(s) + H_p(s)K_{p1}(s)\big)\rho\gamma(s) \\
&u_c(s) = K_{p2}(s)S_p(s)r(s) + K_{p2}(s)S_p(s)H_p(s)b(s) + \\
&\qquad K_{p2}(s)S_p(s)\big(-H_\gamma(s) + H_p(s)K_{p1}(s)\big)\rho\gamma(s) \\
&\varepsilon_\gamma(s) = \rho S_p(s)r(s) - \rho S_p(s)H_p(s)b(s) + \\
&\qquad \rho S_p(s)\big(H_\gamma(s) - H_p(s)K_{p1}(s)\big)\rho\gamma(s) \\
&S_p(s) = \big(I + H_p(s)K_{p2}(s)\big)^{-1}
\end{aligned}
\qquad [3.182]
$$

That is, in matricial form:

$$
\begin{pmatrix} \varepsilon(s) \\ u_c(s) \\ \hline \varepsilon_\Gamma(s) \end{pmatrix} =
$$

$$
\left(\begin{array}{cc|c}
S_p(s) & S_p(s)H_p(s) & S_p(s)\big(H_p(s)K_{p1}(s) - H_\Gamma(s)\big)\rho \\
K_{p2}(s)S_p(s) & K_{p2}(s)S_p(s)H_p(s) & K_{p2}(s)S_p(s)\big(H_p(s)K_{p1}(s) - H_\Gamma(s)\big)\rho \\
\hline
\rho S_p(s) & -\rho S_p(s)H_p(s) & -\rho S_p(s)\big(H_p(s)K_{p1}(s) - H_\Gamma(s)\big)\rho
\end{array}\right)
\begin{pmatrix} r(s) \\ b(s) \\ \hline \Gamma(s) \end{pmatrix}
\qquad [3.183]
$$

We can now look for controllers $K_{p2}(s)$ and $K_{p1}(s)$ that stabilize the previous transfer and minimize γ such that:

$$
\left\|\begin{array}{cc|c}
S_p(s) & S_p(s)H_p(s) & S_p(s)\big(H_p(s)K_{p1}(s) - H_\Gamma(s)\big)\rho \\
K_{p2}(s)S_p(s) & K_{p2}(s)S_p(s)H_p(s) & K_{p2}(s)S_p(s)\big(H_p(s)K_{p1}(s) - H_\Gamma(s)\big)\rho \\
\hline
\rho S_p(s) & \rho S_p(s)H_p(s) & \rho S_p(s)\big(H_p(s)K_{p1}(s) - H_\Gamma(s)\big)\rho
\end{array}\right\|_\infty < \gamma \qquad [3.184]
$$

In order to directly determine the final controllers, note that:

$$
\begin{aligned}
&\left(\begin{array}{cc|c}
S_p(s) & S_p(s)H_p(s) & S_p(s)\big(H_p(s)K_{p1}(s) - H_\Gamma(s)\big)\rho \\
K_{p2}(s)S_p(s) & K_{p2}(s)S_p(s)H_p(s) & K_{p2}(s)S_p(s)\big(H_p(s)K_{p1}(s) - H_\Gamma(s)\big)\rho \\
\hline
\rho S_p(s) & \rho S_p(s)H_p(s) & \rho S_p(s)\big(H_p(s)K_{p1}(s) - H_\Gamma(s)\big)\rho
\end{array}\right) \\
&= \begin{pmatrix} I \\ K_{p2}(s) \\ \hline \rho I \end{pmatrix} S_p(s)\left(\begin{array}{cc|c} I & H_p(s) & \big(H_p(s)K_{p1}(s) - H_\Gamma(s)\big)\rho \end{array}\right)
\end{aligned}
\qquad [3.185]
$$

Through a calculation similar to [3.175], we can easily show that:

$$\begin{pmatrix} I \\ K_{p2}(s) \\ \hdashline \rho I \end{pmatrix} S_p(s)\left(I \quad H_p(s) \;\vdots\; \left(H_p(s)K_{p1}(s)-H_\Gamma(s)\right)\rho\right)$$

$$= \begin{pmatrix} W_o(s) \\ W_i(s)^{-1}K_2(s) \\ \hdashline \rho W_o(s) \end{pmatrix} S(s)\left(W_o(s)^{-1} \quad H(s)W_i(s) \;\vdots\; \left(H(s)K_1(s)W_o(s)^{-1}-W_o(s)^{-1}H_\Gamma(s)\right)\rho\right) \quad [3.186]$$

$$S=(I+HK_2)^{-1}$$

Accordingly, by replacing the frequency weights by static scalings, we can thus resolve problem [3.187].

$$\begin{gathered} \min_{\rho, D_o, D_i, K_1(s),\, K_2(s)} \gamma \\ \text{Subject to:} \\ T(s) \in \mathbf{R}H_\infty \\ Z_1 < \overline{Z}_1,\, Z_2 < \overline{Z}_2, \ldots, Z_m < \overline{Z}_m \end{gathered} \quad [3.187]$$

with:

$$T(s)=\begin{pmatrix} A_{cl} & B_{cl} \\ C_{cl} & D_{cl} \end{pmatrix}$$

$$= \begin{pmatrix} D_o \\ D_i^{-1}K_2(s) \\ \hdashline \rho D_o \end{pmatrix} S(s)\left(D_o^{-1} \quad H(s)D_i \;\vdots\; \left(H(s)K_1(s)D_o^{-1}-D_o^{-1}H_\Gamma(s)\right)\rho\right) \quad [3.188]$$

$$S=(I+HK_2)^{-1}$$

The function to be minimized is identical to [3.181].

3.5.3.1.2. n-steps approach

We can readily extend the structured approach above to the general schematic of feedforward compensation of a disturbance Γ as in Figure 3.47.

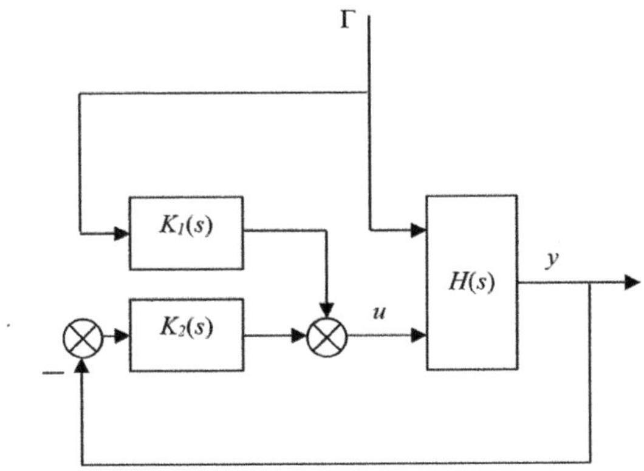

Figure 3.47. *Two-degree-of-freedom correction, general formalism*

In this case, the control law is in the form of [3.189].

$$u = K_1(s)\Gamma - K_2(s)y = \begin{bmatrix} K_1(s) & -K_2(s) \end{bmatrix} \begin{bmatrix} \Gamma \\ y \end{bmatrix} = K(s) \begin{bmatrix} \Gamma \\ y \end{bmatrix} \quad [3.189]$$

In the following expression, $K_1(s)$ represents the feedforward compensator and $K_2(s)$ represents the feedback controller. The $H(s)$ controller is expressed as:

$$y = H_{\Gamma y}(s)\Gamma + K_{uy}(s)u = \begin{bmatrix} H_{\Gamma y}(s) & H_{uy}(s) \end{bmatrix} \begin{bmatrix} \Gamma \\ u \end{bmatrix} = H(s) \begin{bmatrix} \Gamma \\ u \end{bmatrix} \quad [3.190]$$

We now propose to adapt problem [3.133] to problem [3.191]:

$$\begin{array}{c} \min\limits_{D_o, D_i, K_1(s),\, K_2(s)} \gamma \\ \text{Subject to:} \\ T(s) \in \mathbf{R}H_\infty, \\ K_1(s) \in \mathbf{R}H_\infty, \\ Z_1 < \overline{Z}1, Z_2 < \overline{Z}2, \ldots, Z_m < \overline{Z}m \end{array} \quad [3.191]$$

with:

$$\begin{aligned} T(s) &= \left[\begin{array}{c|c} A_{cl} & B_{cl} \\ \hline C_{cl} & D_{cl} \end{array} \right] \\ &= \begin{bmatrix} D_o \\ D_i^{-1} K_2(s) \end{bmatrix} \left(I + H_{uy}(s) K_2(s) \right)^{-1} \begin{bmatrix} H_{uy}(s) D_i & D_o^{-1} \end{bmatrix} \end{aligned} \quad [3.192]$$

Note that the feedforward controller calculation requires an additional constraint, that of the feedforward's stability (which is outside the loop).

For simplicity, we can note that:

$$\Theta_{k_1} = \begin{pmatrix} A_{k_1} & B_{k_1} \\ C_{k_1} & D_{k_1} \end{pmatrix}, \ \Theta_{k_2} = \begin{pmatrix} A_{k_2} & B_{k_2} \\ C_{k_2} & D_{k_2} \end{pmatrix} \quad [3.193]$$

Write:

$$\bar{\lambda} = \max\left[\max_{l}\left(\operatorname{Re}\left(\lambda_l\left(A_{k_1}\right)\right)\right), \ \max_{j}\left(\operatorname{Re}\left(\lambda_j\left(A_{cl}\right)\right)\right)\right] \quad [3.194]$$

Accordingly, we resolve problem [3.191] by considering the following optimization problem:

$$\min_{D_i, D_o, \Theta_{k_1}, \Theta_{k_2}} \ f(D_i, D_o, \Theta_{k_1}, \Theta_{k_2})$$

If $\bar{\lambda} \geq 0$

$$f(D_i, D_o, \Theta_{k_1}, \Theta_{k_2}) = \bar{\lambda} \quad [3.195]$$

Else

$$\gamma = \|T(s)\|_\infty$$

$$f(D_i, D_o, \Theta_{k_1}, \Theta_{k_2}) = -\frac{1}{\gamma + f_c(H, K)}$$

End if

The multiple plant extension is immediate. For this, it should suffice to define a ball of N plants $\wp$ centered on a nominal model $H_0(s)$ such that:

$$\wp = \left\{ H_i(s) = \left[H_{\Gamma yi}(s) \quad H_{uyi}(s) \right], \ i = 0, \ldots, N \right\} \quad [3.196]$$

Writing:

$$\bar{\lambda} = \max\left[\max_{l}\left(\operatorname{Re}\left(\lambda_l\left(A_{k_1}\right)\right)\right), \ \max_{j=0,\ldots,N}\left[\max_{i}\left(\operatorname{Re}\left(\lambda_i\left(A_{cl\,j}\right)\right)\right)\right]\right] \quad [3.197]$$

we can resolve several types of 2-degree-of-freedom robust synthesis problems. For example:

– K_2 stability margin robust, K_1 nominally performing:

$$\min_{D_i, D_o, \Theta_{k_1}, \Theta_{k_2}} f(D_i, D_o, \Theta_{k_1}, \Theta_{k_2})$$

If $\bar{\lambda} \geq 0$

$$f(D_i, D_o, \Theta_{k_1}, \Theta_{k_2}) = \bar{\lambda}$$

Else

$$\gamma = \max_{j=0,\ldots,N} \left\| T_j(s) \right\|_\infty$$

$$f(D_i, D_o, \Theta_{k_1}, \Theta_{k_2}) = -\frac{1}{\gamma + f_c(H_0, K)} \qquad [3.198]$$

End if

– (K_2, K_1) robustly performing:

$$\min_{D_i, D_o, \Theta_{k_1}, \Theta_{k_2}} f(D_i, D_o, \Theta_{k_1}, \Theta_{k_2})$$

If $\bar{\lambda} \geq 0$

$$f(D_i, D_o, \Theta_{k_1}, \Theta_{k_2}) = \bar{\lambda}$$

Else

$$\gamma_j = \left\| T_j(s) \right\|_\infty$$

$$f(D_i, D_o, \Theta_{k_1}, \Theta_{k_2}) = -\frac{1}{\max_{j=0,\ldots,N} \left(\gamma_j + f_c\left(H_j, K\right) \right)} \qquad [3.199]$$

End if

– etc.

While the approach is indeed effective, experience teaches us that the resolution to problem [3.195] takes a tremendous amount of time since it involves determining a stabilizing controller at the same time as a controller dedicated to performance (outside loop). It seems clear that the determination of the controller K_1 is conditioned in advance to that of the feedback controller K_2. A particularly interesting and common case is that among the specified m requirements, n of them (e.g. the first n's) cannot be used to determine K_1 (e.g. we cannot determine K_1 by only optimizing the requirements relative to the internal disturbance rejection).

Write:

$$Z_1 < \overline{Z}_1,\ Z_2 < \overline{Z}_2, \ldots, Z_m < \overline{Z}_m \Rightarrow f_c(H,K) = \frac{\gamma_{obj}}{m} \sum_{j=1}^{m} e^{\frac{\ln(\gamma_{obj})}{0.2|\overline{Z}_j|}(Z_j - \overline{Z}_j)}$$

$$Z_{n+1} < \overline{Z}_{n+1}, \ldots, Z_m < \overline{Z}_m \Rightarrow h_c(H,K) = \sum_{j=1}^{m-n} e^{\frac{\ln(\gamma_{obj})}{0.2|\overline{Z}_j|}(Z_j - \overline{Z}_j)} \qquad [3.200]$$

According to [3.200], we can imagine an optimization in n successive steps:

– Step 1: for fixed K_2, optimization of K_1 is done by the resolution of the following problem.

$$\begin{array}{c} \text{Find } K_1(s) \text{ such that:} \\ K_1(s) \in \mathbf{R}H_\infty, \\ Z_{n+1} < \overline{Z}_{n+1}, \ldots, Z_m < \overline{Z}_m \end{array} \qquad [3.201]$$

The requirements must be evaluated with K_2 fixed and K_1 ongoing.

We write:

$$\overline{\lambda} = \max_l \left(\mathrm{Re}\left(\lambda_l \left(A_{k_1} \right) \right) \right) \qquad [3.202]$$

Accordingly, we determine a feedforward controller by resolving the following problem:

$$\min_{\Theta_{k_1}} f(\Theta_{k_1})$$

If $\overline{\lambda} \geq 0$

$$f(\Theta_{k_1}) = \overline{\lambda}$$

Else

$$f(\Theta_{k_1}) = -\frac{1}{h_c(H,K)} \quad [3.203]$$

End if

From here, either we accept the pair (K_1, K_2) or we can make K_2 robust toward a ball of plants acting on H(s). In this case, we move to Step 2.

– Step 2: with fixed K_1, optimization of K_2 is done by the resolution of the following problem:

$$\begin{gathered} \min_{D_o, D_i, K_2(s)} \gamma \\ \text{Subject to:} \\ T(s) \in \mathbf{R}H_\infty, \\ Z_1 < \overline{Z}_1, Z_2 < \overline{Z}_2, \ldots, Z_m < \overline{Z}_m \end{gathered} \quad [3.204]$$

The requirements must be evaluated with fixed K_1 and K_2 ongoing. This problem has already been developed in [3.136] and later. It is resolved by considering $f_c(H,K)$.

– Start Step 1 again with the previous K_2, etc.

The approach begins with a controller K_2 determined by any method; we can calculate it, for instance, in order to satisfy several of the requirements.

3.5.3.2. *Extension to μ-synthesis and LPV/LFT synthesis*

We have seen in section 2.3 how the structured µ-synthesis can be reformulated for stochastic optimization.

Furthermore, we have an example showing how µ-synthesis can be easily associated with loop-shaping synthesis as in Figure 3.48.

We have seen that by selecting $\alpha_\varepsilon = \alpha_u = \alpha_r = \alpha_b = \gamma_{obj}^{-1/2}$, the problem of structured µ-synthesis can be approached by the resolution of problem [3.205]:

$$\begin{gathered} \min_{D(s),\, K_p(s)} \left\| D(s) F_l(P, K_p) D(s)^{-1} \right\|_\infty \\ \text{Subject to:} \\ F_l(P, K_p) \in \mathbf{R}H_\infty \end{gathered} \quad [3.205]$$

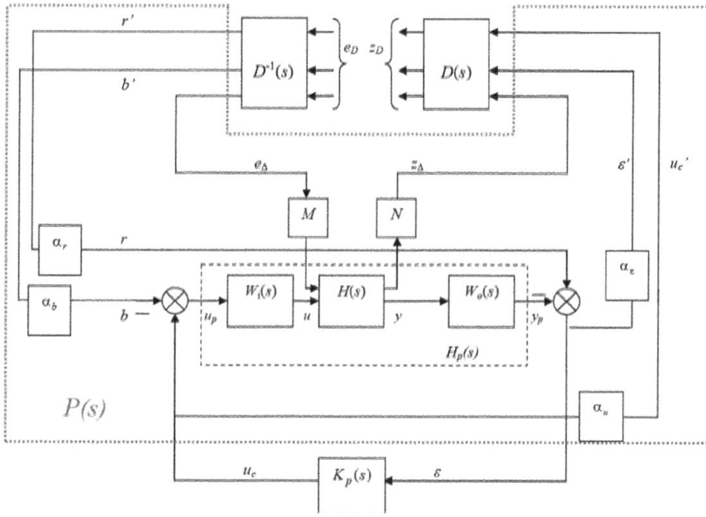

Figure 3.48. *Control scheme for μ-synthesis with scalings (D_o,D_i). For a color version of this figure, see www.iste.co.uk/feyel/optimization.zip*

Thus, the controller K_p will be robust in performance when $\left\| D(s)F_l(P,K_p)D(s)^{-1} \right\|_\infty \le 1$.

By externalizing the frequency weights (which allows to work directly on the final controller) and imposing them as static, there is no reason not to extend the previous techniques of μ-synthesis by adding the frequency scalings to be determined (those of the μ-synthesis) to the decision variables. In this way, we can look to resolve problem [3.206].

$$\begin{gathered} \min_{D_o,D_i,K(s),D(s)} \gamma \\ \text{Subject to:} \\ F_l(P,K) \in \mathbf{R}H_\infty, \\ Z_1 < \overline{Z}_1, Z_2 < \overline{Z}_2, \ldots, Z_m < \overline{Z}_m \end{gathered} \qquad [3.206]$$

with:

$$\gamma = \left\| D(s)F_l(P,K)D(s)^{-1} \right\|_\infty \qquad [3.207]$$

As for the previous problem, we can approach this problem as follows:

$$\min_{\substack{D_o, D_i, K(s), D(s) \\ F_l(P,K) \in RH_\infty}} \gamma + f_c(H,K) \quad [3.208]$$

with:

$$f_c(H,K) = \frac{\gamma_{obj}}{m} \sum_{j=1}^{m} e^{\frac{\ln(\gamma_{obj})}{\varepsilon_j}(z_j - \bar{z}_j)} \quad [3.209]$$

However, solving this problem requires a small change since the final value of γ aimed for at the end of the μ-synthesis is 1, while $f_c(H, K)$ is defined in the loop-shaping synthesis framework with $\gamma_{obj} = 3$ and cannot be equal to 1. In order to recover the feasibility of the optimization problem, it should suffice to complete the uncertainty channels with normalization coefficients $\alpha_M = \alpha_N = \gamma_{obj}^{1/2}$ and choose $\alpha_\varepsilon = \alpha_u = \alpha_r = \alpha_b = 1$ as in Figure 3.49.

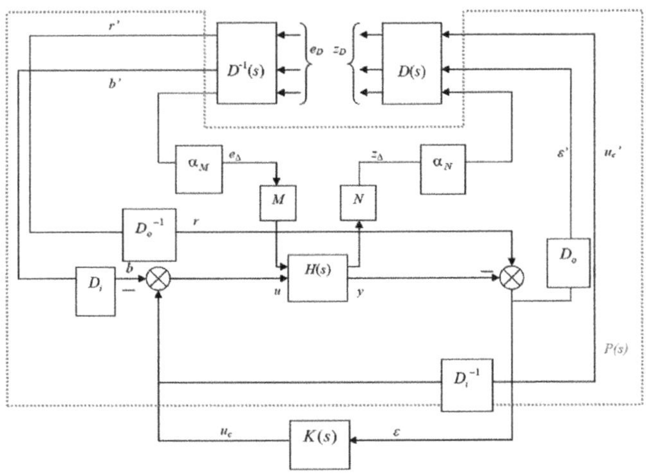

Figure 3.49. *Control schema for μ-synthesis with scalings (D_o,D_i). For a color version of this figure, see www.iste.co.uk/feyel/optimization.zip*

Indeed, as the uncertainties have already been normalized, at the end of optimization, we will have $\left\|D(s)F_l(P,K)D(s)^{-1}\right\|_\infty \rightarrow \gamma_{obj}$, which allows us to resolve problem [3.208].

The extension to structured LPV/LFT synthesis is entirely similar.

3.5.4. *Link with standard control problems*

Let us consider the following general standard form for control $P(s)$:

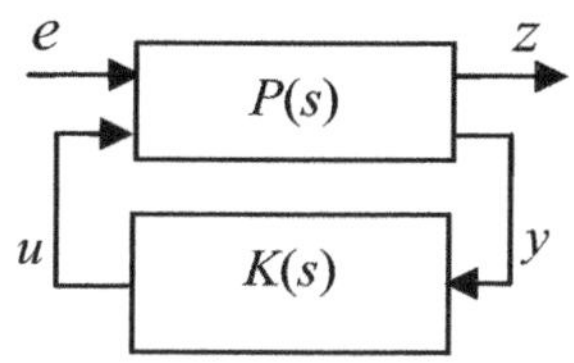

Figure 3.50. *Standard form for control*

The system $P(s)$ is partitioned as follows:

$$P(s) := \left[\begin{array}{c|cc} A & B_1 & B_2 \\ \hline C_1 & D_{11} & D_{12} \\ C_2 & D_{21} & D_{22} \end{array}\right] \quad [3.210]$$

If the problem is not a disturbance feedforward one, the scaled stabilization problem developed in previous sections can be addressed using the plant $H(s) = C_2(sI - A)^{-1}B_2 + D_{22}$ and requirements are evaluated using the closed-loop $F_l(P,K)$.

If a row of C_2 is null and if the corresponding row of D_{21} is not, thus:

$$C_2 = \begin{pmatrix} 0 \\ C_{22} \end{pmatrix}, D_{21} = \begin{pmatrix} D_{211} \\ D_{212} \end{pmatrix}, D_{22} = \begin{pmatrix} D_{221} \\ D_{222} \end{pmatrix} \quad [3.211]$$

In that case $P(s)$ is related to a disturbance feedforward problem and the controller has then the following form:

$$K(s) = \left((K_1(s)\; K_2(s)\right) \quad [3.212]$$

where $K_1(s)$ is a feedforward compensator and $K_2(s)$ is a stabilizing feedback controller. The scaled stabilization problem developed in previous sections is addressed with the system $H(s) = C_{22}(sI - A)^{-1}B_2 + D_{222}$ to compute the stabilizing controller $K_2(s)$, and requirements are evaluated using the global closed loop $F_l(P,K)$ which allows also the optimization of the feedforward part $K_1(s)$.

4

HinfStoch: A Toolbox for Structured and Robust Controller Computation Based on Stochastic Optimization

4.1. Introduction

In this section, we will introduce a *Matlab* tool that illustrates the theoretical developments presented in this chapter. The toolbox is comprised of several routines:

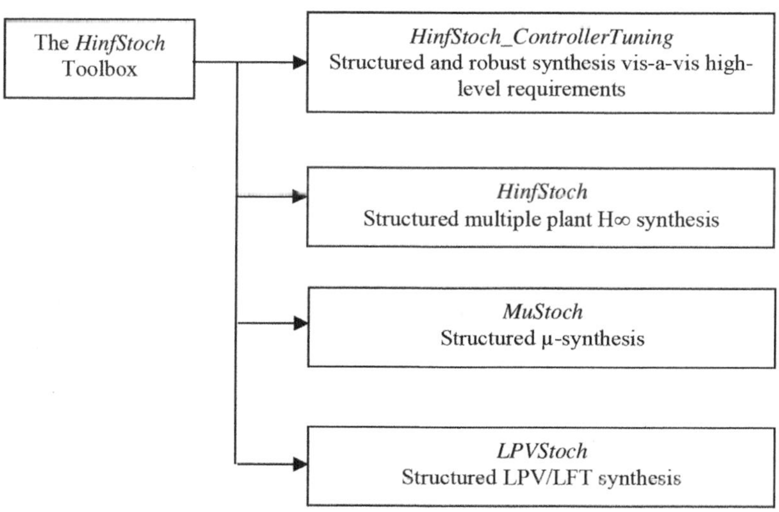

Figure 4.1. *HinfStoch toolbox*

The set of tools comprising *HinfStoch* can be downloaded at https://philippe.feyel.free.fr as well as from the publisher ISTE's website.

4.2. Structured multiple plant H$_\infty$ synthesis

4.2.1. *Principle*

HinfStoch is a numerical tool intended for the computation of structured H∞ controllers in a potentially multiple plant framework.

We consider the following set of m plants:

$$\wp = \{P_i(s), i = 1,\ldots,m\} \qquad [4.1]$$

As in Figure 1.2, the fixed-structure H$_\infty$ synthesis problem in the multiple plant framework consists of determining a controller $K(s)$ of order n_k that:

– internally stabilizes each closed loop $F_l(P_i,K)$;

– minimizes γ such that:

$$\max_{i=1,\ldots,m} \left\| F_l(P_i,K) \right\|_\infty < \gamma \qquad [4.2]$$

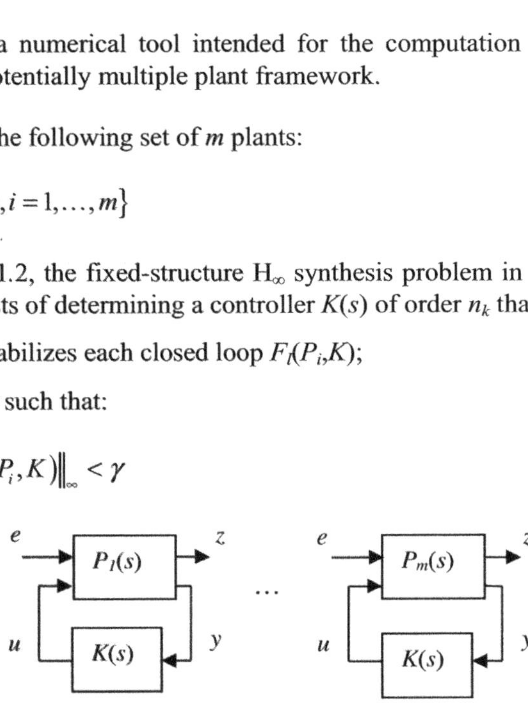

Figure 4.2. *Multiple plant H$_\infty$ synthesis*

The tool acts as a function whose implementation highly resembles that of *hinfsyn* or of *hinfstruct* [MAT 14]:

[K, GAM] = HinfStoch_v1(*P, ny, nu, order, options*)

Below, we will describe the various arguments for the tool's implementation.

4.2.2. *Formalism*

Here we address the set of input parameters necessary for the use of this tool.

4.2.2.1. *System definition*

The definition of the system P follows the same formalism as that of other tools such as *hinfsyn*. We consider the following standard form:

$$\begin{bmatrix} \dot{x} \\ \hline z \\ y \end{bmatrix} = \left[\begin{array}{c|cc} A & B_1 & B_2 \\ \hline C_1 & D_{11} & D_{12} \\ C_2 & D_{21} & D_{22} \end{array}\right] \begin{bmatrix} x \\ \hline e \\ u \end{bmatrix} := \begin{pmatrix} A & B \\ C & D \end{pmatrix} \quad [4.3]$$

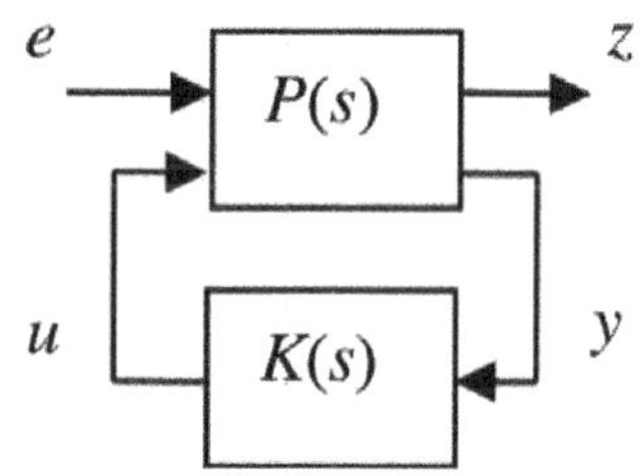

Figure 4.3. *Standard form for control*

where vector e represents the n_e exogenous inputs (disturbances, noise, references, etc.), z is the n_z signals of interest to survey, u is the vector of the n_u control signals and y is the vector of the n_y measurements.

P can be defined as *ss*[1], but the algorithm is faster[2] in execution if P is defined directly by a structure containing the state representation, that is, under the following form:

$$P.a = A; P.b = B; P.c = C; P.d = D; \quad [4.4]$$

In this version, P is a continuous system.

It is possible for the tool to operate in a multiple plant framework as in Figure 4.6.

1 ss = state-space.

2 *ss* manipulation is time consuming and useless if we only proceed to matrix calculation.

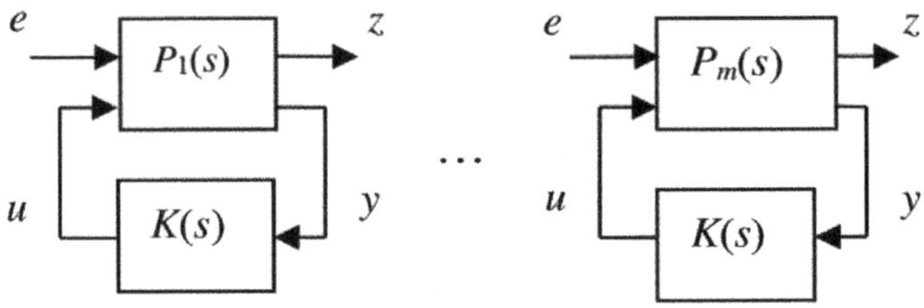

Figure 4.4. *Multiple plant formalism*

With:

$$P_i := \begin{pmatrix} A_i & B_i \\ C_i & D_i \end{pmatrix}, i = 1, \ldots, m \quad [4.5]$$

In this case, the ith plant P_i in the following form:

$$P(i).a = A_i; P(i).b = B_i; P(i).c = C_i; P(i).d = D_i; \quad [4.6]$$

4.2.2.2. *Structure of the searched controller*

It is in the "order" parameter field that we specify the structure and order of the searched controller. Two structural choices are available:

– Centralized controller:

Here, "order" is directly the order of the searched controller, the tool of which will compute and deliver the state representation. If "order" is a unique number, then we directly return to this centralized case.

– Decentralized controller:

Here, "order" is a matrix (of size $n_u \times n_y$) containing the order of each sub-transfer comprising the controller; if a negative order is entered for one of the sub-transfers, then the corresponding sub-transfer is zero.

$$order = \begin{pmatrix} \partial^\circ K_{1,1} & \cdots & \partial^\circ K_{1,n_y} \\ \vdots & \ddots & \vdots \\ \partial^\circ K_{n_u,1} & \cdots & \partial^\circ K_{n_u,n_y} \end{pmatrix} \quad [4.7]$$

$$\partial^\circ K_{i,j} < 0 \Leftrightarrow K_{i,j}(s) = 0$$

4.2.2.3. *Definition of options*

We place the optimization options in the optimization field.

In the following, the default values are positionable by the value [ ] (empty).

The option argument is a structure that covers the following fields:

$$\text{options} \rightarrow \left\{ \begin{array}{c} \text{options.nbr\_run} = \text{strictly positive integer} \\ \text{options.nbr\_iterations\_max} = \text{strictly positive integer} \\ \text{options.convergence\_threshold} = \text{real strictly positive} \\ \text{options.parallel\_computation} = \text{positive integer} \\ \text{options.display} = \{0.1\} \end{array} \right\}$$

The optimization options allow for more precise management of the optimization process. Since the tool is based on evolutionary optimization, it can be worthwhile to launch a certain number of runs successively by setting the *options.nbr_run* field.

Each run is subject to a certain number of iterations to be specified in *options.nbr_iterations_max*. The default value is positioned at 20,000.

It is possible to detect a premature convergence of the optimization process: where this occurs, either we allow the iterations to finish or, using the *options.convergence_threshold* field, we pre-position a threshold from which the algorithm will stop. This threshold is then related to the displayed radius on each iteration (see *options.display*) and is less than 1. The default value is positioned at 1^e-5.

Computation time can be accelerated significantly by making use of Matlab's *Parallel Computing Toolbox*. Accordingly, in the *options.parallel_computation* field, we can specify the number of pools to be used with a positive integer, limited to the number of CPU cores the machine has. The default value is positioned at 0. This is the value to be specified when the *Parallel Computing Toolbox* is not used.

By positioning the *options.display* field to 1, each iteration giving a better result is displayed during the optimization process. In particular, we see the "radius" of the evolutionary algorithm population, a value that announces the similarity of the different individuals with a low value, indicating strong resemblance (so less diversity is often synonymous with premature convergence). Otherwise, only the end result is displayed.

4.3. Structured μ-synthesis

4.3.1. *Principle*

MuStoch is a numerical tool intended for the computation of a structured controller's solution to the μ-synthesis problem.

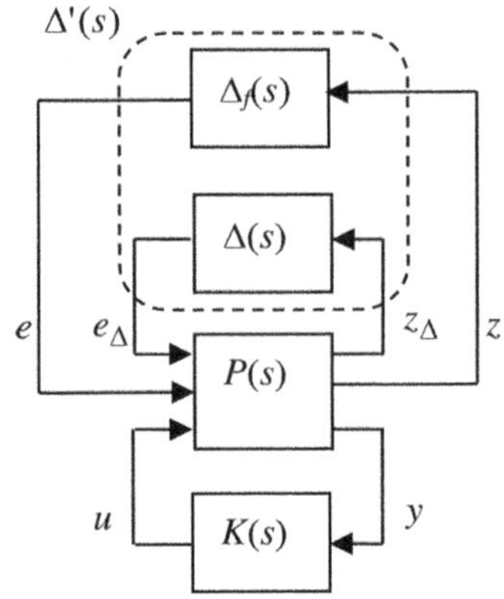

Figure 4.5. *μ-Synthesis*

As in Figure 4.5, the fixed-order structured μ-synthesis problem consists of determining a controller $K(s)$ with order n_k that:

– internally stabilizes the closed loop $F_l(P, K)$;

– that gives[3]:

$$\forall \omega, \mu_{\underline{\Delta}'}\left(F_l\left(P(j\omega), K(j\omega)\right)\right) < 1 \qquad [4.8]$$

With:

$$\underline{\Delta}' := \left\{\Delta' = diag\left(\Delta, \Delta_f\right), \Delta \in \underline{\Delta}, \ \Delta_f \in \mathrm{C}^{n_z \times n_w}\right\} \qquad [4.9]$$

The uncertainty matrix Δ has the following block-diagonal structure:

$$\underline{\Delta} := \left\{ \begin{array}{c} \Delta = diag\left\{\Delta_1, \ldots \Delta_q, \delta_1 I_{r_1}, \ldots, \delta_r I_{r_r}, \varepsilon_1 I_{c_1}, \ldots, \varepsilon_c I_{c_c}\right\} \in \mathbf{C}^{n_\Delta \times n_\Delta} \\ \Delta_i \in \mathbf{C}^{n_i \times n_i}, \ \delta_j \in \mathbf{R}, \ \varepsilon_k \in \mathbf{C} \end{array} \right\} \qquad [4.10]$$

3 Assuming normalized Δ'(s)

We can approach this problem via the following problem: find a stabilizing controller $K(s)$ with order n_k and a transfer matrix $D(s)$ such that:

– $D(s)\Delta'(s) = \Delta'(s)D(s),\ \forall \Delta'(s) \in \underline{\Delta}'$

– $\left\| D(s)F_l(P,K)D(s)^{-1} \right\|_\infty < 1$

It is important to note that $D(s)$ can be chosen without loss of generality under the form [4.11] where each $D_m(s)$ commutes with the corresponding block into $\Delta(s)$, with the latter assumed to be diagonal structured. Therefore:

$$D(s) = \begin{pmatrix} \mathrm{diag}\left(D_m(s), m = 1,\ldots,n_\Delta\right) & 0 \\ 0 & D_{ze}(s) \end{pmatrix} \quad [4.11]$$

Furthermore, for simplicity, $D_{ze}(s) = I$, when $n_z = n_e$.

The tool acts as a function whose implementation greatly resembles that of *hinfsyn* or *hinfstruct* [MAT 14]:

[K, GAM] = MuStoch_v1(*P, ny, nu, nΔ, order_K, order_D, options*)

Below, we describe the various arguments for the tool's implementation.

4.3.2. *Formalism*

Here, we describe the set of input parameters necessary for the use of this tool.

4.3.2.1. *System definition*

The system definition *P* follows the same formalism as that of other tools such as *hinfsyn*. *P* can be obtained by placing the uncertain system into LFT form. Thus, we consider the following standard form:

$$\begin{bmatrix} \dot{x} \\ \overline{Z} \\ y \end{bmatrix} = \left[\begin{array}{c|cc} A & B_1 & B_2 \\ \hline C_1 & D_{11} & D_{12} \\ C_2 & D_{21} & D_{22} \end{array}\right] \begin{bmatrix} x \\ \overline{E} \\ u \end{bmatrix} := \begin{pmatrix} A & B \\ C & D \end{pmatrix} \quad [4.12]$$

With:

$$Z = \begin{pmatrix} z \\ z_\Delta \end{pmatrix},\ E = \begin{pmatrix} e \\ e_\Delta \end{pmatrix},\ e_\Delta = \Delta(s)z_\Delta,\ e = \Delta_f(s)z \quad [4.13]$$

where vector *e* represents the n_e exogenous inputs (disturbances, noise, references, etc.), *z* is the n_z signals to survey, *u* is the vector of the n_u control signals and *y* is the vector of the n_y measurements. The uncertainty matrix $\Delta(s)$ supposed diagonal is of size n_Δ.

P can be defined as *ss*[4], but the algorithm is faster[5] in execution if *P* is directly defined by a structure containing the state representation, that is, in the following form:

$$P.a = A; P.b = B; P.c = C; P.d = D; \quad [4.14]$$

In this version, *P* is a continuous system.

4.3.2.2. *Structure of the searched controller*

It is in the parameter *order_K* that the structure and order of the searched controller are specified. Two structural choices are possible:

– Centralized controller:

Here, *order_K* is directly the order of the searched controller, the tool of which is used to compute and return the state representation. If *order_K* is a unique number, then we directly return to this centralized case.

When *order_K* = [ ], then the computed controller is of full order.

– Decentralized controller:

Here, *order_K* is a matrix (of size $n_u \times n_y$) containing the order of each sub-transfers comprising the controller; if a negative order is entered for one of the sub-transfers, then the corresponding transfer is zero.

$$order\_K = \begin{pmatrix} \partial^\circ K_{1,1} & \cdots & \partial^\circ K_{1,n_y} \\ \vdots & \ddots & \vdots \\ \partial^\circ K_{n_u,1} & \cdots & \partial^\circ K_{n_u,n_y} \end{pmatrix} \quad [4.15]$$
$$\partial^\circ K_{i,j} < 0 \Leftrightarrow K_{i,j}(s) = 0$$

4.3.2.3. *Scaling*

In *order_D* field, we specify the order of the searched scaling $D_m(s)$.

4 ss = state-space.
5 *ss* manipulation is time consuming and useless if we only proceed to matrix calculation.

The tool assumes that all of the $D_m(s)$ have the same order. Therefore, *order_D* is a positive scalar integer.

4.3.2.4. *Definition of options*

We enter the optimization options into the *options* field.

In the following, the default values are positionable by the value [ ] (empty).

The *options* argument is a structure that covers the following fields:

$$\text{options} \rightarrow \left\{ \begin{array}{c} \text{options.nbr\_run} = \text{strictly positive integer} \\ \text{options.nbr\_iterations\_max} = \text{strictly positive integer} \\ \text{options.convergence\_threshold} = \text{strictly positive real} \\ \text{options.parallel\_computation} = \text{positive integer} \\ \text{options.display} = \{0.1\} \end{array} \right\}$$

The optimization options let us manage the optimization process. Since the tool is based on evolutionary optimization, it can be worthwhile to launch a certain number of successive runs by setting the *options.nbr_run* field.

Each run is subject to a certain number of iterations to be specified in *options.nbr_iterations_max*. The default value is positioned at 20,000.

It is possible to detect a premature convergence of the optimization process: in this case, either we allow the iterations to finish or we pre-position a threshold from which the algorithm stops by means of the *options.convergence_threshold* field. This threshold is related to the displayed radius on each iteration (*see options.display*) and is lower than 1. The default value is positioned at 1^e-5.

The computation time can be accelerated significantly by using Matlab's *Parallel Computing Toolbox*. Accordingly, by setting *options.parallel_computation*, we can specify a positive integer, the *pool* number to be used, limited by the machine's number of CPU cores. The default value is positioned at 0. This is the value to be specified when the *Parallel Computing Toolbox* is not used.

By positioning the *options.display* field to 1, each iteration giving a better result is displayed through the optimization process. In particular, we see the population "radius" of the evolutionary algorithm, a value that announces the similarity of the different individuals; a low value indicating strong resemblance (so less diversity is often synonymous with premature convergence). Otherwise, only the end result is displayed.

4.4. Structured LPV/LFT synthesis

4.4.1. *Principle*

LPVStoch is a numerical tool intended for the computation of structured LPV/LFT controllers, stabilizing systems themselves under the form LPV/LFT.

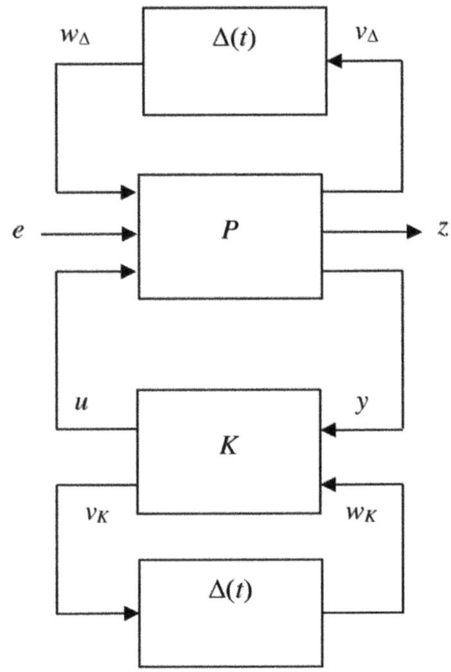

Figure 4.6. *Closed-loop LPV*

As in Figure 4.6, the problem consists of determining an LTI controller K that:

– internally stabilizes the closed loop $T(P, K, \Delta)$ defined by $T(P,K,\Delta) = F_l(F_u(P(s),\Delta(t)),F_l(K(s),\Delta(t)))$ for any possible evolution of $\Delta(t)$ such that $\gamma^2\Delta^T(t)\Delta(t) \leq 1$;

– confers the closed loop $T(P, K, \Delta)$ an L_2 gain between z and e lower than a value of γ, that is:

$$\|T(P,K,\Delta)\|_2 \leq \gamma \qquad [4.16]$$

The tool acts as a function whose implementation greatly resembles that of *hinfsyn* or of *hinfstruct* [MAT 14]:

[K, GAM] = LPVStoch_v1(*P, ny, nu, delta, order, options*)

4.4.2. *Formalism*

Below, we describe the set of input parameters necessary for the tool's use.

4.4.2.1. *System definition*

The definition of the system P follows the same formalism as that of other tools such as *hinfsyn*. It is possible to obtain P by placing the uncertain system under the LFT form. Therefore, we will consider the following form:

$$\begin{bmatrix} \dot{x} \\ \hline Z \\ y \end{bmatrix} = \left[\begin{array}{c|cc} A & B_1 & B_2 \\ \hline C_1 & D_{11} & D_{12} \\ C_2 & D_{21} & D_{22} \end{array}\right] \begin{bmatrix} x \\ \hline E \\ u \end{bmatrix} := \begin{pmatrix} A & B \\ C & D \end{pmatrix} \qquad [4.17]$$

With:

$$Z = \begin{pmatrix} v_\Delta \\ z \end{pmatrix}, \ E = \begin{pmatrix} w_\Delta \\ e \end{pmatrix}, \ w_\Delta = \Delta(t) v_\Delta \qquad [4.18]$$

where vector e represents the exogenous n_e inputs (disturbances, noise, references, etc.), z is the n_z signals to be monitored, u is the vector of the n_u control signals and y is the vector of the n_y measurements. The uncertainty matrix $\Delta(t)$ is assumed diagonal:

$$\underline{\Delta} := \left\{ blockdiag\left(\theta_1(t) I_{r_1}, \ldots, \theta_{n_\Delta}(t) I_{r_{n_\Delta}}\right) : \theta_i(t) \in \mathbf{R} \right\} \qquad [4.19]$$

Its structure has to be entered into the *delta* field (see below).

P can be defined as *ss*[6], but the algorithm is faster[7] in execution if P is directly defined by a structure containing the state representation, that is, under the following form:

$$P.a = A; P.b = B; P.c = C; P.d = D; \qquad [4.20]$$

In this version, P is a continuous system.

6 ss = state-space.

7 Indeed, *ss* manipulation is time consuming and useless if we only proceed to matrix calculation.

4.4.2.2. *Structure of searched controller*

It is important to recall that the controller will have the following form:

$$K(s) := \begin{pmatrix} A_K & B_K \\ C_K & D_K \end{pmatrix} = \left(\begin{array}{c|cc} A_K & B_{K_y} & B_{K_\Delta} \\ \hline C_{K_y} & D_{K_{yy}} & D_{K_{y\Delta}} \\ C_{K_\Delta} & D_{K_{\Delta y}} & D_{K_{\Delta\Delta}} \end{array}\right) := \begin{bmatrix} K_{yy}(s) & K_{y\Delta}(s) \\ K_{\Delta y}(s) & K_{\Delta\Delta}(s) \end{bmatrix} \quad [4.21]$$

Accordingly, the final gain-scheduled controller to be implemented has the following form for its state representation:

$$K(\Delta) = F_l\left(K(s), \Delta(t)\right) := \begin{pmatrix} A_K + B_{K_\Delta} M_\Delta C_{K_\Delta} & B_{K_y} + B_{K_\Delta} M_\Delta D_{K_{\Delta y}} \\ C_{K_y} + D_{K_{y\Delta}} M_\Delta C_{K_\Delta} & D_{K_{yy}} + D_{K_{y\Delta}} M_\Delta D_{K_{\Delta y}} \end{pmatrix} \quad [4.22]$$

$$M_\Delta := \Delta(t)\left(I - D_{K_{\Delta\Delta}} \Delta(t)\right)^{-1}$$

The searched controller's structure and the order are specified in the *order* parameter. Two structural choices are possible:

– Centralized controller

Here, *order* is directly the order of the searched controller, whose tool computes and returns the state representation. If *order* is a unique number, then we directly return to this centralized case.

When *order* = [ ], then the computed controller is of full order.

– Decentralized controller:

Here, *order* is a matrix (of size $(n_u + n_\Delta)\times(n_y + n_\Delta)$, n_Δ being the size of $\Delta(t)$) containing the order of each of the sub-transfers comprising the controller; if a negative order is entered for one of the sub-transfers, then the corresponding sub-transfer is zero.

$$order = \begin{pmatrix} \partial^\circ K_{1,1} & \cdots & \partial^\circ K_{1,n_y+n_\Delta} \\ \vdots & \ddots & \vdots \\ \partial^\circ K_{n_u+n_\Delta,1} & \cdots & \partial^\circ K_{n_u,n_y+n_\Delta} \end{pmatrix} \quad [4.23]$$

$$\partial^\circ K_{i,j} < 0 \Leftrightarrow K_{i,j}(s) = 0$$

4.4.2.3. *Structure of the uncertainty matrix*

The uncertainty matrix $\Delta(t)$ is assumed diagonal:

$$\underline{\Delta} := \left\{ blockdiag\left(\theta_1(t) I_{r_1}, \ldots, \theta_{n_\Delta}(t) I_{r_{n_\Delta}}\right) : \theta_i(t) \in \mathbf{R} \right\} \quad [4.24]$$

Its structure must be entered into the *delta* field, which is a vector containing information on repetition (or not) of the uncertainty blocks considered.

For example, assuming that:

$$\begin{aligned} &\underline{\Delta} := \{blockdiag(\theta_1(t),\ \theta_2(t),\ \theta_3(t))\} \Rightarrow delta = [1 \quad 1 \quad 1] \\ &\underline{\Delta} := \{blockdiag(\theta_1(t),\ \theta_1(t),\ \theta_2(t))\} \Rightarrow delta = [2 \quad 1] \\ &\underline{\Delta} := \{blockdiag(\theta_1(t),\ \theta_2(t),\ \theta_2(t))\} \Rightarrow delta = [1 \quad 2] \\ &\underline{\Delta} := \{blockdiag(\theta_1(t),\ \theta_1(t),\ \theta_1(t))\} \Rightarrow delta = 3 \end{aligned} \quad [4.25]$$

4.4.2.4. *Definition of options*

We position the optimization options by setting the *options* field.

In the following, the default values are positionable by the value [ ] (empty).

The *options* argument is a structure that covers the following fields:

$$\text{options} \rightarrow \left\{ \begin{array}{c} \text{options.nbr\_run} = \text{strictly positive integer} \\ \text{options.nbr\_iterations\_max} = \text{strictly positive integer} \\ \text{options.convergence\_threshold} = \text{strictly positive real} \\ \text{options.parallel\_computation} = \text{positive integer} \\ \text{options.display} = \{0.1\} \end{array} \right\}$$

The optimization options let us manage the optimization process. Since the tool is based on evolutionary optimization, it can be worthwhile to launch a certain number of successive runs using the *options.nbr_run* field.

Each run is subject to a certain number of iterations to be specified in *options.nbr_iterations_max*. The default value is positioned at 20,000.

It is possible to detect a premature convergence of the optimization process: in this case, either we allow the iterations to finish or we pre-position a threshold from which the algorithm stops by means of the *options.convergence_threshold* field. This threshold is related to the displayed radius on each iteration (*see options.display*) and is lower than 1. The default value is positioned at 1^e-5.

The computation time can be accelerated significantly by using Matlab's *Parallel Computing Toolbox*. Accordingly, by setting *options.parallel_computation*, we can specify a positive integer, the *pool* number to be used, limited by the machine's number of CPU cores. The default value is positioned at 0. This is the value to be specified when the *Parallel Computing Toolbox* is not used.

By positioning the *options.display* field to 1, each iteration giving a better result is displayed through the optimization process. In particular, we see the population "radius" of the evolutionary algorithm, a value that announces the similarity of the different individuals; a low value indicating strong resemblance (so less diversity is often synonymous with premature convergence). Otherwise, only the end result is displayed.

4.5. Structured and robust synthesis against high-level requirements with *HinfStoch_ControllerTuning*

4.5.1. *Principle*

HinfStoch_ControllerTuning is a numerical tool for the computation of structured and robust controllers operating under Matlab with the aim of satisfying high-level system requirements related to any given values Z_1, Z_2, etc. These values can be of any nature (frequential, temporal, nonlinear, etc.) so long as the specification is of the form:

$$Z_1 < \overline{Z}_1, Z_2 < \overline{Z}_2, \ldots, Z_m < \overline{Z}_m \qquad [4.26]$$

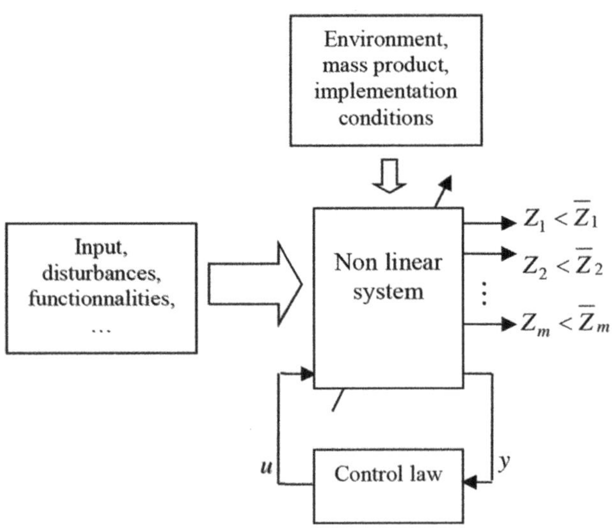

Figure 4.7. *Servo-loop and high-level specifications*

The use of any sort of criteria is made possible by the use of an evolutionary stochastic optimization algorithm dedicated to the problems of structured control (*Rnd2Bst_PDE*), and not requiring gradient or sub-gradient formulation, with the only constraint being the capability of criteria evaluation.

The stability and performance can be ensured here by the use of the tool in a multi-plant framework (see below).

Finally, the computation time can be accelerated significantly by the use of Matlab's *Parallel Computing ToolBox* [MAT 10] on a multi-core computer, as previously mentioned.

This tool is complementary to other commercial tools such as *systune* based on non-smooth optimization [APK 13]. However, *HinfStoch_ControllerTuning* benefits from certain characteristics that make its use particularly worthwhile in the industrial context:

– It can use any type of criteria so long as the requirements under consideration can be evaluated.

– It can work directly on experimental data as well as with "artificial" data (step, sine, etc.).

– Since the requirements are defined using Matlab instructions, nothing prevents the use of this tool with *Simulink* (see examples), thereby allowing for the simulation and evaluation of requirements in a purely nonlinear system framework.

The tool acts as a function whose implementation greatly resembles that of *hinfsyn* or of *hinfstruct* [MAT 14]:

> [K_ss, K_blk] = HinfStoch_ControllerTuning_v2(*P, ny, nu, order, requirements, options*)

Below, we will describe the various arguments for the tool's implementation, before providing some scripts of examples.

A first version of the tool was presented in [FEY 16]. The second version presented here adds functions and options to the first. It is important to note that examples and synthesis performed with the first version are compatible with the second. Here is the description of the usage.

4.5.2. *Formalism*

Below, we describe the set of input parameters necessary for the tool's use.

4.5.2.1. *System definition*

The system definition P follows the same formalism as that of other tools such as *hinfsyn*. We consider the following standard form:

$$\begin{bmatrix} \dot{x} \\ \hline z \\ y \end{bmatrix} = \left[\begin{array}{c|cc} A & B_1 & B_2 \\ \hline C_1 & D_{11} & D_{12} \\ C_2 & D_{21} & D_{22} \end{array}\right] \begin{bmatrix} x \\ \hline e \\ u \end{bmatrix} := \begin{pmatrix} A & B \\ C & D \end{pmatrix} \qquad [4.27]$$

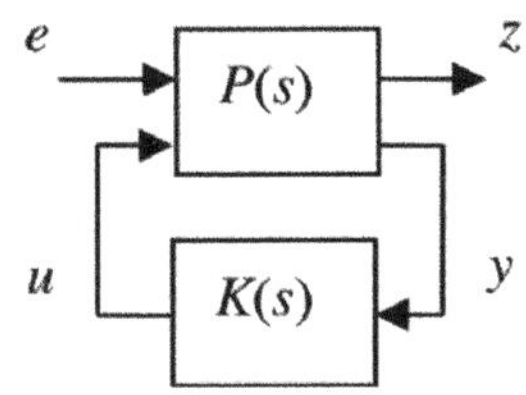

Figure 4.8. *Standard form for control*

where vector e represents the exogenous n_e inputs (disturbances, noise, references, etc.), z is the n_z signals to be monitored, u is the vector of the n_u control signals and y is the vector of the n_y measurements.

P can be defined as *ss*[8], but the algorithm is faster[9] in execution if P is directly defined by a structure containing the state representation, that is, under the following form:

$$P.a = A; P.b = B; P.c = C; P.d = D; \qquad [4.28]$$

In this version, P is a continuous system.

The tool can conceivably be used in the multiple plant framework shown in Figure 4.9.

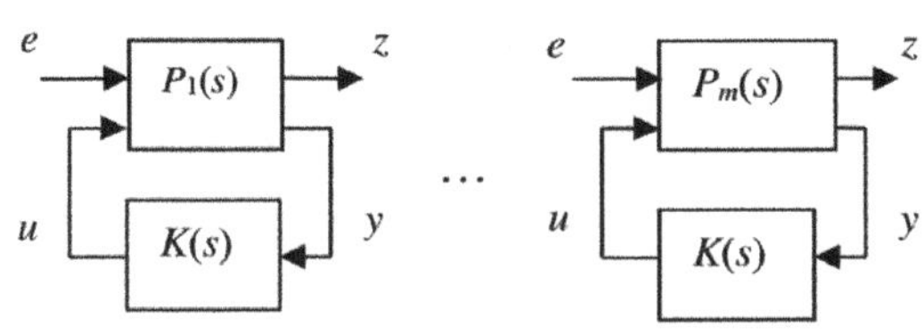

Figure 4.9. *Multiple plant framework*

8 ss = state-space.

9 Indeed, *ss* manipulation is time consuming and useless if we only proceed to matrix calculation.

With:

$$P_i := \begin{pmatrix} A_i & B_i \\ C_i & D_i \end{pmatrix}, i = 1, \ldots, m \qquad [4.29]$$

Here, we define the *i*th plant P_i in the following form:

$$P(i).a = A_i; P(i).b = B_i; P(i).c = C_i; P(i).d = D_i; \qquad [4.30]$$

Where needed, the nominal plant is assumed to be the first element of the array, that is, $P(1)$.

4.5.2.2. *Characteristics of searched controller*

The searched controller's order, structure and degree of robustness are specified in the *order* parameter. Two computation modes are available:

– Direct computation of a structured controller. This mode is denominated mode-K in the following. Here, *order* is defined as follows:

$$\text{order} \rightarrow \left\{ \begin{array}{c} \text{order.mode} = \{\text{'cascade','feedback'}\} \\ \text{order.order\_K} \\ \text{order.robustness} = \{\text{'nominal','stability','margin','performance'}\} \end{array} \right\}$$

It is important to recall that this mode consists of resolving the following problem:

$$\begin{array}{c} \min\limits_{D_o, D_i, K(s)} \gamma, \\ \text{subject to:} \\ T(s) \in \mathbf{R}H_\infty \\ Z_1 < \overline{Z}_1, Z_2 < \overline{Z}_2, \ldots, Z_m < \overline{Z}_m \end{array} \qquad [4.31]$$

With:

$$\begin{array}{l} T(s) = \begin{bmatrix} D_o \\ D_i^{-1} K(s) \end{bmatrix} (I + H(s)K(s))^{-1} \begin{bmatrix} D_o^{-1} & H(s)D_i \end{bmatrix} \\ \gamma = \|T(s)\|_\infty \end{array} \qquad [4.32]$$

– Computation of the structured weights at input and output plant (A, B_2, C_2, D_{22}). The controller is then deduced by the resolution of the loop-shaping synthesis problem on the system, with its order being equal to the sum of that of the system and double of that of the weights in the full-order case. This mode is denominated mode-W in the following. In this case *order* is defined as follows:

$$\text{order} \rightarrow \left\{ \begin{array}{c} \text{order.mode} = \{\text{'weight'}\} \\ \text{order.order\_Wo} \\ \text{order.order\_Wi} \\ \text{order.order\_K} \\ \text{order.robustness} = \{\text{'nominal','stability','margin','performance'}\} \end{array} \right\}$$

It is important to recall that this mode consists of resolving the following problem:

$$\begin{array}{l} \min\limits_{W_o(s),W_i(s)} \gamma \\ \text{subject to:} \\ W_o(s), W_i(s) \in \mathbf{R}H_\infty \\ Z_1 < \overline{Z}_1,\ Z_2 < \overline{Z}_2,\ \ldots, Z_m < \overline{Z}_m \end{array} \qquad [4.33]$$

With:

$$\begin{array}{l} H_p(s) = W_o(s)H(s)W_i(s) \\ K_p(s) = \arg\min \gamma,\ \left\| \begin{bmatrix} I \\ K_p(s) \end{bmatrix} \left(I + H_p(s)K_p(s)\right)^{-1} \begin{bmatrix} I & H_p(s) \end{bmatrix} \right\|_\infty < \gamma \\ K(s) = W_i(s)K_p(s)W_o(s) \end{array} \qquad [4.34]$$

In the two modes:

$$H(s) \stackrel{ss}{=} \begin{pmatrix} A & B_2 \\ C_2 & D_{22} \end{pmatrix} \ or\ H(s) \stackrel{ss}{=} \begin{pmatrix} A & B_2 \\ C_{22} & D_{222} \end{pmatrix} \qquad [4.35]$$

depending on the identified case (see section 3.5.4)

Accordingly, it is the *order.mode* parameter that automatically determines the mode to be selected.

4.5.2.2.1. Mode-K: controller structure and order of the controller

Two structural choices are available:

– Centralized controller:

Here, *order.order_K* is directly the order of the searched feedback controller, the tool of which then computes and returns the state representation. If *order.order_K* is a unique number, then we directly return to this centralized case.

– Decentralized controller:

Here, *order.order_K* is a matrix (of size $n_u \times n_y$), containing the order of each of the sub-transfers comprising the controller; if a negative order is returned for one of the sub-transfers, then the corresponding sub-transfer is zero:

$$order.order\_K = \begin{pmatrix} \partial^\circ K_{1,1} & \cdots & \partial^\circ K_{1,n_y} \\ \vdots & \ddots & \vdots \\ \partial^\circ K_{n_u,1} & \cdots & \partial^\circ K_{n_u,n_y} \end{pmatrix} \quad [4.36]$$

$$\partial^\circ K_{i,j} < 0 \Leftrightarrow K_{i,j}(s) = 0$$

The structure itself of the decentralized control is specified in the *order.mode* field:

– "feedback": the controller is computed directly by feedback implementation as for *hinfsyn* (Figure 4.10);

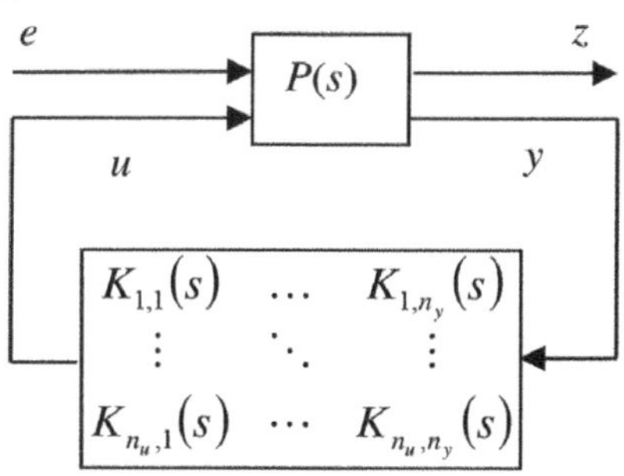

Figure 4.10. *Decentralized feedback controller*

– "cascade": the controller is implemented in cascade form and the matrix order of the controller follows this cascade format (imbrication of outside loops to inside from left to right) as shown in the following scheme:

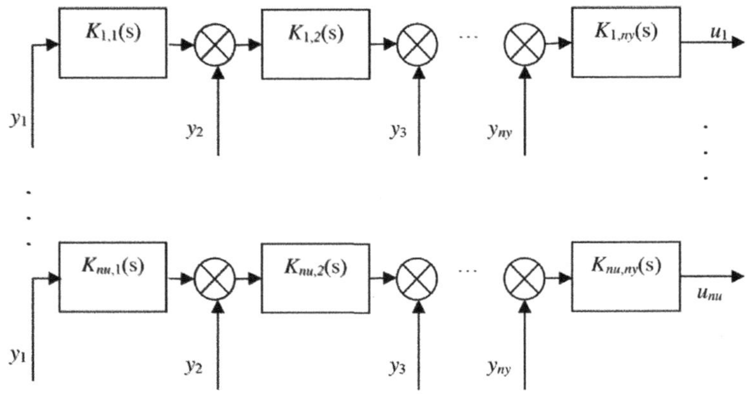

Figure 4.11. *Decentralized cascade controller*

By convention, the indices of measurement entering the controllers go in order from the most external loop to the most internal loop.

4.5.2.2.2. Mode-W: weight structure and order

Input and output plant weights are W_i and W_o, respectively. The structure of each can be either centralized or decentralized.

– Centralized weights:

Here, *order.order_*W_i (resp. *order.order_*W_o) is directly the order of searched weight W_i (resp. W_o), the tool of which computes and returns the state representation. If *order.order_*W_i (resp. *order.order_*W_o) is a unique number, then Wi (resp. W_o) is searched directly as centralized.

– Decentralized weights:

Here, *order.order_*W_i (resp. *order.order_*W_o) is a matrix of size $n_u \times n_u$ (resp. $n_y \times n_y$) containing the order of each sub-transfer comprising the weight W_i (resp. W_o); if a negative order is specified for one of the sub-transfers, then the corresponding sub-transfer is zero.

$$
\begin{aligned}
&order.order\_W_i = \begin{pmatrix} \partial^{\circ}W_{i1,1} & \cdots & \partial^{\circ}W_{i1,n_u} \\ \vdots & \ddots & \vdots \\ \partial^{\circ}W_{in_u,1} & \cdots & \partial^{\circ}W_{in_u,n_u} \end{pmatrix} \\
&\partial^{\circ}W_{im,n} < 0 \Leftrightarrow W_{im,n}(s) = 0 \\
&order.order\_W_o = \begin{pmatrix} \partial^{\circ}W_{o1,1} & \cdots & \partial^{\circ}W_{o1,n_u} \\ \vdots & \ddots & \vdots \\ \partial^{\circ}W_{on_u,1} & \cdots & \partial^{\circ}W_{on_u,n_u} \end{pmatrix} \\
&\partial^{\circ}W_{om,n} < 0 \Leftrightarrow W_{om,n}(s) = 0
\end{aligned} \quad [4.37]
$$

In W-mode, the computed controller can be:

– full-order by specifying *order.order_K* = [ ]. In that case, [MCF 90] is used to compute the precompensator at each iteration;

– fixed-order by specifying the desired order in the field *order.order_K* (a positive integer value). In that case [APK 02] is used to compute the precompensator at each iteration.

4.5.2.2.3. Enforcing frequency components

It is possible to impose *a priori* known components onto the searched controller or weights. In this case, the order matrix is no longer an integer matrix but a cell array matrix in which each element is of dimension 2: the first dimension contains the order of the required controller (not counting the order of the enforced component), the second component contains the expression of the filter that we wish to impose under the form of a character string (the predefined Laplace variable is s):

$$
\left.\begin{aligned}
&M\{1,1,1\} = \partial^{\circ}K'_{1,1},\ M\{1,1,2\} = 'H_{1,1}(s)' \\
&\vdots \\
&M\{n_u,1,1\} = \partial^{\circ}K'_{n_u,1},\ M\{n_u,1,2\} = 'H_{n_u,1}(s)' \\
&\vdots \\
&M\{1,n_y,1\} = \partial^{\circ}K'_{1,n_y},\ M\{1,n_y,2\} = 'H_{1,n_y}(s)' \\
&\vdots \\
&M\{n_u,n_y,1\} = \partial^{\circ}K'_{n_u,n_y},\ M\{n_u,n_y,2\} = 'H_{n_u,n_y}(s)'
\end{aligned}\right\} order.order\_K = M \quad [4.38]
$$

$$
\begin{aligned}
&K_{i,j}(s) \leftarrow K'_{i,j}(s) H_{i,j}(s) \\
&\partial^{\circ}K'_{i,j} < 0 \Leftrightarrow K'_{i,j}(s) = 0
\end{aligned}
$$

4.5.2.2.4. Degree of robustness in a multiple plant framework

We can define the degree of robustness of the searched controller by positioning the *order.robustness* field with the following values: “nominal”, “stability”, “margin” and “performance”. The higher the complexity of the required robustness, the longer the computation time required to achieve it. Definitions of the various robustness modes are given in Table 4.1 (it should be recalled that when using the tool in a multiple plant framework, plant $P(1)$ is the nominal plant):

order.robustness	Definition	Computation time
“Nominal”	– Stabilization of $P_1(s)$ – Minimization of γ for $P_1(s)$ subject to the requirements being satisfied for $P_1(s)$.	+
“Stability”	– Stabilization of $P_i(s)$, $i = 1,\ldots, m$, – Minimization of γ for $P_1(s)$ subject to the requirements being satisfied for $P_1(s)$.	++
“Margin”	– Stabilization of $P_i(s)$, $i = 1,\ldots, m$, – Minimization of worst case γ subject to the requirements being satisfied for $P_1(s)$.	+++
“Performance”	– Stabilization of $P_i(s)$, $i = 1,\ldots, m$, – Minimization of worst case γ subject to the requirements being satisfied for $P_i(s)$, $i = 1,\ldots, m$	++++

Table 4.1. *Robustness of the searched controller*

4.5.2.2.5. Example

Consider the system P composed of m plants, each made up of two measurements for two commands:

1) Searching for a static diagonal output feedback robust in stability. In this case:

```
order.order_K = [[0,-1];[-1,0]] ;
order.mode = 'feedback' ;
order.robustness = 'stability' ;
```

2) The searched controller is now centralized with order 5 :

```
order.order_K = 5 ;
order.robustness = 'stability' ;
```

3) The searched controller is now cascaded:

– for the first command: an order 2 for the external controller and an order 3 for the internal controller;

– for the second command: a static external controller and an order 5 for the internal controller:

```
order.order_K = [[2, 3];[0, 5]] ;
order.mode = ‘cascade’ ;
order.robustness = ‘stability’ ;
```

4) In addition to the above example, the direct sub-transfers of the controller are pure integrators:

```
M{1,1,1} = 2 ; M{1,1,2} = ‘1/s’ ;
M{1,2,1} = 3 ; M{1,2,2} = ‘1’ ;
M{2,1,1} = -1 ; M{2,1,2} = ‘1’ ;
M{2,2,1} = 5 ; M{2,2,2} = ‘1/s’ ;
order.order_K - M ;
order.mode = ‘cascade’ ;
order.robustness = ‘stability’ ;
```

5) Now we are looking for centralized weights W_o and W_i with order 5, solution to the performance robustness problem. In this case:

```
order.order_Wi = 5 ;
order.order_Wo = 5 ;
order.mode = ‘weight’ ;
order.robustness = ‘performance’ ;
```

6) The required searched weight W_o must be a pure integrator. In this case:

```
order.order_Wi = 5 ;
M{1,1,1} = 5 ; M{1,1,2} = ‘1/s*ones(2,2)’ ;
order.order_Wo = M ;
order.mode = ‘weight’ ;
order.robustness = ‘performance’ ;
```

4.5.2.3. *Definition of requirements*

4.5.2.3.1. Formalism

It is in the *requirements* parameter that we define all of the specifications to be met. An advantage of the tool is the possibility of using any Matlab instructions in the *requirements* parameter, which widens the range of possibilities. In particular, this allows us to:

– define requirements of any form or nature. We can also mix many types of requirements: frequential, temporal, etc.;

– use the *Simulink* simulation environment and thereby evaluate the requirements in a potentially nonlinear context;

– use experimental data (with the use of the *loading* instructions from a data file, for instance).

Below, we describe the formalism used for the definition of requirements.

A requirement is composed of:

– a value denoted by Z of any nature;

– its maximal specified value denoted by Z_{max};

– its priority level denoted by L (positive integer, between 0 and *Inf*).

In case of several requirements, one can define an array of Z requirements together with the corresponding arrays for Z_{max} and L.

Accordingly, the *requirements* field has the following formalism:

```
requirements =
{ 'definition of a Z array using all Matlab instructions and
definition of corresponding Zmax and L arrays '; } ;
```

For the sake of readability, we can specify the requirements on several lines:

```
requirements =
'definition of a Z array using all Matlab instructions and
definition of corresponding Zmax and L arrays ';...
'definition of a Z array using all Matlab instructions and
definition of corresponding Zmax and L arrays ';...
 .
 .
'definition of a Z array using all Matlab instructions and
definition of corresponding Zmax and L arrays ';...
```

Or rather:

```
requirements{1,1} =
'definition of a Z array using all Matlab instructions and
definition of corresponding Zmax and L arrays ';
requirements{2,1} =
'definition of a Z array using all Matlab instructions and
definition of corresponding Zmax and L arrays ';
 .
 .
 .
```

In order to define requirements, users should have access to preset variables:

– the state matrices of the $F_l(P,K)$ closed loop denoted by *CL.a, CL.b, CL.c, CL.d*;

– the state matrices of the P system denoted by *P.a, P.b, P.c, P.d*;

– the state matrices of the K controller denoted by *K.a, K.b, K.c, K.d.*

During use in "cascade" mode, users also have access to:

– state matrices of the cascaded sub-transfer of the controller denoted by *K_blk(i,j).a, K_blk(i,j).b, K_blk(i,j).c, K_blk(i,j).d.*

During use in "weight" mode, users also have access to:

– state matrices of weights W_i and W_o denoted by $W_i.a$, $W_i.b$, $W_i.c$, $W_i.d$. et $W_o.a$, $W_o.b$, $W_o.c$, $W_o.d$.

During use in a multiple plant framework, users also have access to:

– state matrices of the *i*th closed loop $F_l(P_i,K)$ denoted by *CL(i).a, CL(i).b, CL(i).c, CL(i).d*;

– state matrices of the *i*th P_i system denoted by *P(i).a, P(i).b, P(i).c, P(i).d.*

4.5.2.3.2. Useful values and associated Matlab instructions

Here, we will review frequently used values together with the associated Matlab instructions. Insofar as any Matlab instruction can be used in the definition of requirements, this list is by no means exhaustive.

– Temporal simulations:

Consider a multi-variable system denoted by *CL*. Here, the temporal response $z(t)$ of *CL* in response to excitation $e(t)$ is obtained by the instruction:

```
z = lsim(CL,e,t)
```

If *CL* is of dimension $n_z \times n_e$, then e must be of size $n_t \times n_e$, n_t being the length of the time vector t and output z will be of size $n_t \times n_z$.

– Response time:

In order to characterize the performance of a closed-loop system, we often excite the latter with a step. A typical response aspect is shown in Figure 4.12.

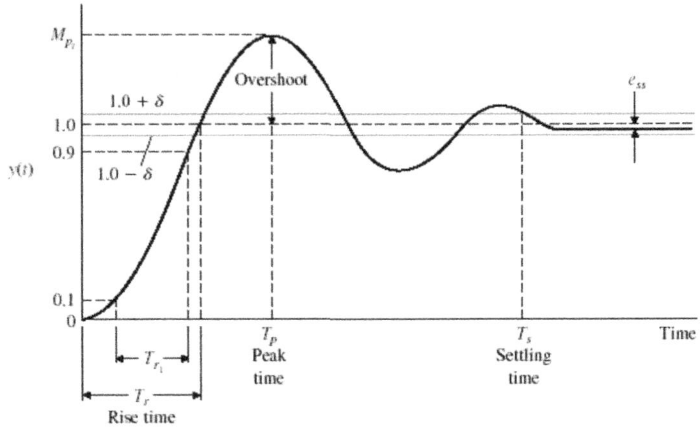

Figure 4.12. *Second-order response. For a color version of this figure, see www.iste.co.uk/feyel/optimization.zip*

We can then characterize the aspect of the temporal response by the following values:

– the rising time T_r at 90%;

– the peak time T_p;

– the settling time T_s typically at ±5% of the final value;

– the percentage overshoot M_p.

Knowing the response value and final value, $y(t)$ and y_{final}, respectively, we can characterize the step response using Matlab with the *stepinfo* instruction and thereby impose constraints on these values.

```
S = stepinfo(y,t,yfinal)
```

The returned value S is consequently a structure including the following fields:

```
S.RiseTime – Rise time
S.SettlingTime – Settling time
S.SettlingMin – Minimum value of y once the response has risen
S.SettlingMax – Maximum value of y once the response has risen
S.Overshoot – Percentage overshoot (relative to yfinal)
S.Undershoot – Percentage undershoot
S.Peak – Peak absolute value of y
S.PeakTime – Time at which this peak is reached
```

It should be recalled that if ξ is the system damping:

$$T_s = T_p \frac{4}{\pi}\sqrt{\frac{1}{\xi^2} - 1} \quad [4.39]$$

For a damping of 0.5, we have:

$$T_s = T_p \frac{4\sqrt{3}}{\pi} \approx 2 \quad [4.40]$$

It is important to note that, unlike with the rising time T_r and the peak time T_p, settling time T_s is only defined if the output reaches y_{final} (to the nearest ±δ in the figure). Accordingly, it is preferable to work with T_s, as imposing a requirement on T_s amounts to also imposing a requirement on the final value of the step response $y(t)$.

– Open-loop characteristics in SISO case:

Where ω_c is the open-loop bandwidth (in rad/s) (i.e. the lowest frequency at which its gain is unitary). In this case, it should be recalled that for a damping $\xi \in$ [0,4 ; 0,7], we have:

$$\omega_c T_p \approx 3 \quad [4.41]$$

By knowing the open loop G, we can retrieve the gain margins (G_m) and phase margins (P_m) (together with the corresponding frequencies ω_g and ω_c) with the *margin* instruction and thereby impose constraints on these values.

```
[Gm,Pm,Wg,Wc] = margin(G);
```

It is important to note that a requirement on the gain margin in dB must be accompanied with a transition to dB from G_m:

```
Gm = 20log10(Gm);
```

A generalized *margin* can be obtained using the instruction *allmargin*:

```
S = allmargin(G);
```

The returned value S is consequently a structure including the following fields:

```
S.GMFrequency – All -180 degree crossover frequencies (in
rad/TimeUnit, where TimeUnit is the time units of the input
dynamic system, specified in the TimeUnit property of sys).
```

S.GainMargin – Corresponding gain margins, defined as 1/G where G is the gain at crossover

S.PMFrequency – All 0 dB crossover frequencies in rad/TimeUnit, where TimeUnit is the time units of the input dynamic system, specified in the TimeUnit property of sys).

S.PhaseMargin – Corresponding phase margins in degrees

S.DMFrequency and DelayMargin – Critical frequencies and the corresponding delay margins. Delay margins are specified in the time units of the system for continuous-time systems and multiples of the sample time for discrete-time systems.

S.Stable – 1 if the nominal closed-loop system is stable, 0 otherwise.

– Open-loop characteristics in MIMO case:

Consider the MIMO plant *P* and the MIMO controller *C*. The generalized stability margins at input and output plant can be computed using the following instruction.

```
[cmi,dmi,mmi,cmo,dmo,mmo,mmio] = loopmargin(P,C)
```

Especially the generalized gain and phase margins (see [FEY 13c] for definition) can be bounded using output arguments *dmi* and *dmo* (respectively at input plant and output plant). The *dm* data is a structure with the following fields:

GainMargin - Smallest gain variation (GM) such that a disk centered at the point -(GM(1) + GM(2))/2 would just touch the loop transfer function
PhaseMargin - Smallest phase variation, in degrees, corresponding to the disk described in the GainMargin field (degrees)
Frequency - Associated with GainMargin/PhaseMargin fields (in radians-per-second)

– Transfer norms and shaping:

For a transfer *H*, we can obtain the H_2 and/or H_∞ norms of *H* with the *norm* instruction and thereby impose constraints on these values.

```
H2 = norm(H,2);
Hinf = norm(H,Inf);
```

It is important to note that the H_2 norm can only be defined if *H* is not strictly proper. If this is not the case, we can return it by adding a small fictitious time constant to the plant.

A frequent use is the shaping of singular values of a transfer. For instance, considering a MIMO system G and a W filter, such that:

$$\bar{\sigma}(G) < |W|^{-1} \Rightarrow \bar{\sigma}(GW) < 1 \Rightarrow \|GW\|_\infty < 1 \qquad [4.42]$$

Another example is the possibility of modulus margin constraint of a servo-loop whose sensitivity function S is known. The modulus margin is defined by:

$$M_d = \frac{1}{\|S\|_\infty} \qquad [4.43]$$

– Transfer stability:

For a transfer H and corresponding state representation (A, B, C, D), H is stable if and only if the eigenvalues of matrix A have their real part strictly negative, that is, if:

$$\max_j \mathrm{Re}\left(\lambda_j(A)\right) < 0 \qquad [4.44]$$

It is possible to retrieve the eigenvalues of a matrix in Matlab using the instruction *eig* and thereby impose constraints on it.

```
temp = ss(H);
vp = eig(temp.a);
```

– Imposing integrators in a transfer:

Where we wish to impose the presence of n integrators in a K transfer whose state representation is (A, B, C, D), we must impose on n the eigenvalues of A have a real part equal to zero (or close to zero). In Matlab, we can retrieve the n eigenvalues in consideration by the following instructions:

```
temp = ss(K);
vp = sort(abs(real(eig(temp.a)))) ;
integ = max(vp(1 : n));
```

4.5.2.4. *Definition of options*

With the options field, we position the optimization *options*.

In the following, the default values are positionable by the value [ ] (empty).

The *options* argument is a structure that covers the following fields:

$$\text{options} \rightarrow \left\{ \begin{array}{c} \text{options.save\_file } = \textit{file\_name} \\ \text{options.debug} = \{0.1\} \\ \text{options.nbr\_run} = \text{strictly positive integer} \\ \text{options.nbr\_iterations\_max} = \text{strictly positive integer} \\ \text{options.convergence\_threshold} = \text{strictly positive real} \\ \text{options.parallel\_computation} = \text{positive integer} \\ \text{options.display} = \{\text{'nominal\_graphic','multimodel\_graphic','iter','end'}\} \end{array} \right\}$$

4.5.2.4.1. Debugging options

By positioning the *options.debug* to 1, we can replay the previous result of a saved optimization in the file specified by the field *options.save_file*.

This assumes that an optimization has already been performed by positioning the *options.debug* to 0 and having taken care to save the result in the file *options.save_file*. If *options.save_file* = [], then a default name is attributed.

4.5.2.4.2. Optimization options

The optimization options allow for more precise management of the optimization process. Since the tool is based on evolutionary optimization, it can be worthwhile to launch a certain number of runs successively by setting the *options.nbr_run* field.

Each run is subject to a certain number of iterations to be specified in *options.nbr_iterations_max*. The default value is positioned at 20,000.

It is possible to detect a premature convergence of the optimization process: where this occurs, either we allow the iterations to finish or, using the *options.convergence_threshold* field, we pre-position a threshold from which the algorithm will stop. This threshold is related to the displayed radius on each iteration (see *options.display*) and is less than 1. The default value is positioned at 1^e-5.

Computation time can be accelerated significantly by using Matlab's *Parallel Computing Toolbox*. Accordingly, in the *options.parallel_computation* field, we can specify the number of pools to be used with a positive integer, limited to the number of CPU cores the machine has. The default value is positioned at 0. This is the value to be specified when the *Parallel Computing Toolbox* is not used.

4.5.2.4.3. Display options

In the interest of visually checking the optimization process, we can position the *options.display* field to the following values:

– "end":

Here, the tool will use the command line only to display any warning messages that occur.

– "iter":

In addition to the warning messages, this displays the ongoing iteration when a better solution is obtained. Here, we find the following information: the number of the ongoing iteration, the cost function value (must be lowest possible), cost function evaluation number and the population radius.

```
itération (génération) n° 925/20000 ; cout = -3.37369e-003 ; nombre devaluations : 17575 ; rayon : 0.235045
itération (génération) n° 999/20000 ; cout = -3.43025e-003 ; nombre devaluations : 18981 ; rayon : 0.295997
itération (génération) n° 1004/20000 ; cout = -3.73616e-003 ; nombre devaluations : 19076 ; rayon : 0.218759
itération (génération) n° 1005/20000 ; cout = -4.54042e-003 ; nombre devaluations : 19095 ; rayon : 0.241254
itération (génération) n° 1032/20000 ; cout = -5.68587e-003 ; nombre devaluations : 19608 ; rayon : 0.231239
```

Figure 4.13. *"iter" display mode*

– "nominal_graphic":

When this value is selected, a graphical window appears and updates as soon as a better solution is obtained. This display is accompanied by the "iter" information in the command line.

The graphical window follows the formalism below:

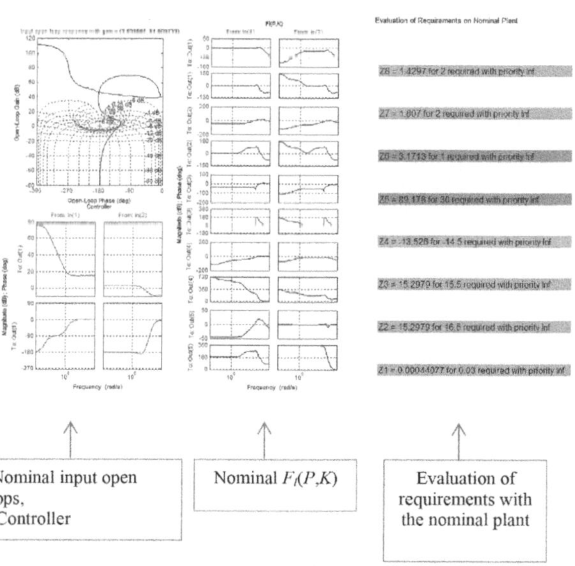

Figure 4.14. *"Nominal_graphic" window mode. For a color version of this figure, see www.iste.co.uk/feyel/optimization.zip*

The requirements are then evaluated with the nominal plant and displayed according to the order (from bottom to top) of their definition. A color code allows for the quick appraisal of the optimization process:

– Green: requirement satisfied;

– Orange: requirement not satisfied but not exceeding 10% of the specified value;

– Red: requirement far from satisfied.

- "multimodel_graphic"

Displays a similar window to the one above, but the frequency responses are plotted for all of the ball of plants, which consequently requires more time. We often use this value in debugging mode to graphically verify the robustness of the controller obtained on all of the ball of plants.

The main options are presented in the following table:

Options	Definition and values
debug	If equal to 1, the tool displays the best result found in the file whose name is specified in the *.save_file* field (see below). Must be positioned to 0 in order to launch the optimization process.
.save_file	Specifies the number of successive runs
nbre_run	Specifies the maximum number of iterations per run. 20,000 by default.
nbre_iteration_max	Threshold used to detect a premature convergence. 1^{e}-5 by default.
convergence_threshold	Sets the pool number to be used during the optimization process. To set to 0 or [ ] if Matlab's *Parallel Computing Toolbox* is not used.
parallel_computation	Sets the pool number to be used during the optimization process. To set to 0 or [ ] if Matlab's *Parallel Computing Toolbox* is not used.
display	Displays information during the optimization process. "end" only displays any warning messages that occur. "iter" displays the value of the cost function when a better value is obtained. Furthermore, "nominal_graphic" displays information as a figure, such as the evolution of requirements, the open loop, etc. when a better value is obtained for the cost function. The use of "multimodel_graphic" resumes the information of "nominal_graphic" but the plotted frequency responses are presented for each system (on the same graphic).

4.5.2.5. *Definition of output*

The tool returns the controller in its state-space form in *K_ss*.

In K-mode, the various sub-transfers of the controller are also retuned in *K_blk*. Therefore, in feedback, these two values are identical. In cascade mode, *K_blk* contains the various transfers of the cascaded controllers (see Figure 4.15).

In W-mode, the weights obtained are also returned in *K_blk*, particularly so in the *K_blk.Wo* and *K_blk.Wi* fields.

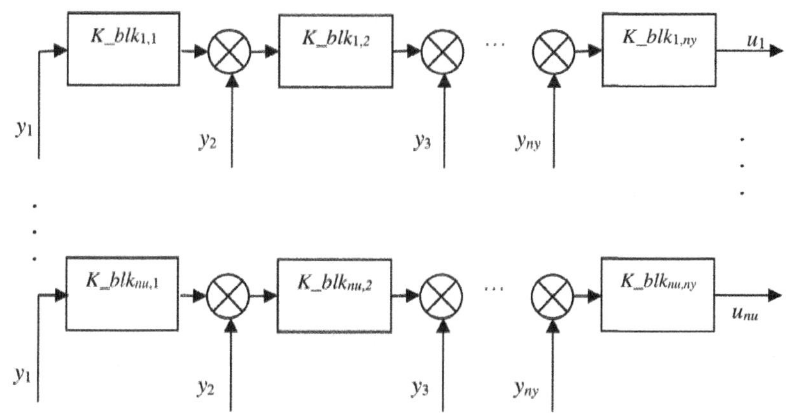

Figure 4.15. *Controller in cascade form*

4.5.3. *Examples*

4.5.3.1. *Example 1 – structured mixed H_2/H_∞ synthesis*

Here, we resume an example presented in [SCH 97]. We consider the following unstable order 3 standard form:

$$\frac{d}{dt}\begin{pmatrix} x_1 \\ x_2 \\ x_3 \end{pmatrix} = \begin{pmatrix} 0 & 10 & 2 \\ -1 & 1 & 0 \\ 0 & 2 & -5 \end{pmatrix}\begin{pmatrix} x_1 \\ x_2 \\ x_3 \end{pmatrix} + \begin{pmatrix} 1 \\ 0 \\ 1 \end{pmatrix} w + \begin{pmatrix} 0 \\ 1 \\ 0 \end{pmatrix} u$$

$$y = x_2 + 2w \qquad [4.45]$$

$$z_\infty = \begin{pmatrix} x_1 \\ u \end{pmatrix}, \quad z_2 = \begin{pmatrix} x_2 \\ x_3 \\ u \end{pmatrix}$$

where u represents the control signal and y represents the measurement. We are interested in the H_∞ norm between w and z_∞ and the H_2 norm between w and z_2. We performed a full-order H_2 synthesis between w and z_2, so the optimal H_2 performance was obtained for $\left\|T_{wz_2}\right\|_2 = 7,748$ with the controller:

$$K_2(s) = \frac{-5.7275(s+5.168)(s-0.2711)}{(s+5.164)(s^2+3.669s+9.933)} \qquad [4.46]$$

For the plant looped by $K_2(s)$, the H_∞ performance returned is $\left\|T_{wz_\infty}\right\|_\infty = 23.586$.

We propose using *HinfStoch_ControllerTuning* to find a stabilizing controller so that:

$$\begin{aligned} &\left\|T_{wz_2}\right\|_2 < 7.75 \\ &\left\|T_{wz_\infty}\right\|_\infty < 23.6 \end{aligned} \qquad [4.47]$$

The order of the searched controller is fixed at 3.

Accordingly, it is a purely frequential use of the tool whose script corresponds to the synthesis below.

```
%
% Taken from Scherer et al. : MULTIOBJECTIF OUTPUT-FEEDBACK
CONTROL VIA LMI OPTIMIZATION
%

clear all;

A = [[0 10 2];...
     [-1 1 0];...
     [0 2 -5]];
B1 = [1;0;1];
B2 = [0;1;0];

% Zinf from 1 to 3, Z2 from 4 to 5
C1 = [[1 0 0];...
      [0 0 0];...
      [0 1 0];...
      [0 0 1];...
      [0 0 0]];
D11 = [0;0;0;0;0];
D12 = [0;1;0;0;1];
```

```
%y
C2 = [0 1 0];
D21 = 2;
D22 = 0;

%
% MO fixed order case
%
P.a = A;
P.b = [B1,B2];
P.c = [C1;C2];
P.d = [[D11,D12];...
       [D21,D22]];

requirements =     {'T2 =
ss(CL.a,CL.b,CL.c(4:5,:),CL.d(4:5,1))*tf([0 1],[1e-10 1]);
Z(1) = norm(T2,2) ; Zmax(1) = 7.75 ; L(1) = Inf ; ' ;...
                   'Tinf =
ss(CL.a,CL.b,CL.c(1:3,:),CL.d(1:3,1)) ; Z(1) = norm(Tinf,inf)
; Zmax(1) = 23.6 ; L(1) = Inf ; ';};

nmeas = 1;
ncon = 1;
order.order_K = 3;
order.mode = 'feedback';
order.robustness = 'nominal';
options.debug = 0;
options.nbr_run = 3;
options.nbr_iteration_max = 2000;
options.display = 'iter';
options.convergence_threshold = 1e-3;
options.parallel_computation = 0;
options.save_file = 'HINFSTOCH_MULTIOBJ_LMI_EXAMPLE.mat';

[K_ss, K_blk] = HinfStoch_ControllerTuning_v2(P,nmeas, ncon,
order, requirements, options);

file = 'CONTROLLER_MULTIOBJ_LMI_EXAMPLE.mat';

save(file,'K_ss', 'K_blk');
```

It is important to note the need to add a very low fictitious time constant in order to ensure that $F_l(P,K)$ is strictly proper, a necessary condition for the evaluation of its H_2 norm.

At the end of the optimization process, we obtain:

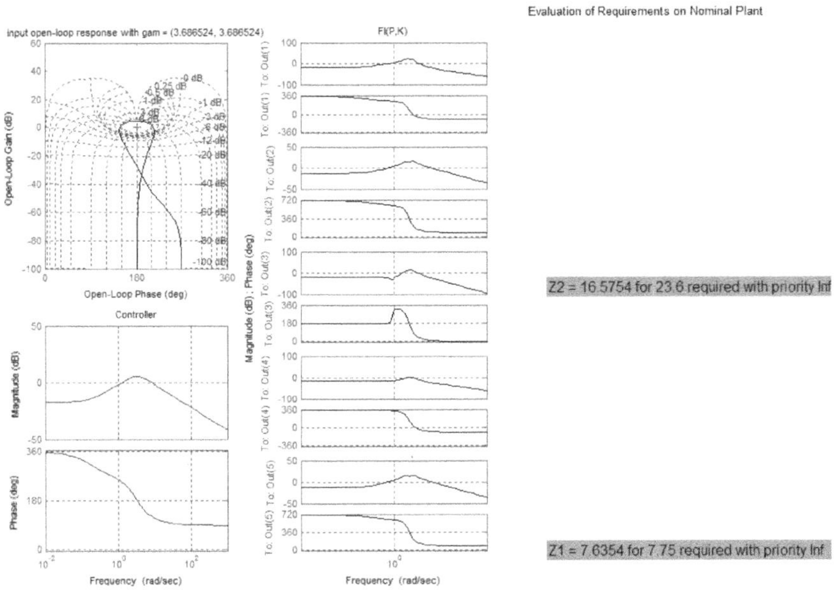

Figure 4.16. *Multi-objective synthesis. For a color version of this figure, see www.iste.co.uk/feyel/optimization.zip*

The controller obtained is:

$$K_{2/\infty}(s) = \frac{-6.5429.10^8\ (s+1.155.10^8)\ (s+5.024)\ (s-0.2383)}{(s+4.299)\ (s^2+4.296s+10.59)} \quad [4.48]$$

Leading to:

$$\begin{aligned} \left\| T_{wz_2} \right\|_2 &= 7,6 \\ \left\| T_{wz_\infty} \right\|_\infty &= 16,6 \end{aligned} \quad [4.49]$$

4.5.3.2. *Example 2 – H_∞ synthesis with a stable controller*

We resume the example taken from [APK 09]. Let us consider the following system:

$$G(s) = \frac{(s^2+0.1s+0.1)}{(s-0.1)(s-1)(s^2+2s+3)} := \begin{pmatrix} A & C \\ B & D \end{pmatrix} \quad [4.50]$$

And the standard form for the control:

$$\begin{pmatrix} \dot{x} \\ z \\ y \end{pmatrix} = \left(\begin{array}{c|ccc} A & B & 0 & B \\ \hline C & 0 & 1 & 0 \\ 0 & 0 & 0 & 1 \\ C & 0 & 1 & 0 \end{array}\right) \begin{pmatrix} x \\ e \\ u \end{pmatrix} := P, \; n_u = 1, n_y = 1 \qquad [4.51]$$

The system is unstable.

We are looking for the controller that minimizes $\gamma = \|F_l(P,K)\|_\infty$. A full-order synthesis with *hinfsyn* leads to $\gamma = 12{,}01$ but the controller is unstable!

Thus, we propose resuming the synthesis but now with the controller's stability constraint. We propose the following script:

```
%
% Taken from "Hinfinity control with stable controllers" in
http://pierre.apkarian.free.fr/EXAMPLES_STRUCTURE/StableContro
llerHTML/StableController.html
%
% Aut : Philippe FEYEL (2015)
%
clear all;

s = tf('s');
G = (s^2+0.1*s+0.1)/((s-0.1)*(s-1)*(s^2+2*s+3)); ny =1; nu =1;
nx = 4;
[a,b,c,d] = ssdata(G) ;
H = ss(a , [b zeros(nx,1) b], [c;zeros(1,nx);c], [0 1 0;0 0 1;
0 1 0]) ;

[K,clp,gamHinfsyn] = hinfsyn(H,ny,nu) ;
gamHinfsyn
if ~isstable(K), disp('K is unstable'); end

%
% MO fixed order case
%
P.a = a;
P.b = [b zeros(nx,1) b];
P.c = [c;zeros(1,nx);c];
P.d = [0 1 0;0 0 1; 0 1 0];
```

```
requirements =     {'Tinf = ss(CL.a,CL.b,CL.c,CL.d) ; Z(1) =
norm(Tinf,inf) ; Zmax(1) =  12.01 ; L(1) = Inf ; ';...
                    'Z(1) = max(real(eig(K.a))) ; Zmax(1) = 0 ;
L(1) = Inf ;' ; };

nmeas = ny;
ncon = nu;
order.order_K = 3;
order.mode = 'feedback';
order.robustness = 'nominal';
options.debug = 0;
options.nbr_run = 10;
options.nbr_iteration_max = 5000;
options.display = 'iter';
options.convergence_threshold = 1e-3;
options.parallel_computation = 0;
options.save_file = 'HINFSTOCH_STABLE_K_EXAMPLE.mat';

[K_ss, K_blk] = HinfStoch_ControllerTuning_v2(P,nmeas, ncon,
order, requirements, options);

file = 'CONTROLLER_STABLE_K_EXAMPLE.mat';

save(file,'K_ss', 'K_blk');
```

At the end of optimization, we find:

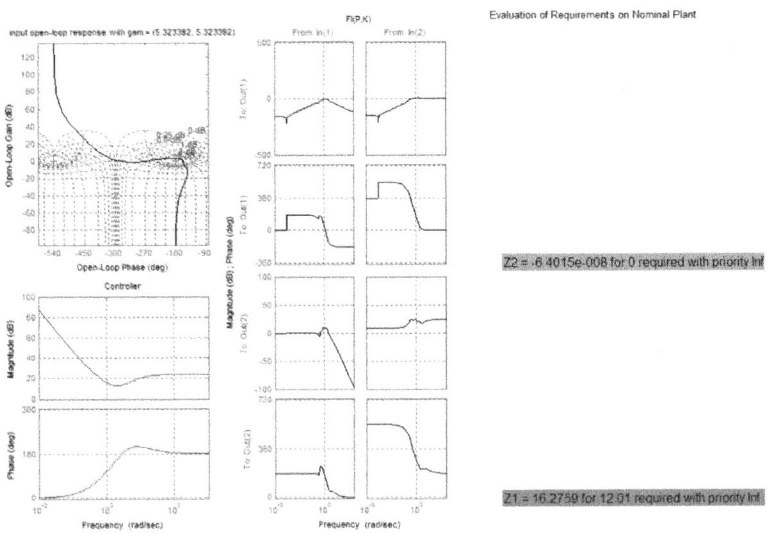

Figure 4.17. *Synthesis with stable controller. For a color version of this figure, see www.iste.co.uk/feyel/optimization.zip*

That is:

$$\gamma = \left\| F_l(P,K) \right\|_\infty = 16.28 \qquad [4.52]$$

And the following controller:

$$K(s) = \frac{-14.8471\ (s+0.4627)\ (s^2+3.168s+3.749)}{(s+10.72)\ (s^2+8.359.10^{-7}s+3.841.10^{-6})} \qquad [4.53]$$

The results are similar to those obtained with *hinfstruct* in [APK 09].

4.5.3.3. *Example 3 – decentralized control of a distillation column*

Here, we resume an example presented in [GAH 11] concerning the distillation column depicted in Figure 4.18.

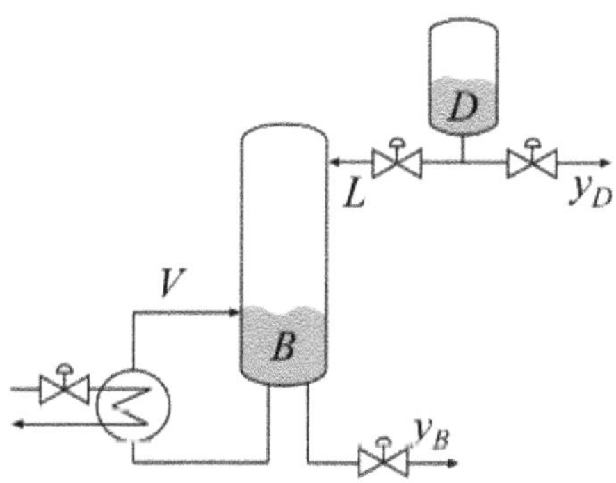

Figure 4.18. *Distillation column*

The transfer function of the process is:

$$G(s) = \frac{1}{75s+1}\begin{pmatrix} 87.8 & -86.4 \\ 108.2 & -109.6 \end{pmatrix} \qquad [4.54]$$

The control structure is as follows:

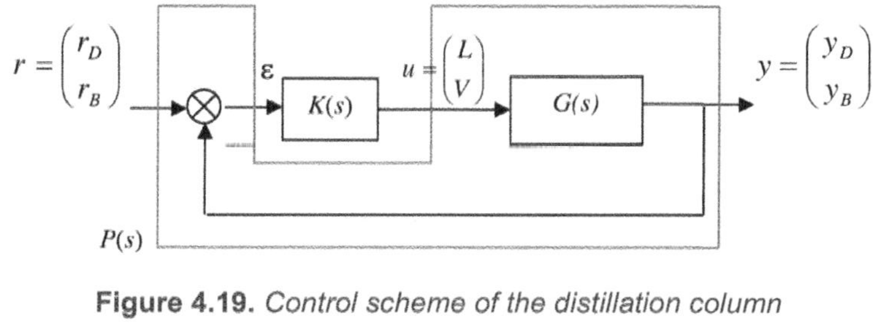

Figure 4.19. *Control scheme of the distillation column*

4.5.3.3.1. Tuning for tracking reference decoupling

Initially, we wish to determine a decentralized MIMO controller with order 1 so as to satisfy the following specifications:

– overshoot of 5% in step regime for direct transfers with time response about 20 min (for a settling time about 30 min);

– in step regime, coupling rate between channels below 40%;

– in order to limit the control energy, the direct open-loop bandwidth must not exceed 0.1 rad/s.

The first stage consists of establishing the standard form for control. Exogenous inputs include the reference vector r and the monitored outputs of vector y. For measurement, the controller takes regulation deviation ε.

If we write:

$$G(s) := \begin{pmatrix} A & B \\ C & D \end{pmatrix} \quad [4.55]$$

Then:

$$\begin{pmatrix} \dot{x} \\ \hline y \\ \varepsilon \end{pmatrix} = \left(\begin{array}{c|cc} A & 0 & B \\ \hline C & 0 & D \\ -C & I & -D \end{array}\right) \begin{pmatrix} x \\ \hline r \\ u \end{pmatrix} := P(s) \quad [4.56]$$

Accordingly, we propose the following script:

```
%
% Taken from "Decentralized and Fixed-Structure Hinf Control
in MATLAB"
%
% Aut : Philippe FEYEL (2015)
%

clear all;

s = tf('s');
G = 1/(75*s+1)*[[87.8, -86.4];[108.2, -109.6]];
```

```
[A,B,C,D] = ssdata(G);
nx = size(A,1);
nu = 2;
ny = 2;

P.a = A;
P.b = [zeros(nx,ny) B B];
P.c = [C;-C];
P.d = [[zeros(ny,ny), D, D];...
       [eye(ny,ny), -D, -D]];

requirements = {'t = transpose(0:1:6000); e =
[ones(size(t,1),1), zeros(size(t,1),3)]; z =
lsim(ss(CL.a,CL.b,CL.c,CL.d),e,t) ; Info =
stepinfo(z(:,1),t,1) ; Z(1) = Info.SettlingTime ; Zmax(1) =
60*30 ; L(1) = Inf ; Z(2) = Info.Peak ; Zmax(2) = 1.05 ; L(2)
= Inf ; Z(3) = max(abs(z(:,2))) ; Zmax(3) = 0.4 ; L(3) = Inf ;
';...
                 't = transpose(0:1:6000); e =
[zeros(size(t,1),1), ones(size(t,1),1), zeros(size(t,1),2)]; z
= lsim(ss(CL.a,CL.b,CL.c,CL.d),e,t) ; Info =
stepinfo(z(:,2),t,1) ; Z(1) = Info.SettlingTime ; Zmax(1) =
60*30 ; L(1) = Inf ; Z(2) = Info.Peak ; Zmax(2) = 1.05 ; L(2)
= Inf ; Z(3) = max(abs(z(:,1))) ; Zmax(3) = 0.4 ; L(3) = Inf ;
';...
                 't = transpose(0:1:6000); e =
[zeros(size(t,1),2), 0.2*ones(size(t,1),1),
zeros(size(t,1),1)]; z = lsim(ss(CL.a,CL.b,CL.c,CL.d),e,t) ;
Info = stepinfo(z(:,1),t,0) ; Z(1) = Info.SettlingTime ;
Zmax(1) = 60*20 ; L(1) = Inf ; Z(2) = max(abs(z(:,1))) ;
Zmax(2) = 4 ; L(2) = Inf ; ';...
                 't = transpose(0:1:6000); e =
[zeros(size(t,1),3), 0.2*ones(size(t,1),1)]; z =
lsim(ss(CL.a,CL.b,CL.c,CL.d),e,t) ; Info =
stepinfo(z(:,2),t,0) ; Z(1) = Info.SettlingTime ; Zmax(1) =
60*20 ; L(1) = Inf ; Z(2) = max(abs(z(:,2))) ; Zmax(2) = 4 ;
L(2) = Inf ; ';...
                 's = tf([1 0],[0 1]); G = 1/(75*s+1)*[[87.8, -
86.4];[108.2, -109.6]] ; H_BO = ss(K.a,K.b,K.c,K.d)*G ; H_BO11
= H_BO(1,1) ; [Gm,Pm,Wg,Wp] = margin(H_BO11) ; Z(1) = Wp ;
Zmax(1) = 0.1 ; L(1) = Inf ; H_BO22 = H_BO(2,2) ;
[Gm,Pm,Wg,Wp] = margin(H_BO22) ; Z(2) = Wp ; Zmax(2) = 0.1 ;
L(2) = Inf ;' ;};

nmeas = ny;
ncon = nu;
order.order_K = [[1 1];...
                 [1 1]];
order.mode = 'feedback';
order.robustness = 'nominal';
```

```
options.debug = 0;
options.nbr_run = 3;
options.nbr_iteration_max = 20000;
options.display = 'iter';
options.convergence_threshold = 1e-3;
options.parallel_computation = 8;
options.save_file = 'HINFSTOCH_DISTILLATION_EXAMPLE.mat';

[K_ss, K_blk] = HinfStoch_ControllerTuning_v2(P,nmeas, ncon,
order, requirements, options);

file = 'CONTROLLER_DISTILLATION_EXAMPLE.mat';

save(file,'K_ss', 'K_blk');
```

At the end of optimization, we obtain the closed-loop responses of Figure 4.20. All of the temporal requirements are satisfied (time response, overshoot and coupling rate). In addition, in Figure 4.21, we find the singular values of the open loop. The bandwidth of the upper singular value does not exceed 0.1 rad/s.

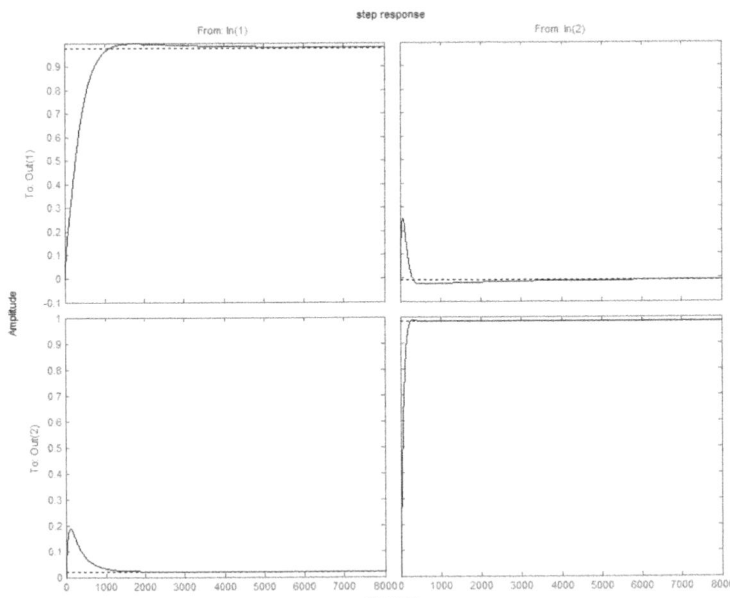

Figure 4.20. *Closed-loop step responses. For a color version of this figure, see www.iste.co.uk/feyel/optimization.zip*

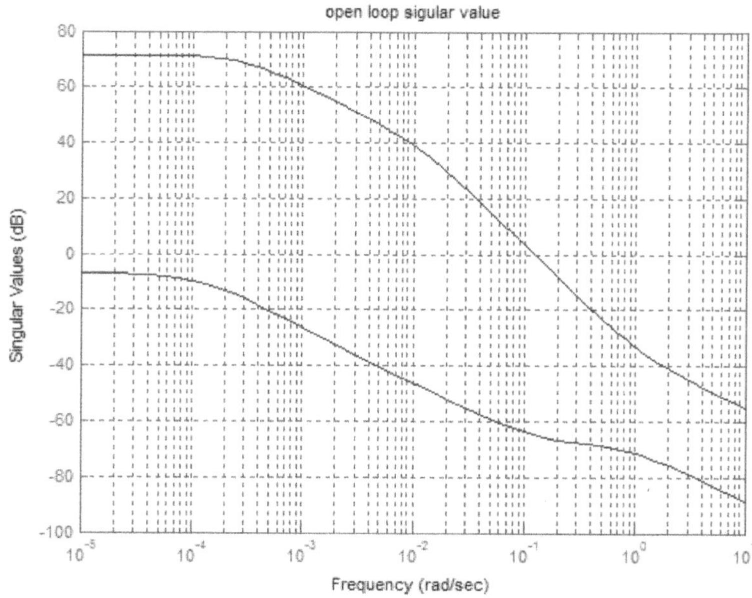

Figure 4.21. *Singular values of the open loop*

4.5.3.3.2. Tuning for reference tracking, decoupling and disturbance rejection

Now we will add an additional constraint of disturbance rejection at model input. This transforms the control scheme into that of Figure 4.22.

For each channel and for a disturbance with magnitude of 0.2:

– the rejection time is about 30 min;

– the corresponding output command must be <4.

The requirements in the above paragraph must remain satisfied. The correction is still decentralized MIMO with order 1.

Here, we have to modify the standard form for control. Exogenous inputs include the reference vector r and the disturbance vector b and the monitored output is vector y. For the sake of measurement, the controller takes the regulation deviation ε.

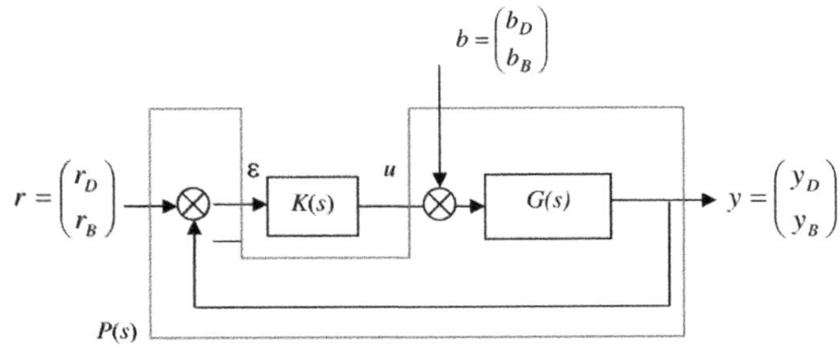

Figure 4.22. *Control scheme of the distillation column with disturbance rejection. For a color version of this figure, see www.iste.co.uk/feyel/optimization.zip*

Thus, it is easy to reach form [4.57].

$$\begin{pmatrix} \dot{x} \\ \hline y \\ \varepsilon \end{pmatrix} = \left(\begin{array}{c|ccc} A & 0 & B & B \\ \hline C & 0 & D & D \\ -C & I & -D & -D \end{array}\right) \begin{pmatrix} x \\ \hline r \\ b \\ u \end{pmatrix} := P(s) \qquad [4.57]$$

Accordingly, we propose the following script:

```
%
% Taken from "Decentralized and Fixed-Structure Hinf Control
in MATLAB"
%
% Aut : Philippe FEYEL (2015)
%

clear all;

s = tf('s');
G = 1/(75*s+1)*[[87.8, -86.4];[108.2, -109.6]];
[A,B,C,D] = ssdata(G);
nx = size(A,1);
nu = 2;
ny = 2;

P.a = A;
P.b = [zeros(nx,ny) B B];
P.c = [C;-C];
P.d = [[zeros(ny,ny), D, D];...
       [eye(ny,ny), -D, -D]];

requirements = {'t = transpose(0:1:6000); e =
[ones(size(t,1),1), zeros(size(t,1),3)]; z =
lsim(ss(CL.a,CL.b,CL.c,CL.d),e,t) ; Info =
```

```
stepinfo(z(:,1),t,1) ; Z(1) = Info.SettlingTime ; Zmax(1) =
60*30 ; L(1) = Inf ; Z(2) = Info.Peak ; Zmax(2) = 1.05 ; L(2)
= Inf ; Z(3) = max(abs(z(:,2))) ; Zmax(3) = 0.4 ; L(3) = Inf ;
';...
                    't = transpose(0:1:6000); e =
[zeros(size(t,1),1), ones(size(t,1),1), zeros(size(t,1),2)]; z
= lsim(ss(CL.a,CL.b,CL.c,CL.d),e,t) ; Info =
stepinfo(z(:,2),t,1) ; Z(1) = Info.SettlingTime ; Zmax(1) =
60*30 ; L(1) = Inf ; Z(2) = Info.Peak ; Zmax(2) = 1.05 ; L(2)
= Inf ; Z(3) = max(abs(z(:,1))) ; Zmax(3) = 0.4 ; L(3) = Inf ;
';...
                    't = transpose(0:1:6000); e =
[zeros(size(t,1),2), 0.2*ones(size(t,1),1),
zeros(size(t,1),1)]; z = lsim(ss(CL.a,CL.b,CL.c,CL.d),e,t) ;
Info = stepinfo(z(:,1),t,0) ; Z(1) = Info.SettlingTime ;
Zmax(1) = 60*20 ; L(1) = Inf ; Z(2) = max(abs(z(:,1))) ;
Zmax(2) = 4 ; L(2) = Inf ; ';...
                    't = transpose(0:1:6000); e =
[zeros(size(t,1),3), 0.2*ones(size(t,1),1)]; z =
lsim(ss(CL.a,CL.b,CL.c,CL.d),e,t) ; Info =
stepinfo(z(:,2),t,0) ; Z(1) = Info.SettlingTime ; Zmax(1) =
60*20 ; L(1) = Inf ; Z(2) = max(abs(z(:,2))) ; Zmax(2) = 4 ;
L(2) = Inf ; ';...
                    's = tf([1 0],[0 1]); G = 1/(75*s+1)*[[87.8, -
86.4];[108.2, -109.6]] ; H_BO = ss(K.a,K.b,K.c,K.d)*G ; H_BO11
= H_BO(1,1) ; [Gm,Pm,Wg,Wp] = margin(H_BO11) ; Z(1) = Wp ;
Zmax(1) = 0.1 ; L(1) = Inf ; H_BO22 = H_BO(2,2) ;
[Gm,Pm,Wg,Wp] = margin(H_BO22) ; Z(2) = Wp ; Zmax(2) = 0.1 ;
L(2) = Inf ;' ;};

nmeas = ny;
ncon = nu;
order.order_K = [[1 1];...
                    [1 1]];
order.mode = 'feedback';
order.robustness = 'nominal';
options.debug = 0;
options.nbr_run = 3;
options.nbr_iteration_max = 20000;
options.display = 'iter';
options.convergence_threshold = 1e-3;
options.parallel_computation = 8;
options.save_file = 'HINFSTOCH_DISTILLATION_EXAMPLE_2.mat';

[K_ss, K_blk] = HinfStoch_ControllerTuning_v2(P,nmeas, ncon,
order, requirements, options);

file = 'CONTROLLER_DISTILLATION_EXAMPLE_2.mat';

save(file,'K_ss', 'K_blk');
```

At the end of optimization, we obtain the closed-loop responses of Figure 4.23.

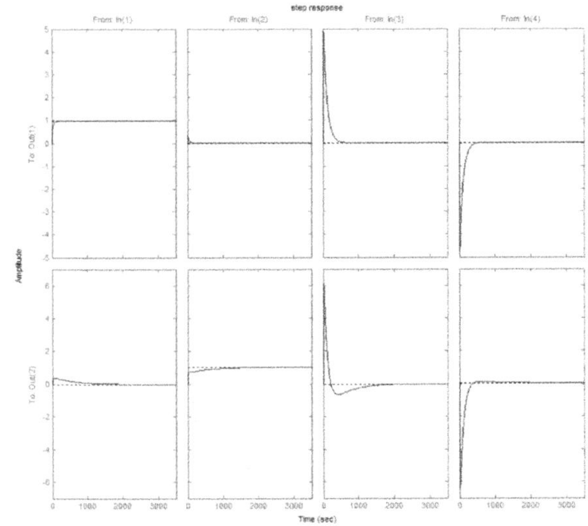

Figure 4.23. *Step responses of references and disturbance rejection (here unitary). For a color version of this figure, see www.iste.co.uk/feyel/optimization.zip*

All of the temporal requirements are satisfied (time response, overshoot and coupling rate). In addition, in Figure 4.21, we find the singular values of the open loop. The bandwidth of the upper singular value is close to 0.1 rad/s

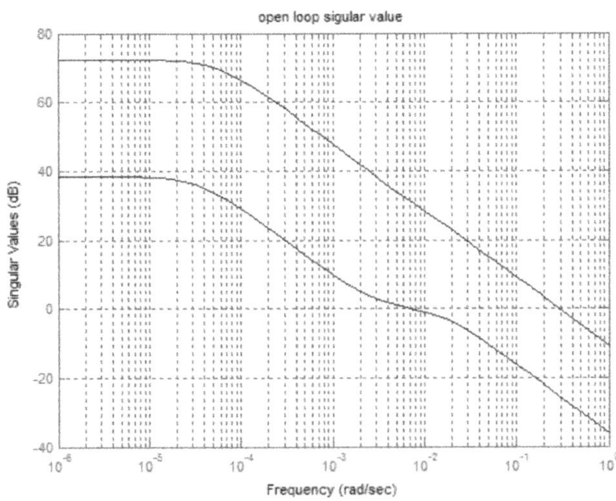

Figure 4.24. *Singular values of the open loop*

4.5.3.4. *Example 4 – HIMAT decentralized correction*

Here, we resume the example of the Himat drawn from [SAF 81]. The objective is to synthesize a control law that controls the "pitching chain" of a highly maneuverable experimental aircraft, called HIMAT[10]. This aircraft has a high number of control surfaces. The two outputs to control are the angle of incidence α and the pitch θ by acting on two control surfaces in particular: the "elevon" and the "canard flap". The angle of incidence α is defined as the angle between the main axis of the aircraft and the projection of the velocity vector in the vertical plane containing the aircraft's main axis. The pitch θ corresponds to the angle between the aircraft's main axis and the horizontal plane of the terrestrial reference.

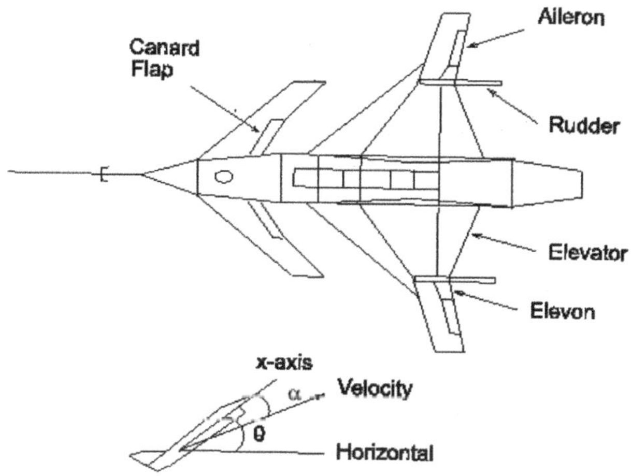

Figure 4.25. *HIMAT*

4.5.3.4.1. Process modelization

The dynamics of the process can be represented by a nonlinear model. The model used for control law synthesis corresponds to the linearization of a nonlinear model around a particular operating point of flight. The model obtained is unstable. The singular values of the process are presented in Figure 4.26 and correspond to the state space model [4.58].

10 This aircraft was tested at the end of the 1970s: it is currently in the National Air and Space Museum in Washington DC.

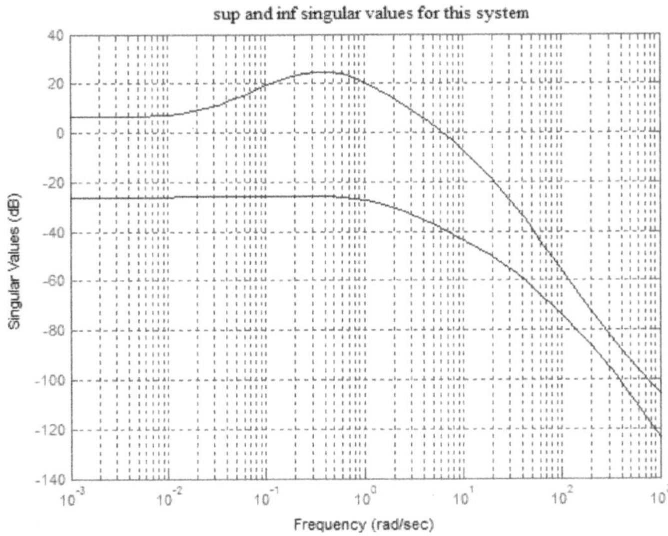

Figure 4.26. *Singular values of the linearized process*

$$\left.\begin{array}{l}\dot{x} = Ax + Bu \\ y = Cx + Du\end{array}\right\} := H(s)$$

$$u = \begin{pmatrix} u_\alpha & u_\theta \end{pmatrix}^T, \; y = \begin{pmatrix} \alpha & \theta \end{pmatrix}^T$$

$$\begin{array}{l}
\mathrm{A} = [\; -2.2567\mathrm{e}-2,\; -3.661\mathrm{e}1,\; -1.8897\mathrm{e}1,\; -3.209\mathrm{e}1,\; 3.2509,\; -7.6257\mathrm{e}-1\,; \\
\quad 9.2572\mathrm{e}-5,\; -1.8997,\; 9.8312\mathrm{e}-1,\; -7.2562\mathrm{e}-4,\; -1.708\mathrm{e}-1,\; -4.9652\mathrm{e}-3\,; \\
\quad 1.2338\mathrm{e}-2,\; 1.172\mathrm{e}1,\; -2.63116,\; 8.7582\mathrm{e}-4,\; -3.1604\mathrm{e}1,\; 2.2396\mathrm{e}1\,; \\
\quad 0,\; 0,\; 1,\; 0,\; 0,\; 0\,; \\
\quad 0,\; 0,\; 0,\; 0,\; -3\mathrm{e}1,\; 0\,; \\
\quad 0,\; 0,\; 0,\; 0,\; 0,\; -3\mathrm{e}1 \;]; \\
\mathrm{B} = [\, 0,\; 0\,; \\
\quad 0,\; 0\,; \\
\quad 0,\; 0\,; \\
\quad 0,\; 0\,; \\
\quad 30,\; 0\,; \\
\quad 0,\; 30 \,]; \\
\mathrm{C} = [\, 0, 1, 0, 0, 0, 0\,; \\
\quad 0, 0, 0, 1, 0, 0 \,]; \\
\mathrm{D} = [\, 0, 0\,; \\
\quad 0, 0 \,];
\end{array} \qquad [4.58]$$

This model does not account for certain flexible modes of the aircraft, which were not included. The existence of these flexible modes can be taken into account nonetheless by the introduction of a multiplicative uncertainty on process output of upper singular value equal to:

$$\bar{\sigma}(\Delta(j\omega)) = 50\left(\frac{\omega^2 + 100^2}{\omega^2 + 10000^2}\right)^{1/2} \qquad [4.59]$$

4.5.3.4.2. High-level specification

4.5.3.4.2.1. Reference tracking

The control law must allow for three types of longitudinal maneuvers to be performed:

– pitch θ is kept constant while the angle of incidence α varies. This is achieved by an altitude that remains constant with a rotation of the velocity vector;

– the $\alpha - \theta$ differential remains constant. The velocity vector does not perform a rotation;

– the angle of incidence α remains constant. The aircraft has vertical acceleration without altering its incidence.

The variations considered will be step-like. The response times required (rising time) *are about 0.3 s, with an overshoot below 20%.* The type of maneuver considered requires a decoupling objective, that is, for example, for a step reference to be sent on pitch with the required angle of attack remaining constant, which is announced in the aircraft's angle of attack by a variation that must be as low as possible. Accordingly, the coupling rate *must not exceed 20%.*

4.5.3.4.2.2. Commands

The commands correspond to angles in degrees of inclination of the ailerons and tailplane. The order of magnitude is of *several decades* (but strictly less than 50°). The objective is to obtain trajectory tracking performances while keeping the lowest possible command amplitudes.

4.5.3.4.2.3. Robustness

The control has to be *stability robustness* relative to the dynamics neglected during process modelization. In addition, it has to have good stability margins (*modulus margins of at least 0.5*).

The searched correction matrix is decentralized with each of its sub-transfers being of order 5.

4.5.3.4.3. Control law computation

We consider the control scheme in Figure 4.27 and establish the standard form for the corresponding control.

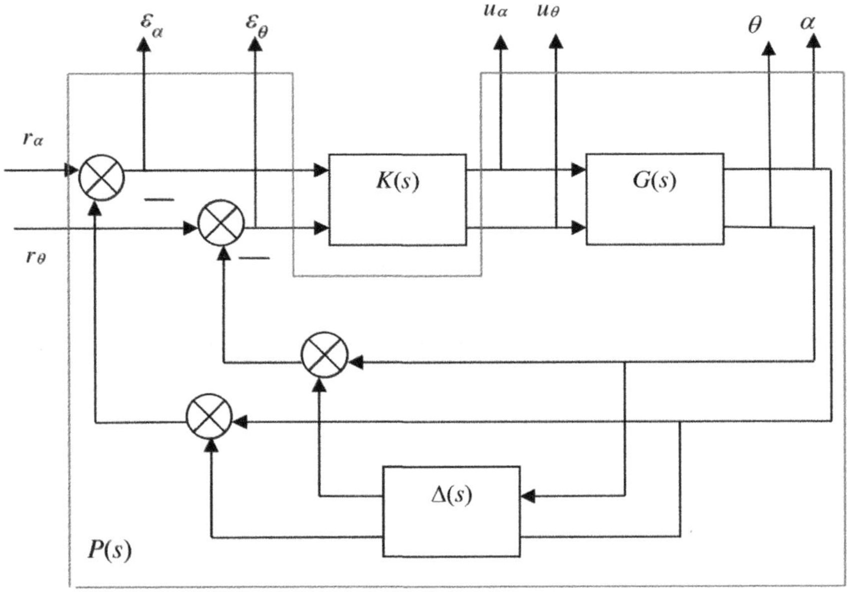

Figure 4.27. *HIMAT controller synthesis scheme. For a color version of this figure, see www.iste.co.uk/feyel/optimization.zip*

According to the small gain theorem, it is easy to show that the closed loop is stable for any uncertainty Δ if:

$$\left\| T_{ry} \Delta \right\|_\infty < 1 \qquad [4.60]$$

According to the scheme, we have:

$$\left.\begin{aligned} \dot{x} &= Ax + 0r + Bu \\ z_1(=y) &= Cx + 0r + Du \\ z_2(=u) &= 0x + 0r + Iu \\ y(=\varepsilon) &= -Cx - Du + Ir \end{aligned}\right\} := P(s) \qquad [4.61]$$

Thus, we can propose the following script for the controller synthesis:

```
%
% HIMAT Example
%
% Aut : Philippe FEYEL (2015)

clear all;

A = [
-2.2567e-02  -3.6617e+01  -1.8897e+01  -3.2090e+01
3.2509e+00  -7.6257e-01;
9.2572e-05  -1.8997e+00   9.8312e-01  -7.2562e-04  -1.7080e-01
-4.9652e-03;
1.2338e-02   1.1720e+01  -2.6316e+00   8.7582e-04  -3.1604e+01
2.2396e+01;
0            0   1.0000e+00            0            0
0;
0            0            0            0  -3.0000e+01
0;
0            0            0            0            0  -
3.0000e+01];

B =   [0     0; 0     0; 0     0; 0     0; 30     0; 0    30];

C = [0     1     0     0     0     0;
     0     0     0     1     0     0];

D = [0     0;
     0     0];

ny = 2;
nu = 2;
nx = size(A,1);

P.a = A;
P.b = [zeros(nx,ny) B];
P.c = [C;...
        zeros(nu,nx);...
       -C];
P.d = [[zeros(ny,ny), D];...
        [zeros(nu,ny), eye(nu,nu)];...
        [eye(ny,ny), -D]];

% temporal requirements for a step on teta
requirements{1,1} = 't = transpose(0:0.001:10) ; FLPK_e1 =
ss(CL.a,CL.b(:,1),CL.c,CL.d(:,1)) ; z = step(FLPK_e1,t) ; Info
= stepinfo(z(:,1),t,1) ; Z(1) = Info.SettlingTime ; Zmax(1) =
0.9 ; L(1) = Inf ; Z(2) = Info.Peak ; Zmax(2) = 1.2 ; L(2) =
Inf ; Z(3) = max(abs(z(:,3))) ; Zmax(3) = 50 ; L(3) = Inf
;Z(4) = max(abs(z(:,4))) ; Zmax(4) = 50 ; L(4) = Inf ; Z(5) =
max(abs(z(:,2))) ; Zmax(5) = 0.4 ; L(5) = Inf ; Info =
```

```
stepinfo(z(:,2),t,0) ; Z(6) = Info.SettlingTime ; Zmax(6) =
0.9 ; L(6) = Inf ;';

% temporal requirements for a step on alpha
requirements{2,1} = 't = transpose(0:0.001:10) ; FLPK_e2 =
ss(CL.a,CL.b(:,2),CL.c,CL.d(:,2)) ; z = step(FLPK_e2,t) ; Info
= stepinfo(z(:,2),t,1) ; Z(1) = Info.SettlingTime ; Zmax(1) =
0.9 ; L(1) = Inf ; Z(2) = Info.Peak ; Zmax(2) = 1.2 ; L(2) =
Inf ; Z(3) = max(abs(z(:,3))) ; Zmax(3) = 50 ; L(3) = Inf
;Z(4) = max(abs(z(:,4))) ; Zmax(4) = 50 ; L(4) = Inf ; Z(5) =
max(abs(z(:,1))) ; Zmax(5) = 0.4 ; L(5) = Inf ; Info =
stepinfo(z(:,1),t,0) ; Z(6) = Info.SettlingTime ; Zmax(6) =
0.9 ; L(6) = Inf ;';

% frequency requirements
requirements{3,1} = 'T = ss(CL.a,CL.b,CL.c,CL.d) ; S =
eye(2,2) - T(1:2,:) ; Z(1) = norm(S,Inf) ; Zmax(1) = 2 ; L(1)
= Inf ; s = tf([1 0],[0 1]); Delta = 50*(s+100)/(s+10000) ;
Z(2) = norm(T(1:2,:)*Delta,Inf) ; Zmax(2) = 1 ; L(2) = Inf;';

nmeas = ny;
ncon = nu;
order.order_K = [[5 5];[5 5]];
order.mode = 'feedback';
order.robustness = 'nominal';
options.debug = 0;
options.nbr_run = 3;
options.nbr_iteration_max = 100000;
options.display = 'iter';
options.convergence_threshold = 1e-3;
options.parallel_computation = 8;
options.save_file = 'HINFSTOCH_HIMAT_EXAMPLE.mat';

[K_ss, K_blk] = HinfStoch_ControllerTuning_v2(P,nmeas, ncon,
order, requirements, options);

file = 'CONTROLLER_HIMAT_EXAMPLE.mat';

save(file,'K_ss', 'K_blk');
```

As we see in Figure 4.28, the majority of the requirements is satisfied or quasi-satisfied, which is confirmed by the step responses of the closed-loop system in Figure 4.29. Figure 4.30 presents the control matrix obtained containing filtered PID networks for components K_{11}, K_{21}, K_{22} of the controller and a low-pass filter for component K_{12}.

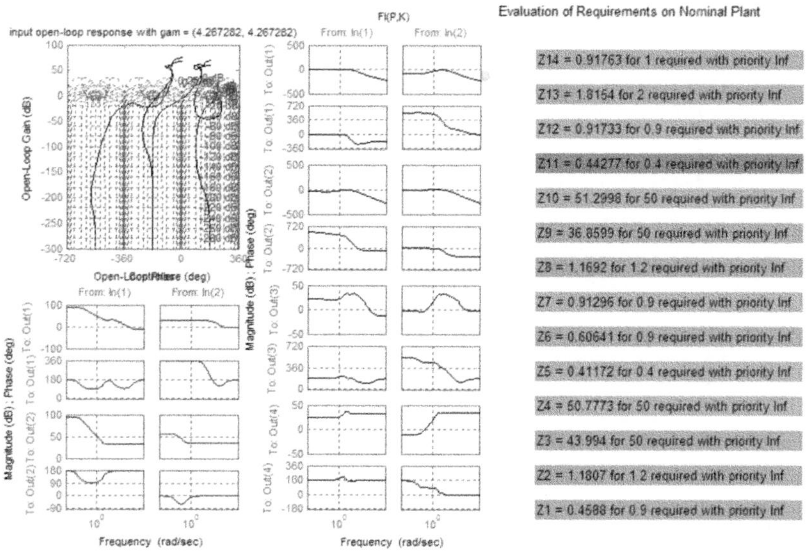

Figure 4.28. *Result of HIMAT controller synthesis. For a color version of this figure, see www.iste.co.uk/feyel/optimization.zip*

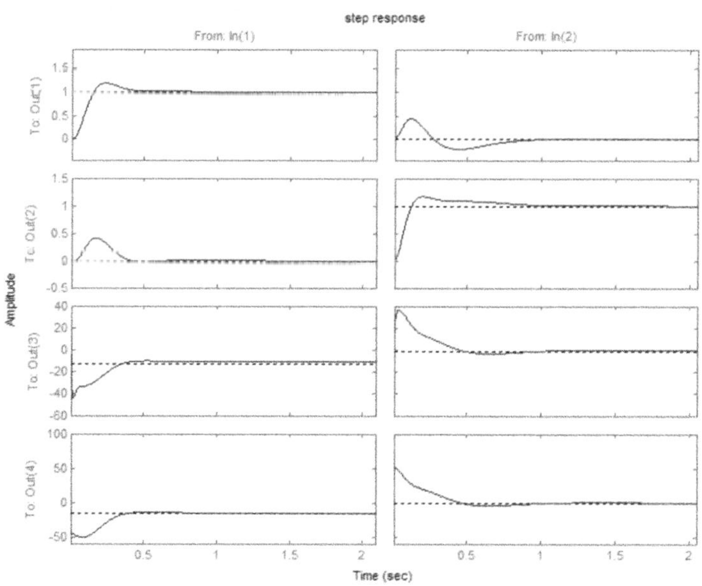

Figure 4.29. *HIMAT step response. For a color version of this figure, see www.iste.co.uk/feyel/optimization.zip*

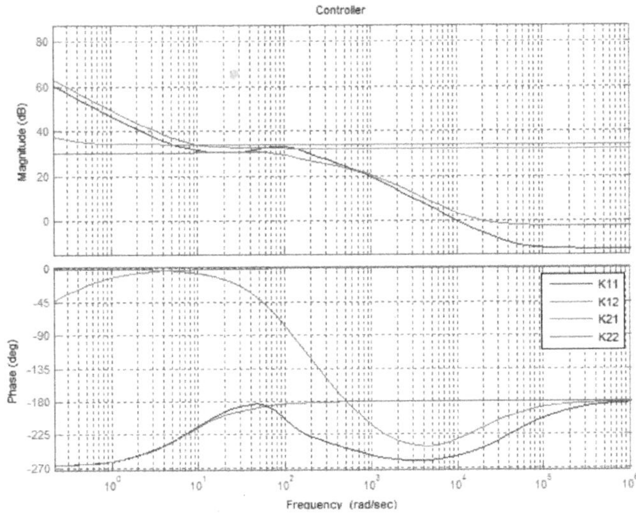

Figure 4.30. *HIMAT controller. For a color version of this figure, see www.iste.co.uk/feyel/optimization.zip*

4.5.3.5. *Example 5 – head disk assembly (HDA) control with 2DoF controller*

Here, we resume the example of the head disk assembly (HDA) control developed in [MAT 14] illustrating the *looptune* tool. The transfer function of the system, denoted by $G(s)$ in the following, is shown in Figure 4.31.

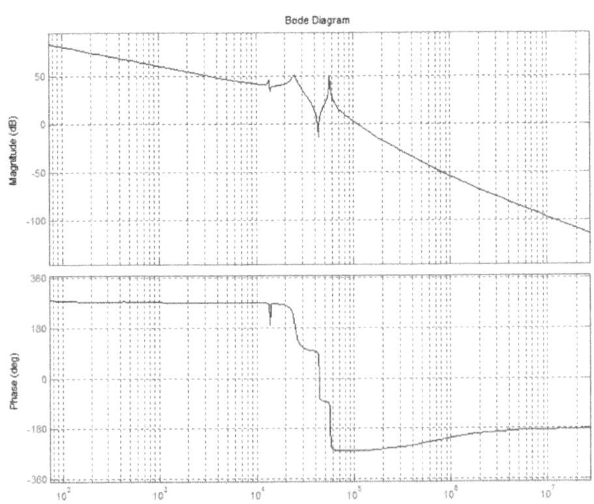

Figure 4.31. *HDA transfer function*

The correction structure retained is a controller with 2 degrees of freedom as in Figure 4.32.

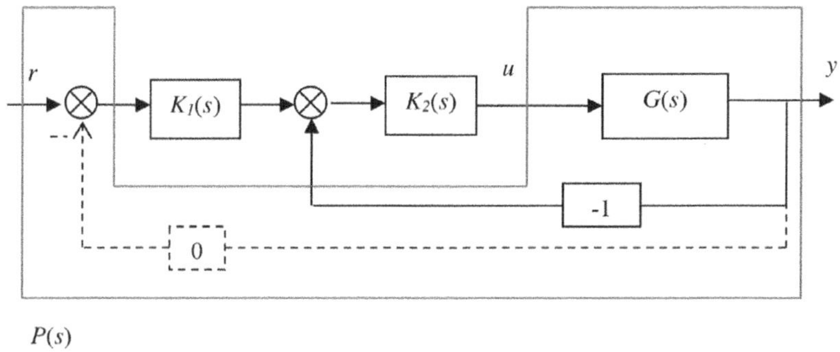

Figure 4.32. *Control scheme of the hard disk assembly. For a color version of this figure, see www.iste.co.uk/feyel/optimization.zip*

We observe that such a structure can be seen as a cascade-type implementation of a loop controller with two inputs (y and $r - 0y$) and one output (u).

The specification to satisfy is the following:

– phase margin $\Delta\Phi > 60°$;

gain margin $\Delta G > 10$ dB;

– response time of y about 1.5 ms without steady-state error for a unitary step of r (i.e. a settling time around 3 ms);

– to ensure disturbance rejection:

- the $K_2(s)$ loop controller must have an integral action;

- the open-loop bandwidth GK_2 must be greater than 1,000 rad/s.

Initially, we construct a standard form $P(s)$ for the posed problem. Without prejudicing the control structure, we will choose:

– for measurements, vector $\begin{pmatrix} r-0y & -y \end{pmatrix}^T$;

– for signals of interest, output y;

– for control signal, value u;

– for exogenous input, reference r.

We can easily define the following standard form:

$$\begin{pmatrix} \dot{x} \\ \hline z(=y) \\ \hline y1(=y-0r) \\ y2(=-y) \end{pmatrix} = \left(\begin{array}{c|c|c} A & 0 & B \\ \hline C & 0 & D \\ \hline 0C & 1 & 0D \\ -C & 0 & -D \end{array}\right) \begin{pmatrix} \dot{x} \\ \hline e(=r) \\ u \end{pmatrix} := P(s) \qquad [4.62]$$

The controller is synthesized by means of the following synthesis script:

```
%
% Taken from "Robust Control Toolbox - Systune" : control of a
head-disk assembly (HDA) in a hard-disk drive.
%
% Aut : Philippe FEYEL (2016)
%
clear all;
close all;

A = 1.0e+05 * [[-0.0001    0.0000    -0.0111          0
0.1717          0       1.2622    0.1235     0.6410 ];...
                         [ 0    -0.0004    -0.0038          0          0
0          0          0          0 ]; ...
                         [ 0     0.0051          0          0          0
0          0          0          0 ];...
                         [ 0          0          0    -0.0014    -0.1166
0          0          0          0 ]; ...
                         [ 0          0          0     0.1638          0
0          0          0          0 ];...
                         [ 0          0          0          0          0
-0.0251    -0.1928          0          0 ];...
                         [ 0          0          0          0          0
0.3277          0          0          0 ];...
                         [ 0          0          0          0          0
0          0    -0.0057    -0.4879 ];...
                         [ 0          0          0          0          0
0          0     0.6554          0 ]];

B = [ 0;2;0;16;0;128;0;128;0];

C = [976.5625  0  0 0 0 0  0  0 0];

D = 0;

nx = size(A,1);
nu = 1;
ny = 2;
```

```
P.a = A;
P.b = [zeros(nx,1) B];
P.c = [C;...
       0*C;...
       -C];
P.d = [[0, D];...
       [1, 0*D];...
       [0, -D]];

requirements =     {'FLPK = ss(CL.a,CL.b,CL.c,CL.d); t =
transpose(0:1e-5:1); e = ones(size(t,1),1); z = lsim(FLPK,e,t)
; Info = stepinfo(z,t,1) ; Z(1) = Info.SettlingTime ; Zmax(1)
= 3e-3 ; L(1) = Inf ; Z(2) = Info.Peak ; Zmax(2) = 1.05 ; L(2)
= Inf ;';...
                    'H = ss(P.a,P.b(:,2),P.c(3,:),P.d(3,2));
Cor = ss(K_blk(1,2).a, K_blk(1,2).b, K_blk(1,2).c,
K_blk(1,2).d) ;  H_BO = -Cor*H ; [Gm,Pm,Wg,Wp] = margin(H_BO)
; Z(1) = -1*Wp ; Zmax(1) = -1000 ; L(1) = Inf; Z(2) = -
abs(20*log10(Gm)) ; Zmax(2) = -10 ; L(2) = Inf ; Z(3) = -
abs(Pm) ; Zmax(3) = -60 ; L(3) = Inf ; ' } ;%S = 1/(1+H_BO) ;
Z(2) = norm(S,Inf) ; Zmax(2) = 2 ; L(2) = Inf ; ' ; };

nmeas = ny;
ncon = nu;
order{1,1,1} = 1; order{1,1,2} = '1';
order{1,2,1} = 3; order{1,2,2} = '1/s';
order.order_K = order;
order.mode = 'cascade';
order.robustness = 'nominal';
options.debug = 0;
options.nbr_run = 3;
options.nbr_iteration_max = 5000;
options.display = 'iter';
options.convergence_threshold = 1e-3;
options.parallel_computation = 8;
options.save_file = 'HINFSTOCH_HDA_EXAMPLE_2DOF.mat';

[K_ss, K_blk] = HinfStoch_ControllerTuning_v2(P,nmeas, ncon,
order, requirements, options);

file = 'CONTROLLER_HDA_EXAMPLE_2DOF.mat';

save(file,'K_ss', 'K_blk');
```

Accordingly, we obtain the results presented in Figure 4.33, in which we see that most requirements are satisfied or quasi-satisfied.

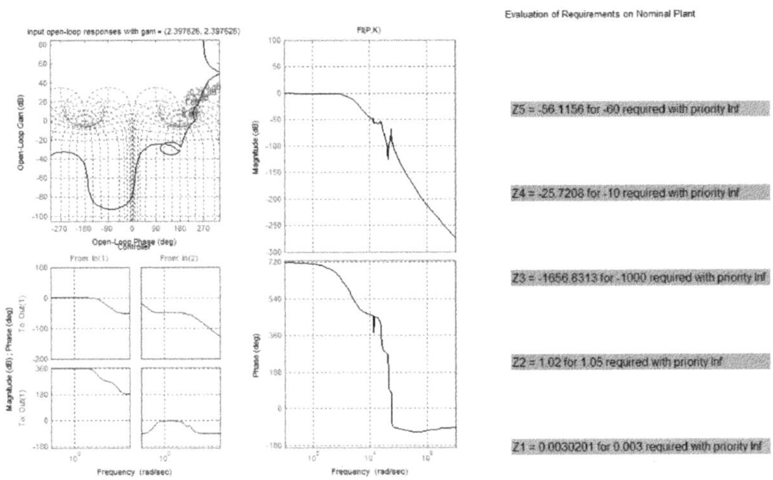

Figure 4.33. *Optimization of HAD controller. For a color version of this figure, see www.iste.co.uk/feyel/optimization.zip*

The K_1 and K_2 controllers are shown in Figure 4.34. K_1 is low-pass filter, while K_2 is a PID. The closed-loop step response is shown in Figure 4.35.

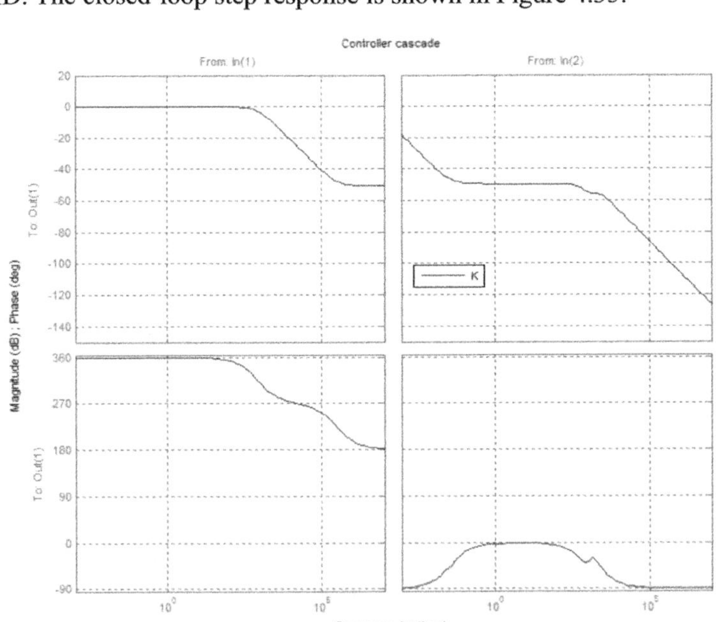

Figure 4.34. *K_1 controllers (left) and K_2 (right)*

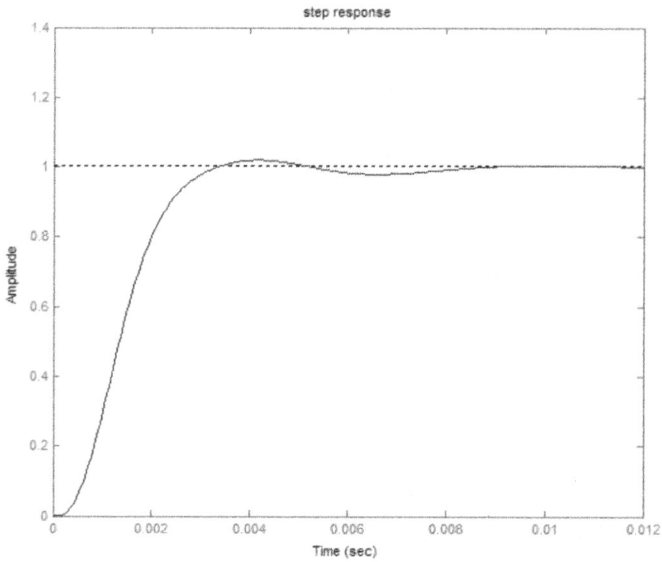

Figure 4.35. *Step HDA response. For a color version of this figure, see www.iste.co.uk/feyel/optimization.zip*

4.5.3.6. *Example 6 – inertial line of sight stabilization*

The issue of line of sight stabilization was evoked in the previous chapter.

Recall the control scheme in Figure 4.36.

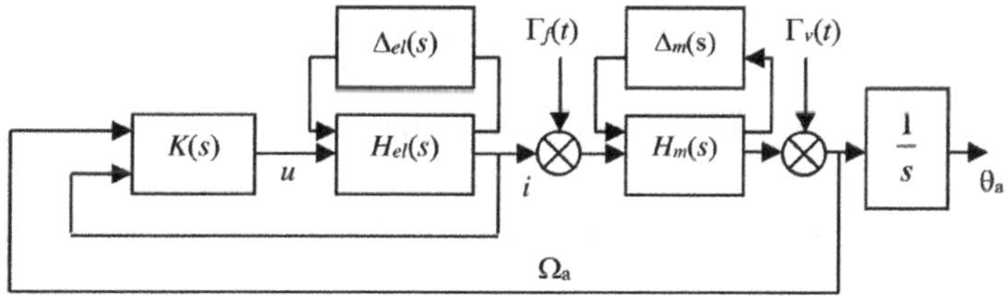

Figure 4.36. *Line of sight stabilization*

This is an electromechanical system controlled by two measures: motor current i and the charge absolute rate Ω_a.

$$H(s) = \begin{pmatrix} H_m(s) \\ 1 \end{pmatrix} H_{el}(s) \qquad [4.63]$$

The system is subject to two external disturbances:

– A friction torque $\Gamma_f(t)$ (returned to electrical current by dividing by the motor's torque constant K_c);

– An absolute rate noise on plant output, of which we have several temporal recordings denoted by $\Gamma_x(t)$, $\Gamma_y(t)$ and $\Gamma_z(t)$.

On the one hand, the specifications act on the consumption of the loop (peak current) and its positioning performance, observable via the standard deviation of the absolute position in response to each excitation noise. On the other hand, they act on the dynamic and static rejection of the friction torque.

To this, we add robustness specifications: classically, the stabilization of a ball of plant, together with a modulus margin of at least 0.5 for the nominal plant. Thus, the specification is the following:

Stability robustness on the ball of plants,

and for the nominal model :

modulus margin > 0.5

$$|\theta_a(t)| \leq \bar{\theta}_f,\ \forall t > 0 \text{ in response to step } \Gamma_f(t)$$

$$\max_{t > T_f} |\theta_a(t)| \leq \varepsilon_f \text{ in response to step } \Gamma_f(t)$$

$$|i(t)| \leq i_{\max},\ \forall t > 0 \text{ in response to step } \Gamma_f(t)$$

$$\sigma_{\theta vx} \leq \bar{\theta}_v \text{ in response to excitation } \Gamma_{vx}(t)$$

$$|i(t)| \leq i_{\max},\ \forall t > 0 \text{ in response to excitation } \Gamma_{vx}(t) \qquad [4.64]$$

$$\sigma_{\theta vy} \leq \bar{\theta}_v \text{ in response to excitation } \Gamma_{vy}(t)$$

$$|i(t)| \leq i_{\max},\ \forall t > 0 \text{ in response to excitation } \Gamma_{vy}(t)$$

$\sigma_{\theta vz} \leq \bar{\theta}_v$ in response to excitation $\Gamma_{vz}(t)$

$|i(t)| \leq i_{\max}$, $\forall t > 0$ in response to excitation $\Gamma_{vz}(t)$

First, we establish the standard form for control by introducing the disturbances as exogenous input (here, $e = \left[\Gamma_f(t), \Gamma_v(t)\right]^T$) and the signals which have temporal requirements as output (here $z = \left[\Omega_a(t), i(t)\right]^T$). The measurements provided to the controllers are the opposite of the absolute rate and motor current (here, $y = \left[-\Omega_a(t), -i(t)\right]^T$) for a negative feedback.

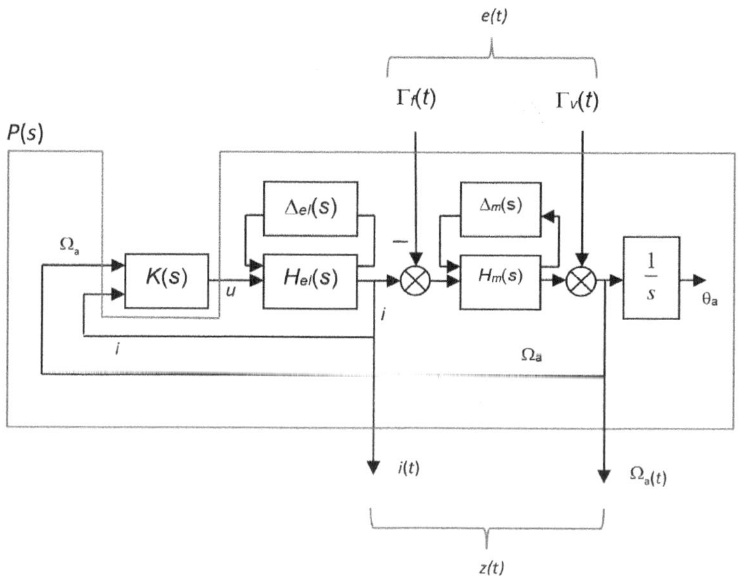

Figure 4.37. *Definition of the standard form. For a color version of this figure, see www.iste.co.uk/feyel/optimization.zip*

Writing:

$$H_m(s) := \begin{pmatrix} A_m & B_m \\ C_m & D_m \end{pmatrix}, \; H_{el}(s) := \begin{pmatrix} A_{el} & B_{el} \\ C_{el} & D_{el} \end{pmatrix} \qquad [4.65]$$

It is easy to establish the state representation of P:

$$\left.\begin{aligned}\frac{dx}{dt} &= Ax(t)+B_1e(t)+B_2u(t)\\ z(t) &= C_1x(t)+D_{11}e(t)+D_{12}u(t)\\ y(t) &= C_2x(t)+D_{21}e(t)+D_{22}u(t)\end{aligned}\right\} \quad [4.66]$$

With:

$$A=\begin{bmatrix}A_m & B_mC_{el}\\ 0 & A_{el}\end{bmatrix},\ B_1=\begin{bmatrix}-\dfrac{B_m}{K_c} & 0\\ 0 & 0\end{bmatrix},\ B_2=\begin{bmatrix}B_mD_{el}\\ B_{el}\end{bmatrix}$$

$$C_1=\begin{bmatrix}C_m & D_mC_{el}\\ 0 & C_{el}\end{bmatrix},\ D_{11}=\begin{bmatrix}-\dfrac{D_m}{K_c} & 1\\ 0 & 0\end{bmatrix},\ D_{12}=\begin{bmatrix}B_mD_{el}\\ D_{el}\end{bmatrix} \quad [4.67]$$

$$C_2=-C_1,\ D_{21}=-D_{11},\ D_{22}=-D_{12}$$

And so:

$$P:=\begin{pmatrix}A & B\\ C & D\end{pmatrix} \quad [4.68]$$

With:

$$\begin{aligned}B &= \begin{pmatrix}B_1 & B_2\end{pmatrix}\\ C &= \begin{pmatrix}C_1\\ C_2\end{pmatrix}\\ D &= \begin{pmatrix}D_{11} & D_{12}\\ D_{21} & D_{22}\end{pmatrix}\end{aligned} \quad [4.69]$$

In a function P = *genere_modele( )*, we can assume that we have thus defined a ball of N plants:

$$\wp=\left\{P_i:=\begin{pmatrix}A_i & B_i\\ C_i & D_i\end{pmatrix},\ i=0,\ldots,N\right\} \quad [4.70]$$

We can also assume to have achieved a temporal scenario generation function. Typically:

```
%
% generates disturbance temporals
%
function [Gv Gf t] = genere_temporel(indice)

switch indice

    case 1

        % load temporel : Gv, t, are in the workspace
        load('data_file_X.mat');

        % disturbance on model input of friction type zero
        Gf = zeros(size(t,1),1);

    case 2

        % load temporel : Gv, t, are in the workspace
        load('data_file_Y.mat');

         % disturbance of friction type zero
        Gf = zeros(size(t,1),1);

    case 3

        % load temporel : Gv, t, are in the workspace
        load('fichier_donnees_Z.mat');

         % disturbance on model input of friction type zero
        Gf = zeros(size(t,1),1);

    case 4

        % disturbance on model input of friction type zero:
step
        Ts = 1/1500;
        t = (0:Ts:3)';
        Gf = G0*ones(size(t,1),1);

        % disturbance on model output zero
        Gv = zeros(size(t,1),1);


end
```

We wish to synthesize a control law respecting the specifications above, implemented in cascade form, where the absolute rate controller is order 12 and the current controller is order 4. In order to accelerate the computations, we propose making use of the available 8 core CPU of our machine.

To this end, we propose the following *HinfStoch_ControllerTuning* implementation script:

```
P = genere_modele();

requirements{1,1} = '[Gv Gf t]=genere_temporel(1); e=[Gf Gv] ;
FLPK = ss(CL(1).a,CL(1).b,CL(1).c,CL(1).d) ;z =
lsim(FLPK,e,t); Z(1) =
max(abs(z(floor(size(t,1)*0.1/max(t)):size(t,1),2 ))); Teta =
lsim(tf(1,[1 0]),z(:,1),t) ; Z(2) = std(Teta); Zmax(1) = 5.5 ;
Zmax(2) = 10e-6; L(1) = Inf; L(2) = Inf;';

requirements{2,1} = '[Gv Gf t]=genere_temporel(2); e=[Gf Gv] ;
FLPK = ss(CL(1).a,CL(1).b,CL(1).c,CL(1).d) ;z =
lsim(FLPK,e,t); Z(1) =
max(abs(z(floor(size(t,1)*0.1/max(t)):size(t,1),2 ))); Teta =
lsim(tf(1,[1 0]),z(:,1),t) ; Z(2) = std(Teta); Zmax(1) = 5.5 ;
Zmax(2) = 10e-6; L(1) = Inf; L(2) = Inf;';

requirements{3,1} = '[Gv Gf t]=genere_temporel(3); e=[Gf Gv] ;
FLPK = ss(CL(1).a,CL(1).b,CL(1).c,CL(1).d) ;z =
lsim(FLPK,e,t); Z(1) =
max(abs(z(floor(size(t,1)*0.1/max(t)):size(t,1),2 ))); Teta =
lsim(tf(1,[1 0]),z(:,1),t) ; Z(2) = std(Teta); Zmax(1) = 5.5 ;
Zmax(2) = 10e-6; L(1) = Inf; L(2) = Inf;';

% Requirements on Frictions rejection
requirements{4,1} = '[Gv Gf t]=genere_temporel(13); e=[Gf Gv]
; FLPK = ss(CL(1).a,CL(1).b,CL(1).c,CL(1).d) ;z =
lsim(FLPK,e,t); Teta = lsim(tf(1,[1 0]),z(:,1),t) ; Z(1) =
max(abs(Teta)); Z(2) =
max(abs(Teta(floor(size(t,1)*0.1/max(t)):size(t,1)))); Zmax(1)
= 10e-6; Zmax(2) = 1e-6; L(1) = Inf; L(2) = Inf;';


nmeas = 2;
ncon = 1;
order.order_K = [12 4];
order.mode = 'cascade';
order.robustness = 'margin';
options.debug = 1;
options.nbr_run = 10;
```

```
options.nbr_iteration_max = 20000;
options.display = 'nominal_graphic';
options.convergence_threshold = 1e-3;
options.parallel_computation = 8;
options.save_file = 'HINFSTOCH_STAB.mat';

[K_ss, K_blk] = HinfStoch_ControllerTuning_v2(P,nmeas, ncon,
order, requirements, options);

file = 'CONTROLLER_STAB.mat';

save(file,'K_ss', 'K_blk');
```

4.5.3.7. *Example 7 – multi-loop control of a helicopter with Simulink*

Here is an example taken from [MAT 14]. This example shows how to use *HinfStoch_ControllerTuning* to tune a multi-loop controller for a rotorcraft using the *Simulink* environment. Note that the plant has been already tuned using *systune* in [MAT 14] and here we want to show the efficiency of the tool in tuning complex structured controllers, against more complex requirements, using *Simulink*.

The model deals with an 8-states helicopter model at the hovering trim condition. The state vector $x = [u, w, q, theta, v, p, phi, r]$ consists of the longitudinal velocity u (m/s), the lateral velocity v (m/s), the normal velocity w (m/s), the pitch angle *theta* (deg), the roll angle *phi* (deg), the roll rate p (deg/s), the pitch rate q (deg/s) and the yaw rate r (deg/s). The controller generates commands *ds*, *dc*, *dT* in degrees for the longitudinal cyclic, lateral cyclic, and tail rotor collective using measurements of *theta*, *phi*, p, q, and r.

The *Simulink* control scheme is depicted in Figure 4.38. It is taken from [MAT 14] and slightly modified to be used directly with *HinfStoch_ControllerTuning*. The major differences are related to:

– adding actuator saturations (± 5°) on command u, and a noisy disturbance in output plant (noisy sine at 17 Hz and 34 Hz due to engine for instance) for more realistic simulations (remember that an advantage of *HinfStoch_ControllerTuning* is to optimize complex temporal requirements using non-linear simulations);

– adding input and output ports that define the connection between the plant and the controller to be tuned;

– adding "*To Workspace*" blocks to record temporal data needed for evaluating requirements during the optimization process.

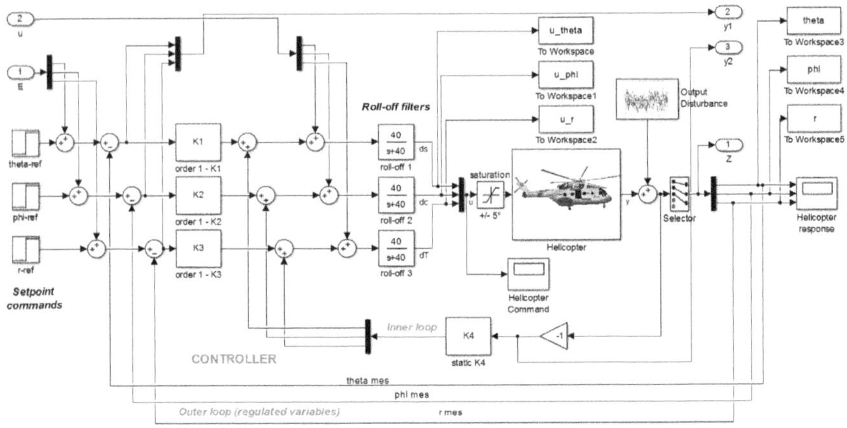

Figure 4.38. *Modified* Simulink *model* 'modified_rct_helico.slx'

The high-level specification is recalled below:

– *Tracking requirement*: the response of *theta*, *phi*, *r* to step commands *theta_ref*, *phi_ref*, *r_ref* must resemble a decoupled first-order response with a one-second time constant: this goal can be achieved by enforcing a step response rise time about 2.2 s, zero over-shoot and a coupling rate lower than 10%. As an additional complex requirement for this paper, each command must be lower than 5° and higher than -5°;

– *Stability margins*: the multivariable gain and phase margins at the plant inputs *u* and plant outputs *y* must be at least 5 dB and 40 degrees;

– *Fast dynamics*: the magnitude of the closed-loop poles must not exceed 25 to prevent fast dynamics and jerky transient. Note the roll-off filter in the model to ensure robustness.

We keep the structuration proposed in [MAT 14] i.e. a 3×5 static inner loop controller and a diagonal structured outer loop controller whose sub transfer is order 1 and which has at least one pure integrator to cancel the steady-state error in step response and enhance the coupling rate.

The 15s temporal scenario used here is composed by:

– A unitary *theta*-step starting at $t = 0$;

– A unitary *phi*-step starting at $t = 5$s;

– A unitary *r*-step starting at $t = 10$s.

As a direct transcription of the high-level specification, we propose the following *Matlab* script to compute the controller with *HinfStoch_ControllerTuning*:

```
% defining a standard form by lineariz. from Simulink
assignin('base', 'K1', 0);
assignin('base', 'K2', 0);
assignin('base', 'K3', 0);
assignin('base', 'K4', zeros(3,5));
ny = 8; nu = 3;
[A,B,C,D]=linmod('modified_rct_helico');
P2 = ss(A,B,C,D); P2 = minreal(P2);
P.a = P2.a; P.b = P2.b; P.c = P2.c; P.d = P2.d;

% running non-linear simulation with simulink...
Simulink_1 = 'assignin(''base'', ''K1'', ss(K_blk(1,1).a,
K_blk(1,1).b, K_blk(1,1).c, K_blk(1,1).d)); ';
Simulink_2 = 'assignin(''base'', ''K2'', ss(K_blk(2,2).a,
K_blk(2,2).b, K_blk(2,2).c, K_blk(2,2).d)); ';
Simulink_3 = 'assignin(''base'', ''K3'', ss(K_blk(3,3).a,
K_blk(3,3).b, K_blk(3,3).c, K_blk(3,3).d)); ';
Simulink_4 = 'A=K.a ; B=K.b ; C=K.c ; D=K.d ;
assignin(''base'', ''K4'', ss(A, B(:,4:8), C(1:3,:),
D(1:3,4:8))); ';
Simulink_5 = 'sim(''modified_rct_helico.slx'',15);';

% Tracking requirements
step_req_theta1 = 'indT = find(theta.time <= 5); T =
theta.time(indT); S = theta.signals.values(indT) ; U =
u_theta.signals.values(indT) ; Info = stepinfo(S,T,1) ; Z(1) =
Info.RiseTime ; Zmax(1) = 2.2 ; L(1) = Inf ; Z(2) = -
Info.RiseTime ; Zmax(2) = -2.2 ; L(2) = Inf ; Z(3) = Info.Peak
; Zmax(3) = 1 ; L(3) = Inf ; Z(4) = max(abs(U)) ; Zmax(4) = 5
; L(4) = Inf ;  ';
step_req_theta2 = 'indT = find(phi.time <= 5); T =
phi.time(indT) ; S = phi.signals.values(indT) ; Z(5) =
max(abs(S)) ; Zmax(5) = 10/100 ; L(5) = Inf ; ';
step_req_theta3 = 'indT = find(r.time <= 5); T = r.time(indT)
; S = r.signals.values(indT) ; Z(6) = max(abs(S)) ; Zmax(6) =
10/100 ; L(6) = Inf ; ';
step_req_phi1 = 'indT = find(phi.time <= 10 & phi.time > 5); T
= phi.time(indT) - 5; S = phi.signals.values(indT) ; U =
u_phi.signals.values(indT) ; Info = stepinfo(S,T,1) ; Z(7) =
Info.RiseTime ; Zmax(7) = 2.2 ; L(7) = Inf ; Z(8) = -
Info.RiseTime ; Zmax(8) = -2.2 ; L(8) = Inf ; Z(9) = Info.Peak
; Zmax(9) = 1 ; L(9) = Inf ; Z(10) = max(abs(U)) ; Zmax(10) =
5 ; L(10) = Inf ; ';
```

```
step_req_phi2 = 'indT = find(theta.time <= 10 & theta.time >
5); T = theta.time(indT) - 5 ; S = 1 -
theta.signals.values(indT) ; Z(11) = max(abs(S)) ; Zmax(11) =
10/100 ; L(11) = Inf ; ';
step_req_phi3 = 'indT = find(r.time <= 10 & r.time > 5); T =
r.time(indT) - 5 ; S = r.signals.values(indT) ; Z(12) =
max(abs(S)) ; Zmax(12) = 10/100 ; L(12) = Inf ; ';
step_req_r1 = 'indT = find(r.time <= 15 & phi.time > 10); T =
r.time(indT) - 10; S = r.signals.values(indT) ; U =
u_r.signals.values(indT) ; Info = stepinfo(S,T,1) ; Z(13) =
Info.RiseTime ; Zmax(13) = 2.2 ; L(13) = Inf ; Z(14) = -
Info.RiseTime ; Zmax(14) = -2.2 ; L(14) = Inf ; Z(15) =
Info.Peak ; Zmax(15) = 1 ; L(15) = Inf ;  Z(16) = max(abs(U))
; Zmax(16) = 5 ; L(16) = Inf ; ';
step_req_r2 = 'indT = find(theta.time <= 15 & theta.time >
10); T = theta.time(indT) - 10 ; S = 1 -
theta.signals.values(indT) ; Z(17) = max(abs(S)) ; Zmax(17) =
10/100 ; L(17) = Inf ; ';
step_req_r3 = 'indT = find(phi.time <= 15 & phi.time > 10); T
= phi.time(indT) - 10 ; S = 1 - phi.signals.values(indT) ;
Z(18) = max(abs(S)) ; Zmax(18) = 10/100 ; L(18) = Inf ; ';
requirements{1,1} = [Simulink_1 Simulink_2 Simulink_3
Simulink_4 Simulink_5 step_req_theta1 step_req_theta2
step_req_theta3 step_req_phi1 step_req_phi2 step_req_phi3
step_req_r1 step_req_r2 step_req_r3];

% Stability margins
loop_margin_req1 = 'Sys = -1*ss(P.a, P.b(:,4:6), P.c(4:11,:),
P.d(4:11,4:6)) ; Cor = ss(K.a,K.b,K.c,K.d) ;
[cmi,dmi,mmi,cmo,dmo,mmo,mmio] = loopmargin(Sys,Cor) ; ';
loop_margin_req2 = '[GM1,GM2,GM3] = dmi.GainMargin ; GM = [GM1
GM2 GM3];  GMdB = 20*log10(GM) ; Z(1) = -min(abs(GMdB));
Zmax(1) = -5 ; L(1) = Inf ; [PM1, PM2, PM3] = dmi.PhaseMargin;
PM = [PM1 PM2 PM3] ; Z(2) = -min(abs(PM)); Zmax(2) = -40 ;
L(2) = Inf ; ';
loop_margin_req3 = '[GM1,GM2,GM3,GM4,GM5,GM6,GM7,GM8] =
dmo.GainMargin ; GM = [GM1 GM2 GM3 GM4 GM5 GM6 GM7 GM8];  GMdB
= 20*log10(GM) ; Z(3) = -min(abs(GMdB)) ; Zmax(3) = -5 ; L(3)
= Inf ; [PM1, PM2, PM3 ,PM4,PM5,PM6,PM7,PM8] =
dmo.PhaseMargin; PM = [PM1 PM2 PM3 PM4 PM5 PM6 PM7 PM8] ; Z(4)
= -min(abs(PM)); Zmax(4) = -40 ; L(4) = Inf ; ';
requirements{2,1} = [loop_margin_req1 loop_margin_req2
loop_margin_req3];

% Fast dynamics
requirements{3,1} = 'Z(1) = max(abs(eig(CL.a))) ; Zmax(1) =
25; L(1) = Inf ; ' ;
```

```
% controller structure
M{1,1,1} = 1; M{1,1,2} = '1/s';
M{1,2,1} = -1; M{1,2,2} = '1';
M{1,3,1} = -1; M{1,3,2} = '1';
M{1,4,1} = 0; M{1,4,2} = '1';
M{1,5,1} = 0; M{1,5,2} = '1';
M{1,6,1} = 0; M{1,6,2} = '1';
M{1,7,1} = 0; M{1,7,2} = '1';
M{1,8,1} = 0; M{1,8,2} = '1';

M{2,1,1} = -1; M{2,1,2} = '1';
M{2,2,1} = 1; M{2,2,2} = '1/s';
M{2,3,1} = -1; M{2,3,2} = '1';
M{2,4,1} = 0; M{2,4,2} = '1';
M{2,5,1} = 0; M{2,5,2} = '1';
M{2,6,1} = 0; M{2,6,2} = '1';
M{2,7,1} = 0; M{2,7,2} = '1';
M{2,8,1} = 0; M{2,8,2} = '1';

M{3,1,1} = -1; M{3,1,2} = '1';
M{3,2,1} = -1; M{3,2,2} = '1';
M{3,3,1} = 1; M{3,3,2} = '1/s';
M{3,4,1} = 0; M{3,4,2} = '1';
M{3,5,1} = 0; M{3,5,2} = '1';
M{3,6,1} = 0; M{3,6,2} = '1';
M{3,7,1} = 0; M{3,7,2} = '1';
M{3,8,1} = 0; M{3,8,2} = '1';

order.order_K = M;
options.debug = 0;
options.nbr_run = 3;
order.mode = 'feedback';
order.robustness = 'nominal';
options.nbr_iteration_max = 2000;
options.display = 'nominal_graphic';
options.convergence_threshold = 1e-3;
options.parallel_computation = 4;
options.save_file = [];

[K_ss, K_blk] = HinfStoch_ControllerTuning_v2(P,ny, nu, order,
requirements, options);
```

Note the use of the `linmod` tool to compute the standard form P to be stabilized. Here the interest is to evaluate requirements using the non-linear *Simulink* environment for a more realistic result.

Computer calculation is done with *MatlabR2014b* on an Intel Core-i7 processor with 8 Gb RAM and *the Parallel Computing Toolbox* with 4 pools used. With a computation time about 10 hours (2000 iterations) and due to the good repeatability of the algorithm, we succeed in computing a suitable controller in one run; as shown in Figure 4.39, temporal and closed-loop pole placement requirements are achieved. Furthermore, as the *HinfStoch_ControllerTuning*'s window shows (Figure 4.40), the generalized gain and phase margins (respectively ΔG_i and $\Delta\phi_i$) at input plant and the generalized gain and phase margins (respectively ΔG_o and $\Delta\phi_o$) at output plant are respectively equal to:

$$\begin{aligned} &\Delta G_i = 7.05\ dB,\ \ \Delta\Phi_i = 42.1\,^\circ \\ &\Delta G_o = 8.15\ dB,\ \ \Delta\Phi_o = 47.3^\circ \end{aligned} \qquad [4.71]$$

Those results show the efficiency of the tool, which is a good alternative to other modern tools in cases of complex high-level requirements being applied to complex systems.

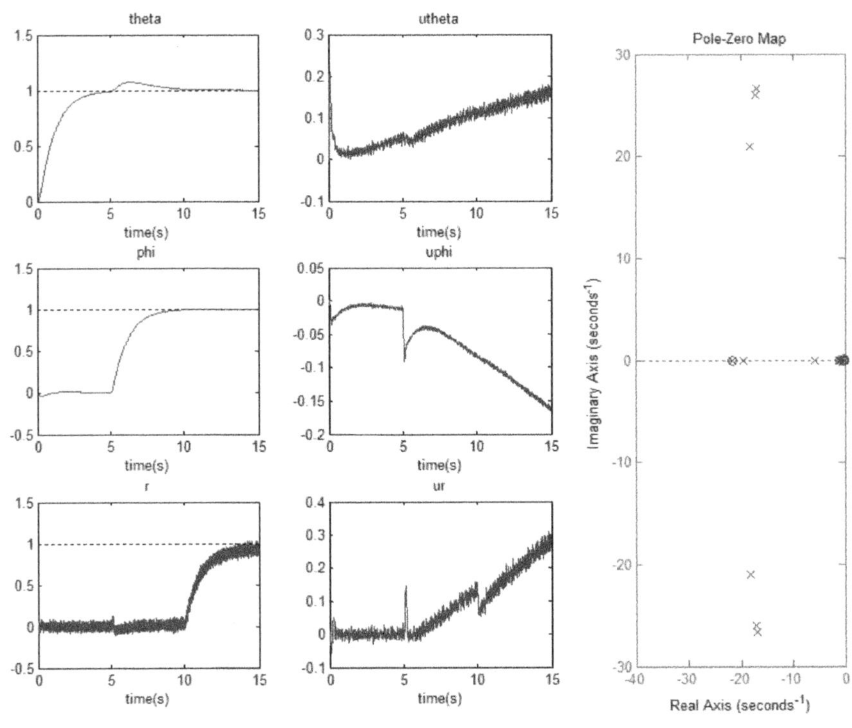

Figure 4.39. *Optimization results*

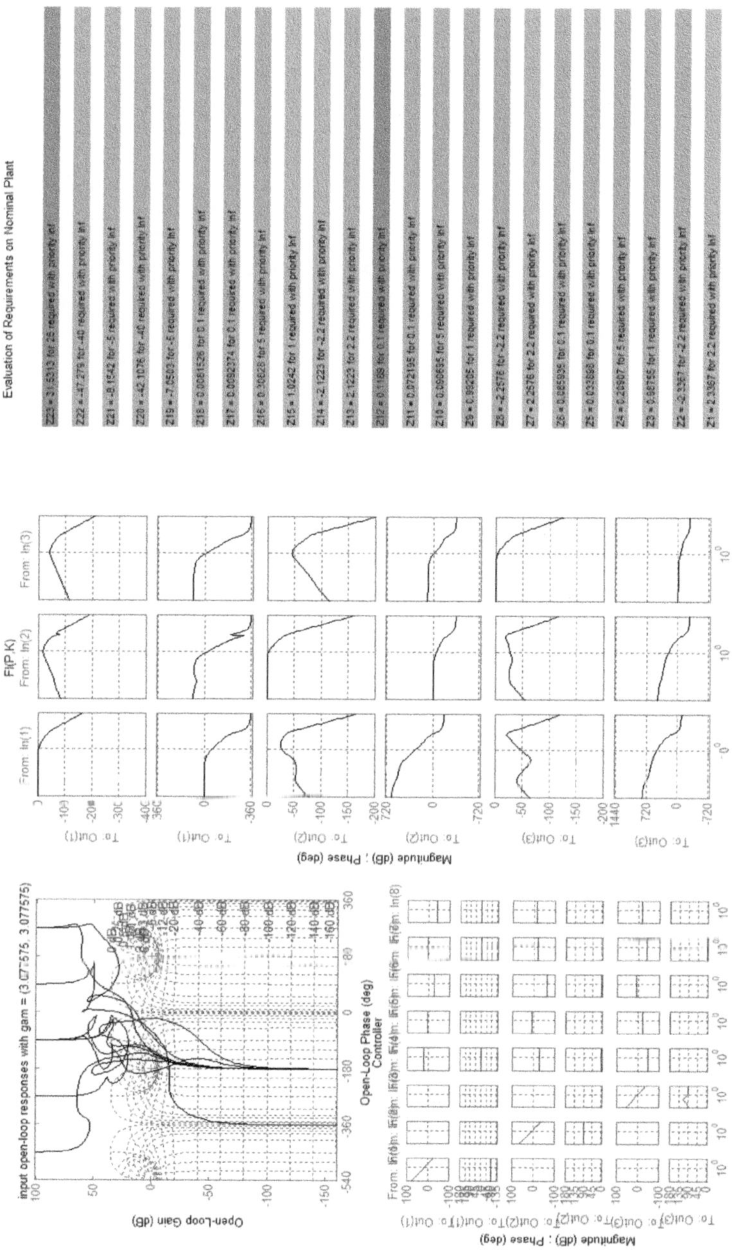

Figure 4.40. *Final* HinfStoch_ControllerTuning's *window for the rotorcraft controller optimization*

APPENDICES

Appendix A

Notions of Coprime Factorizations

A.1. Definitions

Two transfer matrices[1] $M(s) \in \mathbf{R}H_\infty^{m\times m}$ and $N(s) \in \mathbf{R}H_\infty^{p\times m}$ constitute a right coprime factorization of a rational transfer matrix *H*(*s*) of dimensions $p \times m$ if and only if:

– *M*(*s*) is square, and $\det\left(M(s)\right) \neq 0$;

– $\forall s \in \mathbf{C},\ H(s) = N(s)M(s)^{-1}$;

– $\exists \overline{U}(s) \in \mathbf{R}H_\infty^{m\times p}$, $\exists \overline{V}(s) \in \mathbf{R}H_\infty^{m\times m}$ such that:

$$\forall s \in C,\ \overline{V}(s)M(s) + \overline{U}(s)N(s) = I_m \qquad [A.1]$$

Two transfer matrices $\overline{M}(s) \in \mathbf{R}H_\infty^{p\times p}$ and $\overline{N}(s) \in \mathbf{R}H_\infty^{p\times m}$ constitute left coprime factorization of a rational transfer matrix *H*(*s*) of dimensions $p \times m$ if and only if:

– $\overline{M}(s)$ is square, and $\det\left(\overline{M}(s)\right) \neq 0$;

– $\forall s \in \mathrm{C},\ H(s) = \overline{M}(s)^{-1}\overline{N}(s)$;

– $\exists U(s) \in \mathbf{R}H_\infty^{m\times p}$, $\exists V(s) \in \mathbf{R}H_\infty^{p\times p}$ such that:

$$\forall s \in C,\ \overline{M}(s)V(s) + \overline{N}(s)U(s) = I_p \qquad [A.2]$$

---

1 Recall that $\mathbf{R}H_\infty$ is the set of rational, proper and stable matrices with real coefficients.

As an arbitrarily large number of right or left coprime factorizations are defined for the same matrix $H(s)$, we distinguish two normalized factorizations.

We may consider a right coprime factorization of $H(s)$ as normalized if and only if:

$$\forall s \in C,\ N^* N + M^* M = I_m \qquad [A.3]$$

We may consider a left coprime factorization of $H(s)$ as normalized if and only if:

$$\forall s \in C,\ \overline{N}\,\overline{N}^* + \overline{M}\,\overline{M}^* = I_p \qquad [A.4]$$

By definition of the H_∞ norm, a consequence of [A.4] is that:

$$\begin{gathered}\overline{M}\,\overline{M}^* + \overline{N}\,\overline{N}^* = I_p \\ \Leftrightarrow \\ \begin{pmatrix}\overline{M} & \overline{N}\end{pmatrix}\begin{pmatrix}\overline{M}^* \\ \overline{N}^*\end{pmatrix} = I_p \\ \Leftrightarrow \\ \sqrt{\lambda_{\max}\left(\begin{pmatrix}\overline{M} & \overline{N}\end{pmatrix}\begin{pmatrix}\overline{M}^* \\ \overline{N}^*\end{pmatrix}\right)} = 1 \\ \Leftrightarrow \\ \left\| \begin{pmatrix}\overline{M} & \overline{N}\end{pmatrix} \right\|_\infty = 1\end{gathered} \qquad [A.5]$$

where $\lambda_{\max}(\ )$ represents the largest eigenvalue.

Similarly:

$$\left\| \begin{pmatrix} N \\ M \end{pmatrix} \right\|_\infty = 1 \qquad [A.6]$$

Again we find many elements in the formalism of coprime factorization in [VID 85].

A.2. Practical computation of normalized coprime factorizations

Consider that $H(s) = [A, B, C, D]$, then the normalized coprime factorizations are built by the following two theorems.

The right normalized coprime factorization of $H(s)$ is given by:

$$N(s) = \left[\begin{array}{c|c} A+BF & BS^{-1/2} \\ \hline C+DF & DS^{-1/2} \end{array}\right], \quad M(s) = \left[\begin{array}{c|c} A+BF & BS^{-1/2} \\ \hline F & S^{-1/2} \end{array}\right] \qquad [A.7]$$

with:

$$\begin{aligned} &F = -S^{-1}\left(D^T C + B^T X\right) \\ &S = I_m + D^T D = S^{1/2T} S^{1/2},\ S^{1/2} \text{ squared} \end{aligned} \qquad [A.8]$$

Note that a $S^{1/2T} S^{1/2}$-type factorization can be obtained by an upper Cholesky factorization.

In addition, $X > 0$, solution of the Generalized Control Riccati Algebraic equation (GCARE):

$$\begin{aligned} &\left(A - BS^{-1}D^T C\right)^T X + X\left(A - BS^{-1}D^T C\right) \\ &-XBS^{-1}B^T X + C^T\left(I_p + DD^T\right)^{-1} C = 0 \end{aligned} \qquad [A.9]$$

The left normalized coprime of $H(s)$ is given by:

$$\overline{N}(s) = \left[\begin{array}{c|c} A+LC & B+LD \\ \hline R^{-1/2}C & R^{-1/2}D \end{array}\right], \quad \overline{M}(s) = \left[\begin{array}{c|c} A+LC & L \\ \hline R^{-1/2}C & R^{-1/2} \end{array}\right] \qquad [A.10]$$

with:

$$\begin{aligned} &L = -\left(YC^T + BD^T\right)R^{-1} \\ &R = I_p + DD^T = R^{1/2} R^{1/2T},\ R^{1/2} \text{ squared} \end{aligned} \qquad [A.11]$$

Note that a $R^{1/2} R^{1/2T}$-type factorization can be obtained by a lower Cholesky factorization.

In addition, $Y > 0$, solution of the Generalized Filter Algebraic Riccati Equation (GFARE):

$$\begin{aligned}&\left(A-BD^T R^{-1}C\right)Y+Y\left(A-BD^T R^{-1}C\right)^T\\&-YC^T R^{-1}CY+B\left(I_m+D^T D\right)^{-1}B^T=0\end{aligned} \qquad [A.12]$$

Moreover, consider a square transfer matrix $G(s) = [A, B, C, D]$, then, if D^{-1} exists, the state-space representation of the inverse of $G(s)$ is:

$$G^{-1}(s)=\left[\begin{array}{c|c}A-BD^{-1}C & -BD^{-1}\\\hline D^{-1}C & D^{-1}\end{array}\right] \qquad [A.13]$$

As the matrices $\overline{M}(s)$ and $M(s)$ are square, we can deduce that:

$$M(s)^{-1}=\left[\begin{array}{c|c}A & -B\\\hline S^{1/2}F & S^{1/2}\end{array}\right],\quad \overline{M}(s)^{-1}=\left[\begin{array}{c|c}A & -LR^{1/2}\\\hline C & R^{1/2}\end{array}\right] \qquad [A.14]$$

A.3. Reconstruction of a transfer function by its coprime factors

Consider M and N, whose state-space representation is:

$$\begin{pmatrix}N\\M\end{pmatrix}=\left[\begin{array}{c|c}A_{NM} & B_{NM}\\\hline C_N & D_N\\C_M & D_M\end{array}\right] \qquad [A.15]$$

Then, if the D_M matrix is invertible, a state-space representation of the system $H(s)=N(s)M(s)^{-1}$ is written as:

$$H(s)=\left[\begin{array}{c|c}A_{NM}-B_{NM}D_M{}^{-1}C_M & B_{NM}D_M{}^{-1}\\\hline C_N-D_N D_M{}^{-1}C_M & D_N D_M{}^{-1}\end{array}\right] \qquad [A.16]$$

Consider $\overline{N}$ and $\overline{M}$, whose state-space representation is:

$$\begin{pmatrix}\overline{N} & \overline{M}\end{pmatrix}=\left[\begin{array}{c|cc}A_{\overline{NM}} & B_{\overline{N}} & B_{\overline{M}}\\\hline C_{\overline{NM}} & D_{\overline{N}} & D_{\overline{M}}\end{array}\right] \qquad [A.17]$$

Then, if the matrix $D_{\overline{M}}$ is invertible, a state-space representation of the system $H(s) = \overline{M}(s)^{-1}\overline{N}(s)$ is written as:

$$H(s) = \left[\begin{array}{c|c} A_{\overline{NM}} - B_{\overline{M}} D_{\overline{M}}^{-1} C_{\overline{NM}} & B_{\overline{N}} - B_{\overline{M}} D_{\overline{M}}^{-1} D_{\overline{N}} \\ \hline D_{\overline{M}}^{-1} C_{\overline{NM}} & D_{\overline{M}}^{-1} D_{\overline{N}} \end{array} \right] \qquad \text{[A.18]}$$

Appendix B

Examples of LFT Form Used for Uncertain Systems

B.1. Example of LFT form use for a neglected dynamic

Considering a nominal system $G_0(s)$ in which a dynamic has been neglected:

$$G(s) = G_0(s)\frac{1}{1+\tau s},\ \tau < \tau_{\max} \qquad [B.1]$$

We observe that:

$$\left|\frac{G(j\omega)}{G_0(j\omega)} - 1\right| = \left|\frac{\tau j\omega}{1+\tau j\omega}\right| \leq \left|\frac{\tau_{\max} j\omega}{1+\tau_{\max} j\omega}\right| \qquad [B.2]$$

Thus, we can account for the neglected dynamic under the form of a multiplicative uncertainty on the nominal system $G_0(s)$:

$$\begin{cases} \dfrac{G(s)}{G_0(s)} - 1 = w_1(s)\Delta_1(s) \Leftrightarrow G(s) = G_0(s)\left(1 + w_1(s)\Delta_1(s)\right) \\ w_1(s) = \dfrac{\tau_{\max s}}{1+\tau_{\max s}} \text{ et } \forall \omega, \left|\Delta_1(j\omega)\right| < 1 \Leftrightarrow \left\|\Delta_1(s)\right\|_\infty < 1 \end{cases} \qquad [B.3]$$

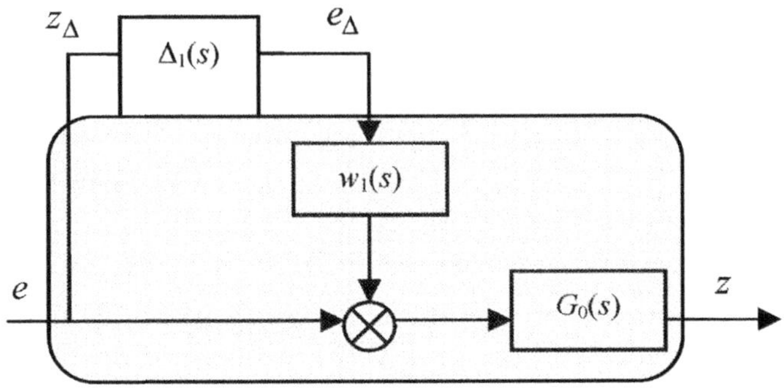

Figure B.1. *Use of LFT form for a neglected dynamic*

The LFT form (Figure B.1) is written as:

$$\begin{pmatrix} z_\Delta(s) \\ z(s) \end{pmatrix} = \begin{pmatrix} 0 & 1 \\ G_0(s)w_1(s) & G_0(s) \end{pmatrix} \begin{pmatrix} e_\Delta(s) \\ e(s) \end{pmatrix}; \ \Delta(s) = \Delta_1(s) \quad \text{[B.4]}$$

The matrix is $\Delta(s)$ diagonal.

B.2. Example of LFT form use for a parametric uncertainty on a transfer function

Considering a nominal system $G_0(s)$ in which a parameter is uncertain:

$$G_0(s) = \frac{1}{(s+a)^2}, \ a_0 - b < a < a_0 + b \quad \text{[B.5]}$$

We observe that:

$$\frac{1}{s+a} = \frac{1}{s+a_0}\left(1+\delta\frac{b}{s+a_0}\right)^{-1}, \ a = a_0 + \delta b, \ -1 < \delta < 1 \quad \text{[B.6]}$$

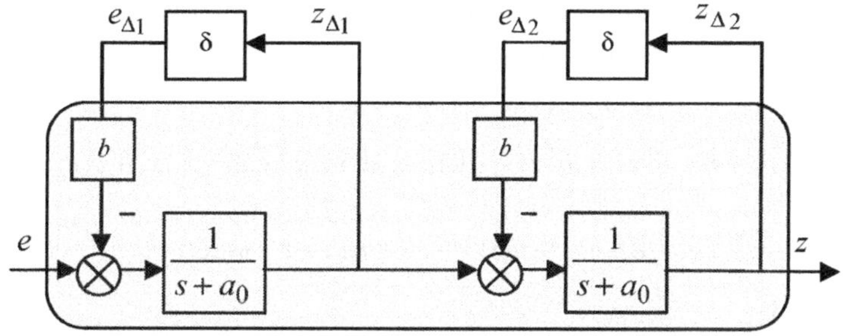

Figure B.2. *LFT form use for a parametric uncertainty on a transfer function*

The LFT form in Figure B.2 is written as:

$$\begin{pmatrix} z_{\Delta 1}(s) \\ z_{\Delta 2}(s) \\ \hline z(s) \end{pmatrix} = \left(\begin{array}{cc|c} \dfrac{-b}{s+a_0} & 0 & \dfrac{1}{s+a_0} \\ \dfrac{-b}{(s+a_0)^2} & \dfrac{-b}{s+a_0} & \dfrac{1}{(s+a_0)^2} \\ \hline \dfrac{-b}{(s+a_0)^2} & \dfrac{-b}{s+a_0} & \dfrac{1}{(s+a_0)^2} \end{array}\right) \begin{pmatrix} e_{\Delta 1}(s) \\ e_{\Delta 2}(s) \\ \hline e(s) \end{pmatrix}, \quad \Delta(s) = \begin{pmatrix} \delta & 0 \\ 0 & \delta \end{pmatrix} \qquad [B.7]$$

The matrix $\Delta(s)$ is diagonal.

B.3. Example of LFT form use for a parametric uncertainty in a state-space representation

Considering the state-space representation:

$$\frac{d}{dt}\begin{pmatrix} x_1 \\ x_2 \end{pmatrix} = \begin{pmatrix} 0 & a^2 \\ -1 & -a \end{pmatrix}\begin{pmatrix} x_1 \\ x_2 \end{pmatrix} + \begin{pmatrix} 0 \\ 1 \end{pmatrix} u(t), \quad a = a_0 + \delta a_1, \ -1 < \delta < 1 \qquad [B.8]$$

Reforming the uncertain expression of a in the state equation and introducing new inputs and outputs in order to model the uncertainty channels gives us:

$$\left.\begin{array}{l}
\dfrac{dx_1}{dt}=(a_0+\delta a_1)^2x_2=a_0{}^2x_2+\delta\left(2a_1a_0x_2+\delta a_1{}^2x_2\right)=a_0{}^2x_2+e_{\Delta 2}\\
\dfrac{dx_2}{dt}=-x_1-(a_0+\delta a_1)x_2+u(t)=-x_1-a_0x_2-e_{\Delta 1}+u(t)\\
e_{\Delta 1}=\delta z_{\Delta 1},\ z_{\Delta 1}=a_1x_2\\
e_{\Delta 2}=\delta\left(2a_0a_1x_2+a_1\delta a_1x_2\right)=\delta z_{\Delta 2},z_{\Delta 2}=2a_0a_1x_2+a_1e_{\Delta 1}
\end{array}\right\}\quad [B.9]$$

The use of LFT form in the state space is then:

$$\left.\begin{array}{l}
\dfrac{d}{dt}\begin{pmatrix}x_1\\x_2\end{pmatrix}=\begin{pmatrix}0 & a_0{}^2\\-1 & -a_0\end{pmatrix}\begin{pmatrix}x_1\\x_2\end{pmatrix}+\left(\begin{array}{cc|c}0 & 1 & 0\\-1 & 0 & -1\end{array}\right)\begin{pmatrix}e_{\Delta 1}\\e_{\Delta 2}\\\hline u(t)\end{pmatrix}\\
\begin{pmatrix}z_{\Delta 1}\\z_{\Delta 2}\end{pmatrix}=\begin{pmatrix}0 & a_1\\0 & 2a_0a_1\end{pmatrix}\begin{pmatrix}x_1\\x_2\end{pmatrix}+\left(\begin{array}{cc|c}0 & 0 & 0\\a_1 & 0 & 0\end{array}\right)\begin{pmatrix}e_{\Delta 1}\\e_{\Delta 2}\\\hline u(t)\end{pmatrix}\\
\begin{pmatrix}e_{\Delta 1}\\e_{\Delta 2}\end{pmatrix}=\begin{pmatrix}\delta & 0\\0 & \delta\end{pmatrix}\begin{pmatrix}z_{\Delta 1}\\z_{\Delta 2}\end{pmatrix}
\end{array}\right\}\Leftrightarrow\begin{array}{l}\dfrac{dX}{dt}=A_0X+Bu+Ne_\Delta\\z_\Delta=MX+Oe_\Delta\\e_\Delta=\Delta z_\Delta,\ \Delta=diag(\delta_i)\end{array}\quad [B.10]$$

The matrix $\Delta(s)$ is diagonal.

B.4. Example of LFT form use for a complex uncertainty

This type of uncertainty can allow for modeling the gain and phase notion margins.

$$G(s)=G_0(s)ke^{i\varphi},\ ke^{i\varphi}=1+s_c\varepsilon_c,\ s_c\in\mathbf{R},\ \varepsilon_c\in\mathbf{C},\ |\varepsilon_c|<1\quad [B.11]$$

This is a special instance of Figure B.1.

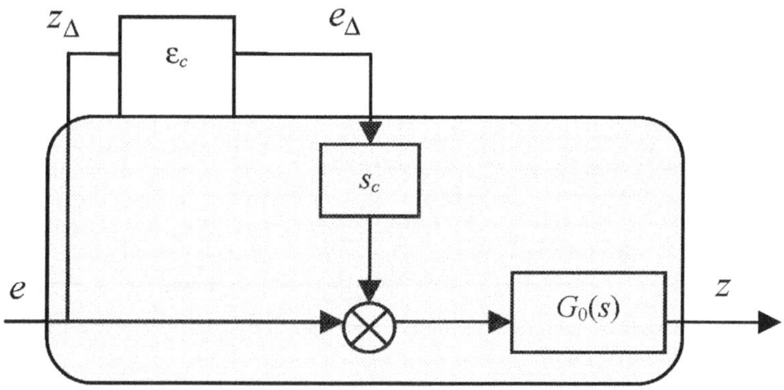

Figure B.3. *LFT form use for a neglected dynamic*

B.5. Example of LFT form use for a closed-loop system

Now, let us consider the following system:

$$G_0(s) = \frac{1}{(s+a)^2} \qquad \text{[B.12]}$$

This system:

– has a neglected dynamic $\frac{1}{(1+\tau s)}$, $\tau < \tau_{\max}$;

– has a parametric uncertainty on a: $a_0 - b < a < a_0 + b$;

– has a complex uncertainty: $ke^{i\varphi} = 1 + s_c\varepsilon_c$, $s_c \in \mathbf{R}$, $\varepsilon_c \in \mathbf{C}$, $|\varepsilon_c| < 1$

– is looped by a controller $K(s)$.

The association of several LFTs giving one LFT is shown in Figure B.4.

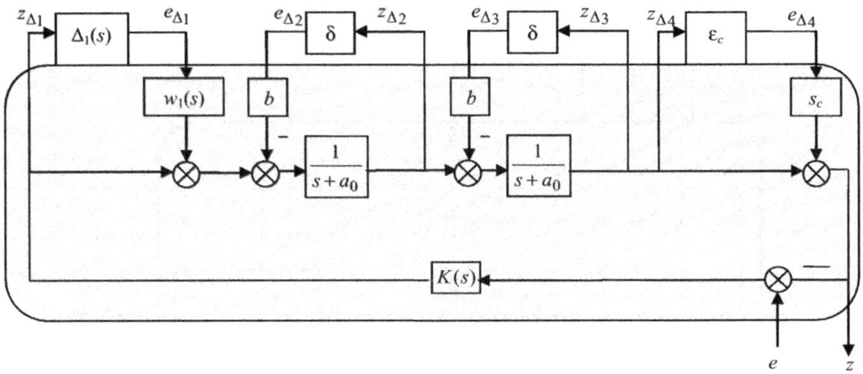

Figure B.4. *LFT form use of an uncertain loop*

We can easily show that:

$$\begin{pmatrix} z_{\Delta 1}(s) \\ z_{\Delta 2}(s) \\ z_{\Delta 3}(s) \\ z_{\Delta 4}(s) \\ \hline z(s) \end{pmatrix} = \left(\begin{array}{cccc|c} -Tw_1 & Tb & \dfrac{KSb}{s+a_0} & KSs_c & KS \\ -\dfrac{Sw_1}{s+a_0} & -\dfrac{Sb}{s+a_0} & Tb & \dfrac{KSs_c}{s+a_0} & \dfrac{KS}{s+a_0} \\ -\dfrac{Sw_1}{(s+a_0)^2} & -\dfrac{Sb}{(s+a_0)^2} & \dfrac{-Sb}{s+a_0} & -Ts_c & T \\ -\dfrac{Sw_1}{(s+a_0)^2} & -\dfrac{Sb}{(s+a_0)^2} & \dfrac{-Sb}{s+a_0} & -Ts_c & T \\ \hline -\dfrac{Sw_1}{(s+a_0)^2} & -\dfrac{Sb}{(s+a_0)^2} & \dfrac{-Sb}{s+a_0} & Ss_c & T \end{array}\right) \begin{pmatrix} e_{\Delta 1}(s) \\ e_{\Delta 2}(s) \\ e_{\Delta 3}(s) \\ e_{\Delta 4}(s) \\ \hline e(s) \end{pmatrix} \quad [B.13]$$

$$S = \frac{1}{1+\dfrac{K}{(s+a_0)^2}}, \; T = 1-S$$

$$\Delta(s) = diag\left(\Delta_1(s), \delta, \delta, \varepsilon_c\right), \; \|\Delta(s)\|_\infty < 1$$

The matrix $\Delta(s)$ is diagonal.

Appendix C

LFT Form Use of an Electromechanical System with Uncertain Flexible Modes

C.1. General model and objective

The system is assumed to have an input noted u (homogeneous to a voltage reference or a current reference according to the case) and a measured output noted y (homogeneous to a position or a velocity according to the case).

The general model used for the study is that of Figure C.1. It includes:

a "flexible n modes" type of mechanical model;

$$\frac{\theta}{C_{em}} = \left(\frac{1}{Js^2} + \sum_{i=1}^{n} \frac{a_i}{\frac{s^2}{\omega_i^2} + \frac{2\xi}{\omega_i} s + 1} \right) = H_m(s) \qquad [C.1]$$

– associated with a DC motor:

- either current piloted via a current loop, in this case:

$$\frac{C_{em}}{u} = \frac{K_c}{1+\tau_{el} s} = H_{el}(s) \qquad [C.2]$$

- or directly supplied in voltage[1], in this case:

$$\frac{C_{em}}{u} \approx \frac{K_c}{R+Ls} = H_{el}(s) \qquad [C.3]$$

1 f.c.e.m is assumed to be negligible here.

– associated with a measurement chain:

- either in position, in this case:

$$\frac{y}{\theta} = H_r(s) = H_{mes}(s) \qquad [C.4]$$

- or in velocity, in this case:

$$\frac{y}{\theta} = sH_r(s) = H_{mes}(s) \qquad [C.5]$$

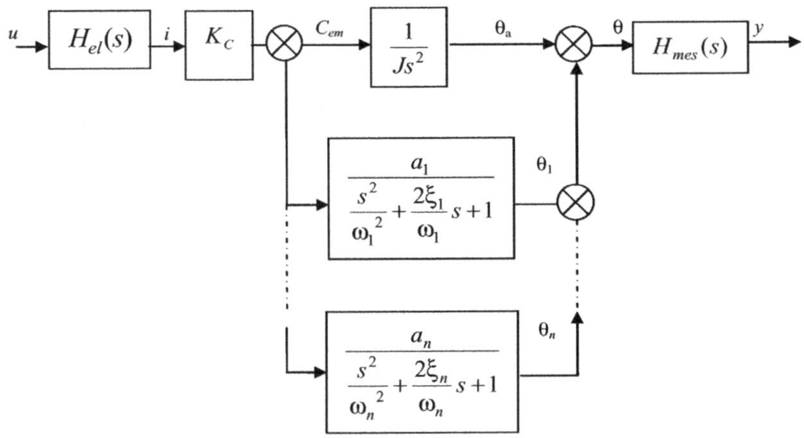

Figure C.1. *Electromechanical system with flexible modes*

For such systems, experience shows that the following parameters vary predominantly:

– the time constant τ_{el} of the current loop or the motor resistance R according to the piloting mode considered in $H_{el}(s)$;

– the contribution factors a_i of the flexible modes;

– the damping of flexible modes ξ_i;

– the flexible mode proper pulsation ω_i.

The objective here is to place this system and its uncertainties under LFT form for the μ-analysis (since the uncertainty matrix is diagonal), that is, under the form of Figure C.2

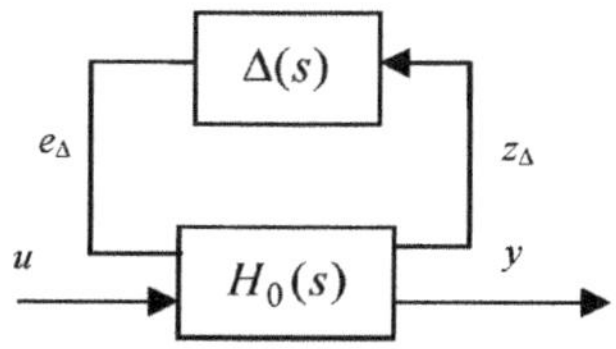

Figure C.2. *LFT form*

C.2. State-space representation of the nominal model

Initially, we establish a parametric state-space representation of the nominal model.

C.2.1. *Motorization*

– Case of current piloted motor (current loop):

$$
\begin{aligned}
&\frac{C_{em}}{u} = \frac{K_c}{1+\tau_{el}s} = H_{el}(s) \\
&\Leftrightarrow \\
&\frac{di}{dt} = -\frac{i(t)}{\tau_{el}} + \frac{i_c}{\tau_{el}} = -\omega_{el}i(t) + \omega_{el}u \\
&C_{em} = K_c i(t) \\
&\Leftrightarrow \\
&\frac{dX_{el}}{dt} = A_{el}X_{el} + B_{el}u \\
&C_{em} = C_{el}X_{el}
\end{aligned}
\qquad [C.6]
$$

– case of voltage piloted motor:

$$
\begin{aligned}
&\frac{C_{em}}{u} = \frac{K_c}{R+Ls} = H_{el}(s) \\
&\Leftrightarrow \\
&\frac{di}{dt} = -\frac{R}{L}i(t) + \frac{u}{L} \\
&C_{em} = K_c i(t) \\
&\Leftrightarrow \\
&\frac{dX_{el}}{dt} = A_{el}X_{el} + B_{el}u \\
&C_{em} = C_{el}X_{el}
\end{aligned}
\qquad [C.7]
$$

C.2.2. *Mechanics*

– Pure inertia:

$$\theta_a = \frac{C_{em}}{Js^2}$$

$$\Leftrightarrow$$

$$\frac{d}{dt}\begin{pmatrix}\theta_a \\ \dot{\theta}_a\end{pmatrix} = \begin{pmatrix}0 & 1 \\ 0 & 0\end{pmatrix}\begin{pmatrix}\theta_a \\ \dot{\theta}_a\end{pmatrix} + \begin{pmatrix}0 \\ J^{-1}\end{pmatrix} C_{em}$$

$$\theta_a = \begin{pmatrix}1 & 0\end{pmatrix}\begin{pmatrix}\theta_a \\ \dot{\theta}_a\end{pmatrix} \quad \text{[C.8]}$$

$$\Leftrightarrow$$

$$\frac{dX_a}{dt} = A_a X_a + B_a C_{em}$$

$$\theta_a = C_a X_a$$

– *i*th flexible mode:

$$\theta_i = \frac{a_i}{\dfrac{s^2}{\omega_i^{\,2}} + \dfrac{2\xi_i}{\omega_i} s + 1} C_{em}$$

$$\Leftrightarrow$$

$$\frac{d}{dt}\begin{pmatrix}\theta_i \\ \dot{\theta}_i\end{pmatrix} = \begin{pmatrix}0 & 1 \\ -\omega_i^{\,2} & -2\xi_i\omega_i\end{pmatrix}\begin{pmatrix}\theta_i \\ \dot{\theta}_i\end{pmatrix} + \begin{pmatrix}0 \\ \omega_i^{\,2} a_i\end{pmatrix} C_{em}$$

$$\theta_i = \begin{pmatrix}1 & 0\end{pmatrix}\begin{pmatrix}\theta_i \\ \dot{\theta}_i\end{pmatrix}$$

$$\Leftrightarrow$$

$$\frac{dX_i}{dt} = A_i X_i + B_i C_{em}$$

$$\theta_i = C_i X_i \quad \text{[C.9]}$$

– State equation of the mechanical model with n flexible modes and pure inertia:

$$\left.\begin{array}{l}\dfrac{dX_a}{dt}=A_aX_a+B_aC_{em}\\ \dfrac{dX_1}{dt}=A_1X_1+B_1C_{em}\\ \qquad\ldots\\ \dfrac{dX_i}{dt}=A_iX_i+B_iC_{em}\\ \qquad\ldots\\ \dfrac{dX_n}{dt}=A_nX_n+B_nC_{em}\end{array}\right\}$$

$$\Leftrightarrow \qquad [C.10]$$

$$\frac{d}{dt}\begin{pmatrix}X_a\\X_1\\\ldots\\X_i\\\ldots\\X_n\end{pmatrix}=\begin{pmatrix}A_a&0&0&0&0&0\\0&A_1&0&0&0&0\\0&0&\ldots&0&0&0\\0&0&0&A_i&0&0\\0&0&0&0&\ldots&0\\0&0&0&0&0&A_n\end{pmatrix}\begin{pmatrix}X_a\\X_1\\\ldots\\X_i\\\ldots\\X_n\end{pmatrix}+\begin{pmatrix}B_a\\B_1\\\ldots\\B_i\\\ldots\\B_n\end{pmatrix}C_{em}\Leftrightarrow\frac{dX_m}{dt}=A_mX_m+B_mC_{em}$$

C.2.3. *Acquisition chain*

$$\frac{y}{\theta}=H_r(s)$$

$$\Leftrightarrow \qquad [C.11]$$

$$\frac{dX_r}{dt}=A_rX_r+B_r\theta=A_rX_r+B_r\left(\theta_a+\sum_i\theta_i\right)$$

$$y-C_rX_r+D_r\theta$$

C.2.4. *Global state-space form*

We have:

$$\left.\begin{array}{l}\dfrac{dX_{el}}{dt}=A_{el}X_{el}+B_{el}U\\ \dfrac{dX_m}{dt}=A_mX_m+B_mC_{em}=B_mC_{el}.X_{el}+A_mX_m\\ \dfrac{dX_r}{dt}=A_rX_r+B_r\theta=B_rC_mX_m+A_rX_r\\ y=C_r.X_r+D_r\theta=C_r.X_r+D_rC_mX_m\end{array}\right\} \qquad [C.12]$$

C.3. LFT form use for uncertainties

The previous nominal state-space representation is subject to parametric variations. The aim is to place the uncertain model under standard form for the μ-analysis.

C.3.1. *Motorization*

– Case of a current piloted motor (current loop) :

We have:

$$\begin{aligned} \frac{di}{dt} &= -\frac{i(t)}{\tau_{el}} + \frac{i_c}{\tau_{el}} = -\omega_{el} i(t) + \omega_{el} u \\ C_{em} &= K_c i(t) \end{aligned} \quad [C.13]$$

For this block, the predominantly varying parameter is the electrical time constant of the current servo-loop, so that[2]:

$$\omega_{el} = \omega_{el0} + \delta\omega_{el}\omega_{el1},\ |\delta\omega_{el}| \leq 1,\ \omega_{el1} = \frac{\omega_{el\max} - \omega_{el\min}}{2} \quad [C.14]$$

such that:

$$\begin{aligned} \frac{di}{dt} &= -\omega_{el} i(t) + \omega_{el} u \\ &= -(\omega_{el0} + \delta\omega_{el}\omega_{el1})i(t) + (\omega_{el0} + \delta\omega_{el}\omega_{el1})u \\ &= -\omega_{el0} i(t) + \omega_{el0} U - \delta\omega_{el}\omega_{el1} i(t) + \delta\omega_{el}\omega_{el1} u \end{aligned} \quad [C.15]$$

or rather[3]:

$$\frac{di}{dt} = -\omega_{el0} i(t) + \omega_{el0} u + \sqrt{\omega_{el1}}\, e_{\Delta el_i} + \sqrt{\omega_{el1}}\, e_{\Delta el_u} \quad [C.16]$$

2 ω_{el0} is assumed to be the center of the variation interval $\omega_{el0} = \frac{\omega_{el\min} + \omega_{el\max}}{2}$.

3 We chose to spread the uncertainty weight equally on $z_{\Delta el}$ and on $e_{\Delta el}$.

with:

$$\begin{aligned} e_{\Delta el_i} &= \delta\omega_{el} z_{\Delta el_i},\ |\delta\omega_{el}| \le 1 \\ z_{\Delta el_i} &= -\sqrt{\omega_{el1}}\, i(t) \\ e_{\Delta el_u} &= \delta\omega_{el} z_{\Delta el_u},\ |\delta\omega_{el}| \le 1 \\ z_{\Delta el_u} &= \sqrt{\omega_{el1}}\, u \end{aligned} \quad [C.17]$$

That is:

$$\frac{dX_{el}}{dt} = A_{el}X_{el} + B_{el}u + \begin{pmatrix} \sqrt{\omega_{el1}} & \sqrt{\omega_{el1}} \end{pmatrix} \begin{pmatrix} e_{\Delta el_i} \\ e_{\Delta el_u} \end{pmatrix}$$

$$\begin{pmatrix} z_{\Delta el_i} \\ z_{\Delta el_u} \end{pmatrix} = \begin{pmatrix} -\sqrt{\omega_{el1}} \\ 0 \end{pmatrix} X_{el} + \begin{pmatrix} 0 \\ \sqrt{\omega_{el1}} \end{pmatrix} u \quad [C.18]$$

$$C_{em} = C_{el}X_{el}$$

$$\begin{aligned} e_{\Delta el_i} &= \delta\omega_{el} z_{\Delta el_i},\ |\delta\omega_{el}| \le 1 \\ e_{\Delta el_u} &= \delta\omega_{el} z_{\Delta el_u},\ |\delta\omega_{el}| \le 1 \end{aligned}$$

or rather noted:

$$\begin{aligned} \frac{dX_{el}}{dt} &= A_{el}X_{el} + B_{el}u + M_{el}e_{\Delta el} \\ z_{\Delta el} &= N_{el}X_{el} + O_{el}u \\ C_{em} &= C_{el}X_{el} \end{aligned} \quad [C.19]$$

$$e_{\Delta el} = \Delta_{el} z_{\Delta el},\ \|\Delta_{el}\|_\infty \le 1$$

– Case of voltage piloted motor:

Recall that:

$$\begin{aligned} \frac{di}{dt} &= -\frac{R}{L}i(t) + \frac{u}{L} \\ C_{em} &= K_c i(t) \end{aligned} \quad [C.20]$$

For this block, the predominantly varying parameter is the electrical resistance of the motor, such that $R \in [R_{\min}; R_{\max}]$ and thus [4]

$$R = R_0 + \delta\omega_R R_1, |\delta R| \leq 1, \; R_1 = \frac{R_{\max} - R_{\min}}{2} \qquad [C.21]$$

We reach:

$$\begin{aligned} \frac{di}{dt} &= -\frac{R}{L} i(t) + \frac{u}{L} \\ &= -\frac{1}{L}(R_0 + \delta\omega_R R_1) i(t) + \frac{u}{L} \\ &= -\frac{R_0}{L} i(t) - \delta\omega_R \frac{R_1}{L} i(t) + \frac{u}{L} \end{aligned} \qquad [C.22]$$

or rather[5]:

$$\frac{di}{dt} = -\frac{R_0}{L} i(t) + \frac{u}{L} + \sqrt{\frac{R_1}{L}} e_{\Delta el_i} \qquad [C.23]$$

with:

$$\begin{aligned} e_{\Delta el_i} &= \delta R z_{\Delta el_i}, \; |\delta R| \leq 1 \\ z_{\Delta el_i} &= -\sqrt{\frac{R_1}{L}} i(t) \end{aligned} \qquad [C.24]$$

That is:

$$\begin{aligned} \frac{dX_{el}}{dt} &= A_{el} X_{el} + B_{el} u + \sqrt{\frac{R_1}{L}} e_{\Delta el_i} \\ z_{\Delta el_i} &= -\sqrt{\frac{R_1}{L}} X_{el} \\ C_{em} &= C_{el} X_{el} \\ e_{\Delta el_i} &= \delta R z_{\Delta el_i}, \; |\delta R| \leq 1 \end{aligned} \qquad [C.25]$$

4 R_0 is assumed to be the center of the variation interval $R_0 = \frac{R_{\min} + R_{\max}}{2}$.

5 We chose to spread the uncertainty weight equally on $w_{\delta el}$ and on $v_{\delta el}$.

or rather, noted:

$$\begin{aligned} &\frac{dX_{el}}{dt} = A_{el}X_{el} + B_{el}u + M_{el}e_{\Delta el} \\ &z_{\Delta el} = N_{el}X_{el} \\ &C_{em} = C_{el}X_{el} \\ &e_{\Delta el} = \Delta_{el}z_{\Delta el},\ \left\|\Delta_{el}\right\|_{\infty} \leq 1 \end{aligned} \qquad [C.26]$$

The input u is homogeneous to a control voltage

C.3.2. *Mechanics*

– Pure inertia:

For pure inertia, the predominantly varying parameter is pure inertia J, such that $J \in [J_{\min}; J_{\max}]$ and consequently $\beta = J^{-1} \in \left[J_{\max}{}^{-1}; J_{\min}{}^{-1}\right] = [\beta_{\min}; \beta_{\max}]$ and thus $\beta = \beta_0 + \delta\beta\,\beta_1$, $|\delta\beta| \leq 1$, $\beta_1 = \dfrac{\beta_{\max} - \beta_{\min}}{2}$, $\beta_0 = \dfrac{\beta_{\max} + \beta_{\min}}{2}$

$$\begin{aligned} \frac{d}{dt}\begin{pmatrix}\theta_a \\ \dot{\theta}_a\end{pmatrix} &= \begin{pmatrix}0 & 1 \\ 0 & 0\end{pmatrix}\begin{pmatrix}\theta_a \\ \dot{\theta}_a\end{pmatrix} + \begin{pmatrix}0 \\ \beta_0 + \delta\beta\,\beta_1\end{pmatrix}C_{em} \\ &= \begin{pmatrix}0 & 1 \\ 0 & 0\end{pmatrix}\begin{pmatrix}\theta_a \\ \dot{\theta}_a\end{pmatrix} + \begin{pmatrix}0 \\ \beta_0\end{pmatrix}C_{em} + \delta\beta\begin{pmatrix}0 \\ \beta_1\end{pmatrix}C_{em} \\ \theta_a &= \begin{pmatrix}1 & 0\end{pmatrix}\begin{pmatrix}\theta_a \\ \dot{\theta}_a\end{pmatrix} \end{aligned} \qquad [C.27]$$

We write $e_{\Delta J} = \begin{pmatrix}0 \\ \sqrt{\beta_1}\end{pmatrix}\delta\beta.z_{\Delta J}$, $z_{\Delta J} = \sqrt{\beta_1}C_{em} = O_aC_{em}$

and so:

$$\begin{aligned} &\frac{dX_a}{dt} = A_aX_a + B_aC_{em} + M_ae_{\Delta J}, M_a = \begin{pmatrix}0 \\ \sqrt{\beta_1}\end{pmatrix} \\ &z_{\Delta J} = N_aX_a + O_aC_{em},\ O_a = \sqrt{\beta_1},\ N_a = 0 \\ &\theta_a = C_aX_a \\ &e_{\Delta J} = \delta\beta.z_{\Delta J} \end{aligned} \qquad [C.28]$$

– ith flexible mode:

For the flexible modes, the predominantly varying parameters are:

– ξ_i such that $\xi_i \in [\xi_{i\min};\xi_{i\max}]$ and:

$$\xi_i = \xi_{i0} + \delta\xi_i\,\xi_{i1},\ |\delta\xi_i| \le 1,\ \xi_{i1} = \frac{\xi_{i\max} - \xi_{i\min}}{2}$$

– a_i such that $a_i \in [a_{i\min};a_{i\max}]$ and:

$$a_i = a_{i0} + \delta a_i a_{i1},\ |\delta a_i| \le 1,\ a_{i1} = \frac{a_{i\max} - a_{i\min}}{2}$$

– ω_i such that $\omega_i \in [\omega_{i\min};\omega_{i\max}]$ and:

$$\omega_i = \omega_{i0} + \delta\omega_i\omega_{i1},\ |\delta\omega_i| \le 1,\ \omega_{i1} = \frac{\omega_{i\max} - \omega_{i\min}}{2}$$

According to the state equation:

$$\begin{aligned}
\frac{d\dot{\theta}_i}{dt} &= -\omega_i^2\theta_i - 2\xi_i\omega_i\dot{\theta}_i + a_i\omega_i^2 C_{em} \\
&= -(\omega_{i0} + \delta\omega_i\omega_{i1})^2\theta_i - 2(\xi_{i0} + \delta\xi_i\xi_{i1})(\omega_{i0} + \delta\omega_i\omega_{i1})\dot{\theta}_i \\
&\quad + (\omega_{i0} + \delta\omega_i\omega_{i1})^2(a_{i0} + \delta a_i a_{i1})C_{em} \\
&= -\left(\omega_{i0}^2 + 2.\delta\omega_i.\omega_{i0}\omega_{i1} + (\delta\omega_i\omega_{i1})^2\right)\theta_i \\
&\quad -2(\xi_{i0}\omega_{i0} + \xi_{i0}\delta\omega_i\omega_{i1} + \delta\xi_i\xi_{i1}\omega_{i0} + \delta\xi_i\xi_{i1}\delta\omega_i\omega_{i1})\dot{\theta}_i \\
&\quad +\left(\omega_{i0}^2 + 2.\delta\omega_i.\omega_{i0}\omega_{i1} + (\delta\omega_i\omega_{i1})^2\right)a_{i0}C_{em} \\
&\quad +\left(\omega_{i0}^2 + 2.\delta\omega_i.\omega_{i0}\omega_{i1} + (\delta\omega_i\omega_{i1})^2\right)\delta a_i a_{i1}C_{em}
\end{aligned} \qquad [C.29]$$

Accordingly, assuming that the terms of order 2 relative to the parametric variations are negligible:

$$\begin{cases} \delta\xi_i < 1 \\ \delta\omega_i < 1 \\ \delta a_i < 1 \end{cases} \Rightarrow \begin{cases} \delta\xi_i\delta\omega_i << 1 \\ \delta\omega_i^2 << 1 \\ \delta a_i\delta\omega_i << 1 \end{cases} \qquad [C.30]$$

Therefore, we can proceed to the following simplification:

$$\begin{aligned}\frac{d\dot{\theta}_i}{dt} &\approx -\left(\omega_{i0}{}^2 + 2\delta\omega_i\omega_{i0}\omega_{i1}\right)\theta_i \\ &-2\left(\xi_{i0}\omega_{i0} + \xi_{i0}\delta\omega_i\omega_{i1} + \delta\xi_i\xi_{i1}\omega_{i0}\right)\dot{\theta}_i \\ &+\left(\omega_{i0}{}^2 a_{i0} + 2a_{i0}\delta\omega_i\omega_{i0}\omega_{i1} + \omega_{i0}\delta a_i a_{i1}\right)C_{em}\end{aligned} \quad [C.31]$$

That is:

$$\begin{aligned}\frac{d\dot{\theta}_i}{dt} &\approx -\omega_{i0}{}^2\theta_i - 2\xi_{i0}\omega_{i0}\,\dot{\theta}_i + a_{i0}\omega_{i0}{}^2 C_{em} \\ &-\delta\omega_i 2\omega_{i1}\left(\omega_{i0}\theta_i + \xi_{i0}\,\dot{\theta}_i\right) - \delta\xi_i\left(2\xi_{i1}\omega_{i0}\right)\dot{\theta}_i \\ &+\delta\omega_i\left(a_{i0}\omega_{i0}2\omega_{i1}\right)C_{em} + \delta a_i\omega_{i0}a_{i1}C_{em}\end{aligned} \quad [C.32]$$

We then write[6]:

$$\begin{aligned} e_{\Delta\omega_i\Theta} &= \delta\omega_i z_{\Delta\omega_i\Theta} \\ z_{\Delta\omega_i\Theta} &= -\sqrt{2\omega_{i1}}\left(\omega_{i0}\theta_i + \xi_{i0}\,\dot{\theta}_i\right) \\[1em] e_{\Delta\omega_i C} &= \delta\omega_i z_{\Delta\omega_i C} \\ z_{\Delta\omega_i C} &= \sqrt{a_{i0}\omega_{i0}2\omega_{i1}}\,C_{em} \\[1em] e_{\Delta\xi_i} &= \delta\xi_i z_{\Delta\xi_i} \\ z_{\Delta\xi_i} &= -\sqrt{\left(2\xi_{i1}\omega_{i0}\right)}\,\dot{\theta}_i \\[1em] e_{\Delta a_i} &= \delta a_i z_{\Delta a_i} \\ z_{\Delta a_i} &= \sqrt{\omega_{i0}a_{i1}}\,C_{em}\end{aligned} \quad [C.33]$$

6 We chose to spread the uncertainty weight equally on z_Δ and on e_Δ .

such that:

$$\begin{aligned}\frac{d\dot{\theta}_i}{dt} &\approx -\omega_{i0}{}^2\theta_i - 2\xi_{i0}\omega_{i0}\dot{\theta}_i + \omega_{i0}{}^2 C_{em} \\ &+\sqrt{2\omega_{i1}}e_{\Delta\omega_i\Theta} + \sqrt{(2\xi_{i1}\omega_{i0})}e_{\Delta\xi_i} \\ &+\sqrt{a_{i0}\omega_{i0}2\omega_{i1}}e_{\Delta\omega_i C} + \sqrt{\omega_{i0}a_{i1}}e_{\Delta a_i}\end{aligned} \quad [C.34]$$

and, finally, under matricial form:

$$\frac{dX_i}{dt} = A_i X_i + B_i C_{em} + \begin{pmatrix} 0 & 0 & 0 & 0 \\ \sqrt{2\omega_{i1}} & \sqrt{a_{i0}\omega_{i0}2\omega_{i1}} & \sqrt{2\xi_{i1}\omega_{i0}} & \sqrt{\omega_{i0}a_{i1}} \end{pmatrix} \begin{pmatrix} e_{\Delta\omega_i\Theta} \\ e_{\Delta\omega_i C} \\ e_{\Delta\xi_i} \\ e_{\Delta a_i} \end{pmatrix}$$

$$\begin{pmatrix} z_{\Delta\omega_i\Theta} \\ z_{\Delta\omega_i C} \\ z_{\Delta\xi_i} \\ z_{\Delta a_i} \end{pmatrix} = \begin{pmatrix} -\sqrt{2\omega_{i1}}\omega_{i0} & -\sqrt{2\omega_{i1}}\xi_{i0} \\ 0 & 0 \\ 0 & -\sqrt{2\xi_{i1}\omega_{i0}} \\ 0 & 0 \end{pmatrix} X_i + \begin{pmatrix} 0 \\ \sqrt{a_{i0}\omega_{i0}2\omega_{i1}} \\ 0 \\ \sqrt{\omega_{i0}a_{i1}} \end{pmatrix} C_{em}$$

$$\theta_i = C_i X_i \quad [C.35]$$

$$\begin{aligned} e_{\Delta\xi_i} &= \delta\xi_i z_{\Delta\xi_i}, & |\delta\xi_i| \le 1 \\ e_{\Delta a_i} &= \delta a_i z_{\Delta a_i}, & |\delta a_i| \le 1 \\ z_{\Delta\omega_i\Theta} &= \delta\omega_i e_{\Delta\omega_i\Theta}, & |\delta\omega_i| \le 1 \\ z_{\Delta\omega_i C} &= \delta\omega_i e_{\Delta\omega_i C}, & |\delta\omega_i| \le 1 \end{aligned}$$

or rather:

$$\frac{dX_i}{dt} = A_i X_i + B_i C_{em} + M_i \begin{pmatrix} e_{\Delta\omega_i\Theta} \\ e_{\Delta\omega_i C} \\ e_{\Delta\xi_i} \\ e_{\Delta a_i} \end{pmatrix}$$

$$\theta_i = C_i X_i$$

$$\begin{pmatrix} z_{\Delta\omega_i\Theta} \\ z_{\Delta\omega_i C} \\ z_{\Delta\xi_i} \\ z_{\Delta a_i} \end{pmatrix} = N_i X_i + O_i C_{em}$$

$$\begin{aligned} e_{\Delta\xi_i} &= \delta\xi_i z_{\Delta\xi_i}, \quad |\delta\xi_i| \le 1 \\ e_{\Delta a_i} &= \delta a_i z_{\Delta a_i}, \quad |\delta a_i| \le 1 \\ z_{\Delta\omega_i\Theta} &= \delta\omega_i e_{\Delta\omega_i\Theta}, \quad |\delta\omega_i| \le 1 \\ z_{\Delta\omega_i C} &= \delta\omega_i e_{\Delta\omega_i C}, \quad |\delta\omega_i| \le 1 \end{aligned} \qquad [C.36]$$

$$N_i = \begin{pmatrix} -\sqrt{2\omega_{i1}}\,\omega_{i0} & -\sqrt{2\omega_{i1}}\,\xi_{i0} \\ 0 & 0 \\ 0 & -\sqrt{2\xi_{i1}\omega_{i0}} \\ 0 & 0 \end{pmatrix}, \quad O_i = \begin{pmatrix} 0 \\ \sqrt{a_{i0}\omega_{i0}2\omega_{i1}} \\ 0 \\ \sqrt{\omega_{i0}a_{i1}} \end{pmatrix},$$

$$M_i = \begin{pmatrix} 0 & 0 & 0 & 0 \\ \sqrt{2\omega_{i1}} & \sqrt{a_{i0}\omega_{i0}2\omega_{i1}} & \sqrt{2\xi_{i1}\omega_{i0}} & \sqrt{\omega_{i0}a_{i1}} \end{pmatrix}$$

– State-space equation of the uncertain mechanical model with n flexible modes and pure inertia:

$$\frac{d}{dt}\begin{pmatrix} X_a \\ X_1 \\ X_i \\ X_n \end{pmatrix} = \begin{pmatrix} A_a & 0 & 0 & 0 \\ 0 & A_1 & 0 & 0 \\ 0 & 0 & A_i & 0 \\ 0 & 0 & 0 & A_n \end{pmatrix}\begin{pmatrix} X_a \\ X_1 \\ X_i \\ X_n \end{pmatrix} + \begin{pmatrix} B_a \\ B_1 \\ B_i \\ B_n \end{pmatrix} C_{em} + \begin{pmatrix} M_a & 0 & 0 & 0 \\ 0 & M_1 & 0 & 0 \\ 0 & 0 & M_i & 0 \\ 0 & 0 & 0 & M_n \end{pmatrix}\begin{pmatrix} e_{\Delta J} \\ \begin{bmatrix} e_{\Delta\omega_1\Theta} \\ e_{\Delta\omega_1 C} \\ e_{\Delta\xi_1} \\ e_{\Delta a_1} \end{bmatrix} \\ \begin{bmatrix} e_{\Delta\omega_i\Theta} \\ e_{\Delta\omega_i C} \\ e_{\Delta\xi_i} \\ e_{\Delta a_i} \end{bmatrix} \\ \begin{bmatrix} e_{\Delta\omega_n\Theta} \\ e_{\Delta\omega_n C} \\ e_{\Delta\xi_n} \\ e_{\Delta a_n} \end{bmatrix} \end{pmatrix}$$

$$\begin{pmatrix} z_{\Delta J} \\ \begin{bmatrix} z_{\Delta\omega_1\Theta} \\ z_{\Delta\omega_1 C} \\ z_{\Delta\xi_1} \\ z_{\Delta a_1} \end{bmatrix} \\ \begin{bmatrix} z_{\Delta\omega_i\Theta} \\ z_{\Delta\omega_i C} \\ z_{\Delta\xi_i} \\ z_{\Delta a_i} \end{bmatrix} \\ \begin{bmatrix} z_{\Delta\omega_n\Theta} \\ z_{\Delta\omega_n C} \\ z_{\Delta\xi_n} \\ z_{\Delta a_n} \end{bmatrix} \end{pmatrix} = \begin{pmatrix} N_a & 0 & 0 & 0 \\ 0 & N_1 & 0 & 0 \\ 0 & 0 & N_1 & 0 \\ 0 & 0 & 0 & N_n \end{pmatrix}\begin{pmatrix} X_a \\ X_1 \\ X_i \\ X_n \end{pmatrix} + \begin{pmatrix} O_a \\ O_1 \\ O_i \\ O_n \end{pmatrix} C_{em} \qquad [C.37]$$

$$
\begin{pmatrix} e_{\Delta J} \\ \begin{bmatrix} e_{\Delta\omega_1\Theta} \\ e_{\Delta\omega_1 C} \\ e_{\Delta\xi_1} \\ e_{\Delta a_1} \end{bmatrix} \\ \begin{bmatrix} e_{\Delta\omega_i\Theta} \\ e_{\Delta\omega_i C} \\ e_{\Delta\xi_i} \\ e_{\Delta a_i} \end{bmatrix} \\ \begin{bmatrix} e_{\Delta\omega_n\Theta} \\ e_{\Delta\omega_n C} \\ e_{\Delta\xi_n} \\ e_{\Delta a_n} \end{bmatrix} \end{pmatrix} = diag\left(\delta\beta, \begin{pmatrix}
\delta\omega_1 & 0 & 0 & 0 & 0 & 0 & 0 & 0 & 0 & 0 & 0 & 0 \\
0 & \delta\omega_1 & 0 & 0 & 0 & 0 & 0 & 0 & 0 & 0 & 0 & 0 \\
0 & 0 & \delta\xi_1 & 0 & 0 & 0 & 0 & 0 & 0 & 0 & 0 & 0 \\
0 & 0 & 0 & \delta a_1 & 0 & 0 & 0 & 0 & 0 & 0 & 0 & 0 \\
0 & 0 & 0 & 0 & \delta\omega_i & 0 & 0 & 0 & 0 & 0 & 0 & 0 \\
0 & 0 & 0 & 0 & 0 & \delta\omega_i & 0 & 0 & 0 & 0 & 0 & 0 \\
0 & 0 & 0 & 0 & 0 & 0 & \delta\xi_i & 0 & 0 & 0 & 0 & 0 \\
0 & 0 & 0 & 0 & 0 & 0 & 0 & \delta a_i & 0 & 0 & 0 & 0 \\
0 & 0 & 0 & 0 & 0 & 0 & 0 & 0 & \delta\omega_n & 0 & 0 & 0 \\
0 & 0 & 0 & 0 & 0 & 0 & 0 & 0 & 0 & \delta\omega_n & 0 & 0 \\
0 & 0 & 0 & 0 & 0 & 0 & 0 & 0 & 0 & 0 & \delta\xi_n & 0 \\
0 & 0 & 0 & 0 & 0 & 0 & 0 & 0 & 0 & 0 & 0 & \delta a_n
\end{pmatrix} \right) \begin{pmatrix} z_{\Delta J} \\ \begin{bmatrix} z_{\Delta\omega_1\Theta} \\ z_{\Delta\omega_1 C} \\ z_{\Delta\xi_1} \\ z_{\Delta a_1} \end{bmatrix} \\ \begin{bmatrix} z_{\Delta\omega_i\Theta} \\ z_{\Delta\omega_i C} \\ z_{\Delta\xi_i} \\ z_{\Delta a_i} \end{bmatrix} \\ \begin{bmatrix} z_{\Delta\omega_n\Theta} \\ z_{\Delta\omega_n C} \\ z_{\Delta\xi_n} \\ z_{\Delta a_n} \end{bmatrix} \end{pmatrix}
$$

$$
\begin{aligned}
&|\delta\beta| \le 1, \\
&|\delta\xi_i| \le 1, \quad i = 1 \ldots n \\
&|\delta a_i| \le 1, \quad i = 1 \ldots n \\
&|\delta\omega_i| \le 1, \quad i = 1 \ldots n
\end{aligned}
$$

which, for simplicity, we note as:

$$\frac{dX_m}{dt} = A_m X_m + B_m C_{em} + M_m e_{\Delta m}$$

$$z_{\Delta m} = N_m . X_m + O_m C_{em}$$

$$e_{\Delta m} = \Delta_m z_{\Delta m}, \quad \|\Delta_m\|_\infty \le 1 \tag{C.38}$$

$$
\begin{aligned}
\theta &= \theta_a + \theta_1 + \ldots . + \theta_n \\
&= C_m X_m
\end{aligned}
$$

C.3.3. *Global LFT form*

By collecting all of the blocks and the acquisition chain block, in the general case (that of the current piloted motor), we have:

$$\begin{aligned}
\frac{dX_{el}}{dt} &= A_{el}X_{el} + B_{el}u + M_{el}e_{\Delta el} \\
\frac{dX_m}{dt} &= A_m X_m + B_m C_{el} X_{el} + M_m e_{\Delta m} \\
\frac{dX_r}{dt} &= A_r X_r + B_r C_m X_m
\end{aligned} \quad [C.39]$$

$$\begin{aligned}
z_{\Delta el} &= N_{el}X_{el} + O_{el}u \\
z_{\Delta m} &= N_m X_m + O_m C_{em} = N_m X_m + O_m C_{el} X_{el}
\end{aligned}$$

$$y = C_r X_r + D_r \theta = C_r X_r + D_r C_m X_m$$

For the case of the voltage piloted motor, we can take $O_{el} = 0$.

That is, ultimately:

$$\frac{d}{dt}\begin{pmatrix} X_{el} \\ X_m \\ X_r \end{pmatrix} = \begin{pmatrix} A_{el} & 0 & 0 \\ B_m C_{el} & A_m & 0 \\ 0 & B_r C_m & A_r \end{pmatrix}\begin{pmatrix} X_{el} \\ X_m \\ X_r \end{pmatrix} + \begin{pmatrix} M_{el} & 0 \\ 0 & M_m \\ 0 & 0 \end{pmatrix}\begin{pmatrix} e_{\Delta el} \\ e_{\Delta m} \end{pmatrix} + \begin{pmatrix} B_{el} \\ 0 \\ 0 \end{pmatrix} u$$

$$y = C_r X_r + D_r C_m X_m$$

$$\begin{pmatrix} z_{\Delta el} \\ z_{\Delta m} \end{pmatrix} = \begin{pmatrix} N_{el} & 0 & 0 \\ O_m C_{el} & N_m & 0 \end{pmatrix}\begin{pmatrix} X_{el} \\ X_m \\ X_r \end{pmatrix} + \begin{pmatrix} O_{el} \\ 0 \end{pmatrix} u \quad [C.40]$$

$$\begin{pmatrix} e_{\Delta el} \\ e_{\Delta m} \end{pmatrix} = \begin{pmatrix} \Delta_{el} & 0 \\ 0 & \Delta_m \end{pmatrix}\begin{pmatrix} z_{\Delta el} \\ z_{\Delta m} \end{pmatrix}$$

which, for simplicity, we note as:

$$\begin{aligned} \frac{dX}{dt} &= AX + Bu + Me_\Delta \\ y &= CX \\ z_\Delta &= NX + Ou \\ e_\Delta &= \Delta z_\Delta, \quad \|\Delta\|_\infty \leq 1 \end{aligned} \qquad [C.41]$$

We observe that the uncertain output w_δ cannot be strictly proper as it is dependent on u. To make this uncertain system strictly proper, we must add a high-frequency fictitious input dynamic at input plant, such that:

$$u = \frac{u'}{1+\tau_F s} \Rightarrow u + \tau_F \frac{du}{dt} = u' \Rightarrow \frac{du}{dt} = \frac{u'}{\tau_F} - \frac{u}{\tau_F} \qquad [C.42]$$

which allows us to augment the state-space representation of the new state variable u':

$$\begin{aligned} \frac{dX}{dt} &= AX + Bu + Me_\Delta \\ \frac{du}{dt} &= -\frac{u}{\tau_F} + \frac{u'}{\tau_F} \\ y &= CX \\ z_\Delta &= NX + Ou \\ e_\Delta &= \Delta z_\Delta, \quad \|\Delta\|_\infty \leq 1 \end{aligned} \qquad [C.43]$$

or rather:

$$\begin{aligned} \frac{d}{dt}\begin{pmatrix} X \\ u \end{pmatrix} &= \begin{pmatrix} A & B \\ 0 & -\tau_F^{-1} \end{pmatrix}\begin{pmatrix} X \\ u \end{pmatrix} + \begin{pmatrix} M \\ 0 \end{pmatrix} e_\Delta + \begin{pmatrix} 0 \\ \tau_F^{-1} \end{pmatrix} u' \\ y &= \begin{pmatrix} C & 0 \end{pmatrix}\begin{pmatrix} X \\ u \end{pmatrix} \\ z_\Delta &= \begin{pmatrix} N & O \end{pmatrix}\begin{pmatrix} X \\ u \end{pmatrix} \\ e_\Delta &= \Delta z_\Delta, \quad \|\Delta\|_\infty \leq 1 \end{aligned} \qquad [C.44]$$

Appendix D

FTM (1D) Computation from a Time Signal

D.1. Computation principle

Let us consider having a time signal from residual stabilization $\theta_v(t)$ generating blur. The aim is to determine an FTM (at one dimension[1]) resulting from this time signal for a given integration time T_i. If the time acquisitions have a global duration of t_{max}, then we can temporally segment it into N_f segments from each of which an FTM can be computed.

$$N_f = floor\left(\frac{t_{\max}}{T_i}\right) + 1 \qquad [D.1]$$

From this sample of N_f FTMs, we can deduce a mean FTM, an FTM above 70% of the sample FTMs, etc., according to the optronic needs.

By knowing that the time acquisitions are sampled with sample time T_e, each segment of duration T_i will contain N_p points:

$$N_p = floor\left(\frac{T_i}{T_e}\right) + 1 \qquad [D.2]$$

The resulting FTM (1D) blur computation consists of performing the following two steps:

– Computation of the 1D PSF.

1 We can show that the calculation of a 2D FTM from elevation and azimuth data followed by x and y projection in order to obtain two FTMs (one for the elevation and one for the azimuth) is strictly equivalent to the calculations of a 1D FTM from the elevation axis and the azimuth axis data.

PSF is to optronics engineers what the impulse response is to automation engineers.

We now consider one of the N_f segments of residual stabilization (so of size N_p). We write:

$$\bar{\varepsilon} = \max(\theta_v(t)), \underline{\varepsilon} = \min(\theta_v(t)) \quad [D.3]$$

We consider a PSF vector of size n. We construct n points of reference, with spatial coordinates uniformly spaced and covering the interval $\left[\bar{\varepsilon};\underline{\varepsilon}\right]$:

$$x_{ref} = \left[\underline{\varepsilon} - b\left(\bar{\varepsilon} - \underline{\varepsilon}\right), \ldots, \bar{\varepsilon} - b\left(\bar{\varepsilon} - \underline{\varepsilon}\right)\right] \quad [D.4]$$

where b is an integer coefficient allowing an increase of the FTM resolution (principle of time-frequency uncertainty). We can choose $b = 4$, for instance. Usually, we choose $n = 100$ points.

We then compute the coordinate of the closest PSF matrix to that of the measured point:

$$x_{ref}(i) < \theta_v[k] \leq x_{ref}(i+1),\ k \in \left[1, N_p\right] \Rightarrow PSF_{1D}(i) = PSF_{1D}(i) + 1,\ i \in [1, n] \quad [D.5]$$

The PSF is finally normalized:

$$PSF_{1D}(i) = \frac{PSF_{1D}(i)}{\sum_{k=1}^{n} PSF_{1D}(k)},\ i \in [1, n] \quad [D.6]$$

– Computation of the 1D FTM.

FTM is for optronics engineers what the transmission is for automation engineers

FTM is obtained by the gain of the Fourier 1D transform from the PSF, i.e.:

$$FTM_{1D}(\nu) = \left|\sum_{k=1}^{n} e^{-2\pi j\nu} PSF_{1D}(k)\right| \quad [D.7]$$

Appendix E

Choice of Iteration Number for CompLeib Tests

Test name	*K degree*	*Number of iterations*
AC1	0	250
AC1	2	250
AC2	0	125
AC2	2	250
AC3	0	500
AC3	2	1,000
AC4	0	250
AC4	2	1,000
AC5	0	250
AC5	2	1,000
AC6	0	500
AC6	2	2,000
AC6	4	2,000
AC7	0	250
AC7	3	1,500
AC7	5	1,500
AC8	0	500
AC8	3	1,500
AC8	5	1,500
AC9	0	1,000

AC9	3	1,500
AC9	5	2,000
AC9	8	3,000
AC10	0	500
AC10	3	10,000
AC10	10	10,000
AC10	20	10,000
AC11	0	500
AC11	2	500
AC12	0	500
AC12	2	1,000
AC13	0	500
AC13	2	500
AC13	7	1,000
AC13	10	1,500
AC13	20	2,000
AC14	0	500
AC14	2	500
AC14	10	1,000
AC15	0	250
AC15	2	500
AC16	0	250
AC16	2	250
AC17	0	250
AC17	2	250
AC18	0	1,000
AC18	2	1,500
AC18	5	500
HE1	0	4,000
HE1	2	4,000
HE2	0	4,000
HE2	2	4,000
HE3	0	4,000

HE3	2	4,000
HE3	4	4,000
HE4	0	4,000
HE4	2	4,000
HE4	4	4,000
HE5	0	4,000
HE5	2	4,000
HE5	4	4,000
HE6	0	4,000
HE6	2	4,000
HE6	8	4,000
HE6	15	4,000
HE7	0	4,000
HE7	2	4,000
HE7	8	4,000
HE7	10	4,000
HE7	15	4,000
JE1	0	500
JE1	2	3,000
JE1	8	3,000
JE2	0	2,000
JE2	2	2,000
JE2	8	2,000
JE2	10	2,000
JE2	15	2,000
JE3	0	250
JE3	2	1,000
JE3	8	2,000
DIS1	0	500
DIS1	2	1,000
DIS1	4	2,000
DIS2	0	1,000
DIS2	2	2,000

DIS3	0	1,000
DIS3	2	2,000
DIS4	0	500
DIS4	2	1,000
DIS5	0	500
DIS5	2	1,000
EB1	0	250
EB1	2	250
EB1	4	250
EB2	0	250
EB2	2	250
EB2	4	250
EB3	0	250
EB3	2	250
EB3	4	250
NN1	0	250
NN1	2	250
NN2	0	250
NN2	2	250
NN3	0	10,000
NN3	2	10,000
NN4	0	250
NN4	2	250
NN5	0	500
NN5	2	1,000
NN5	4	2,000
NN6	0	500
NN6	2	1,000
NN6	4	2,000
NN7	0	500
NN7	3	1,000
NN7	5	2,000
NN8	0	250

NN8	3	250
NN9	0	250
NN9	3	2,000
NN11	0	250
NN11	2	1,000
NN11	6	4,000
NN12	0	1,000
NN12	2	2,000
NN13	0	500
NN13	2	1,000
NN14	0	1,000
NN14	2	2,000
NN15	0	250
NN15	2	250
NN16	0	250
NN16	2	1,500
NN16	4	3,000
NN17	0	250
NN17	2	250
ROC1	1	500
ROC1	3	1,000
ROC1	5	2,000
ROC2	1	500
ROC2	3	1,000
ROC2	5	2,000
ROC3	1	10,000
ROC3	3	10,000
ROC3	5	10,000
ROC4	1	500
ROC4	3	1,000
ROC4	5	2,000
ROC5	1	500
ROC5	3	1,000

ROC5	5	2,000
HF2D10	0	250
HF2D10	2	500
HF2D11	0	250
HF2D11	2	500
HF2D12	0	250
HF2D12	2	500
HF2D13	0	250
HF2D13	2	500
JE3	10	3,000
WEC1	0	500
WEC1	2	1,000
WEC1	4	2,000
WEC1	9	4,000
WEC2	0	500
WEC2	2	1,000
WEC2	4	2,000
TG1	0	250
TG1	2	500
TG1	4	1,000
AGS	0	250
AGS	2	250
AGS	4	250
PAS	0	500
PAS	2	1,000
TMD	0	250
TMD	2	250
TMD	4	500
CM1	0	250
CM1	2	250
CM1	7	250
IH	0	10,000
IH	2	10,000

IH	8	10,000
IH	10	10,000
IH	15	10,000
TF1	0	250
TF1	2	1,000
TF1	4	2,000
TF2	0	250
TF2	2	250
TF2	4	250
TF3	0	500
TF3	2	1,000
TF3	4	2,000
CSE1	0	250
CSE1	2	250
CSE1	8	250
CSE2	0	250
CSE2	2	250
CSE2	10	250
UWV	0	250
UWV	2	250
UWV	4	250
EB1	0	250
EB1	2	250
EB1	4	250
EB2	0	250
EB2	2	250
EB2	4	250
EB3	0	250
EB3	2	250
EB3	4	250
CM1	0	250
CM1	3	250
CM1	8	250

CM2	0	250
CM2	4	250
CM2	11	250
CM3	0	250
CM3	5	250
CM3	14	500
CM4	0	250
REA1	0	250
REA1	2	250
REA2	0	250
REA2	2	250
REA3	0	250
REA3	2	250
REA3	5	250
REA3	10	250

Appendix F

PDE versus DE

The following tables compare DE and PDE for the DE/Rand1/, DE/Rand-to-best/1 and DE/Best/1 mutation schemes for several problems posed in [SUG 05]. The settings used are:

– dimension $n = 10$, population size $N = 5n$;

– each test comprises 50 successive runs;

– the only stop criterion envisaged is:

$$\exists X_{best} : \frac{\max_{i=1,\ldots,P} \left\| X_i^{(k)} - X_{best} \right\|}{\left\| X_{max} - X_{min} \right\|} < 1.10^{-5} \qquad \text{[F.1]}$$

– $F = 0.75$, $C_r = 0.8$, $F' = 0.25$, $p_m = 0.025$

For the Rand/1 mutation scheme, we see that:

– The lowest value reached by PDE and by DE at the end of 50 runs is identical for the set of tested functions.

– The mean values obtained by PDE are noticeably better for functions f1, f2, f4, f6 and f12. PDE and DE behave similarly for other functions.

– The standard deviation values (Std) reached by PDE are noticeable better for functions f1, f2, f4, f6, f18 and f20. PDE and DE behave similarly for the other functions.

	Smallest value reached		***Mean value***		***Standard deviation***		***Mean computation time per run (min)***	
	DE	PDE	DE	PDE	DE	PDE	DE	PDE
F1	– 450	– 450	– 289.2	– **436.2**	267.0	**58.2**	1.1	1.3
F2	– 450	– 450	14.3	– **421.0**	1,030	**60.4**	2.7	3.5
F3	– 450	– 450	82,097	86,019	66,579	62,702	14.5	15.2
F4	– 450	– 450	– 218.7	– **450.0**	343.5	**0.0**	118.5	**7.2**
F5	– 310	– 310	– 310.0	– 310.0	0.0	0.0	34.0	57.4
F6	390	390.0	5,620,755.7	**390.4**	18,310,470.4	**1.2**	3.0	3.4
F7	1,087.3	1,087.3	1,087.3	1,087.3	0.0	0.0	0.2	0.3
F8	– 120	– 120	– 119.7	– 119.8	0.1	0.1	11.4	25.9
F9	– 330	– 330	– 327.5	– 327.0	1.9	2.0	1.6	3.6
F10	– 319.1	– 320.3	– 307.7	– 308.6	7.5	6.5	8.2	17.3
F11	92.6	93.2	95.5	95.1	1.3	1.1	2.7	7.0
F12	– 460	– 460	2,377.3	**1,793.1**	3,804.5	4,410.9	1.7	4.0
F13	– 129.8	– 129.9	– 129.6	– 129.6	0.1	0.1	6.2	12.6
F14	– 296.6	– 296.6	– 296.4	– 296.4	0.1	0.1	6	22.1
F15	123	123.0	127.3	126.0	3.4	3.2	0.3	0.5
F16	122.6	122.6	124.8	124.5	3.2	2.8	0.7	0.6
F18	64.8	64.8	71.3	68.6	16.2	**10.6**	0.5	0.5
F19	64.8	64.8	70.0	72.1	12.5	13.8	0.5	0.5
F20	233.8	233.6	252.7	254.2	14.3	**8.3**	2.7	2.1
F21	457.4	457.4	457.7	460.3	2.3	9.0	0.6	0.6
F22	432.9	434	439.7	440.1	7.1	6.8	3.2	2.1

Table F.1. *DE versus PDE. DE/Rand/1 mutation scheme*

	Smallest value reached		*Mean value*		*Standard deviation*		*Mean computation time per run (min)*	
	DE	PDE	DE	PDE	DE	PDE	DE	PDE
F1	– 450	– 450	– 121.7	**– 383.1**	339.4	**181.2**	0.3	0.4
F2	– 450	– 450	181.8	**– 170.9**	972.3	**847.8**	0.6	0.8
F3	– 450	– 450	742,229.8	**689,418.4**	2,511,349.6	**1,250,885.0**	2.5	2.6
F4	– 450	– 450	407.6	**– 450**	1,376.3	**0.0**	137.8	**12.0**
F5	– 310	– 310	– 309.3	– 309.1	1.8	2.0	16.8	30.2
F6	390	390	28,505,198.8	**315,651.7**	48,634,956.8	**2,185,994.6**	0.6	1.2
F7	1,087.3	1,087.3	1,087.2	1,087.2	0.0	0.0	0.1	0.1
F8	– 120	– 120	– 119.6	– 119.6	0.1	0.1	1.4	2.0
F9	– 326	– 326	– 315.3	– 314.9	7.8	**6.4**	0.3	0.6
F10	– 322	– 321	– 288.9	**– 291.7**	18.9	**16.9**	0.6	1.1
F11	91.4	91.6	95.8	95.6	1.7	1.6	0.8	1.2
F12	– 460	– 460	5,618.1	5,730.5	**6,157.5**	10,155	0.5	0.9
F13	– 130	– 130	– 128.9	– 129	1.5	**1.1**	1.2	2.2
F14	– 296.6	– 296.6	– 296.3	– 296.4	0.1	0.1	6.6	22.2
F15	123	123	132.4	132.6	6.0	**5.1**	0.3	0.5
F16	122.6	122.6	130.3	129	5.1	**4.1**	0.4	0.6
F18	64.8	64.8	149	**132.9**	71.6	**61.8**	0.4	0.4
F19	64.8	64.8	128.1	123.4	62.3	**53.6**	0.2	0.4
F20	233.7	235	274.6	277	34.0	32.2	0.7	1.0
F21	457.4	457.4	516.8	523.4	82.5	88.1	0.5	0.8
F22	434.4	434.4	465.3	461.6	30.1	**22.1**	0.5	0.7

Table F.2. *DE versus PDE. DE/Best/1 mutation scheme*

	Smallest value reached		***Mean value***		***Standard deviation***		***Mean computation time per run (min)***	
	DE	PDE	DE	PDE	DE	PDE	DE	PDE
F1	– 450	– 450	– 424.5	**– 447.9**	79.2	**14.6**	0.3	0.3
F2	– 450	– 450	– 230.7	**– 420.7**	665.2	**62.5**	0.8	0.9
F3	– 450	– 450	65,602	**31,488**	116,612	**75,175**	4.5	5
F4	– 450	– 450	48.1	**– 450**	1,060.1	**0.0**	96.6	**3.5**
F5	– 310	– 310	– 310.0	– 310.0	0.0	0.0	30.9	37.1
F6	390	390	1,508,300.2	**391.1**	10,662,525.2	**1.8**	2.0	2.2
F7	1,087.3	1,087.3	1,087.3	1,087.3	0.0	0.0	0.3	0.4
F8	– 120	– 120	– 119.8	– 119.8	0.1	0.1	52.2	58.9
F9	– 327	– 328	– 323.0	– 323.2	2.3	2.1	1.4	1.8
F10	– 323	– 324	– 310.9	– 311.6	6.7	7.9	20.3	31.9
F11	92.3	92.1	94.9	95.0	1.3	**1.0**	3.3	5.0
F12	– 460	– 460	3,636	**2,554**	5,534.8	**3,985.4**	1.1	2.1
F13	– 130	– 130	– 129.6	– 129.6	0.1	**0.1**	5.4	9.1
F14	– 296.6	– 296.5	– 296.5	– 296.5	0.1	0.1	35	32
F15	123	123.0	128.8	128.8	3.8	3.7	0.5	0.7
F16	122.6	122.6	127.1	127.3	3.7	4.0	0.6	1.1
F18	64.8	64.8	91.6	99.7	34.8	34.8	0.6	0.9
F19	64.8	64.8	94.2	93.5	40.0	**33.4**	0.5	0.8
F20	232.5	233.1	264.3	262.1	28.4	**23.2**	3.0	3.5
F21	457.4	457.4	463.7	464.8	14.2	16.3	0.7	0.9
F22	431.9	432.7	441.4	443.0	9.1	9.3	3.4	4.1

Table F.3. *DE versus PDE. DE/Rand– to– Best/1 mutation scheme*

For the Best/1 mutation scheme, we see that:

– The lowest value reached by PDE and by DE at the end of 50 runs is identical for the set of tested functions.

– The mean values obtained by PDE are noticeably better for functions f1, f2, f3, f4, f6 and f18. PDE and DE behave similarly for other functions.

– The standard deviation values (Std) reached by PDE are noticeable better for functions f1, f2, f3, f4, f6, f9, f13, f15, f16, f18 and f19. PDE and DE behave similarly for the other functions.

For the Rand– to– Best/1 mutation scheme (Table F.3), we see that:

– The lowest value reached by PDE and by DE at the end of 50 runs is identical for the set of tested functions.

– The mean values obtained by PDE are noticeably better for functions f1, f2, f3, f4, f6 and f12. PDE and DE behave similarly for other functions.

– The standard deviation values (Std) reached by PDE are noticeable better for functions f1, f2, f3, f4, f6, f12, f13, f19 and f20. PDE and DE behave similarly for other functions.

For most of the functions tested and for the three mutation schemes, the mean computation time is of course greater for PDE than for DE (indeed, PDE does not converge prematurely), but this increase remains reasonable. However, note one exception for function f4, where PDE is faster than DE.

For the three mutation schemes tested, PDE noticeably improves DE in terms of the mean values and standard deviations for several benchmark functions. This means that PDE is more repeatable than DE from one run to the next. Furthermore, the effectiveness of DE is retained for the other functions. As the only stop criterion here is the detection of premature convergence with [F.1], we can say that the extension proposed here is an interesting improvement on the original DE algorithm to limit its premature convergence. In addition, it seems reasonable to conclude similarly for other known mutation schemes of DE (see [DAS 11]).

Bibliography

[ADE 12] ADEGAS F.D., SOUSTRUP J., "Structured control of LPV systems with application to wind turbines", *Proceedings of the 2012 American Control Conference*, pp. 756–761, 2012.

[APK 93] APKARIAN P., BIANNIC J.M., GAHINET P., "Gain scheduled H_∞ control of a missile via linear matrix inequalities", *Journal of Guidance Control and Dynamics*, vol. 18, no. 3, pp. 532–538, 1993.

[APK 95] APKARIAN P., GAHINET P., "A convex characterization of gain-scheduled H_∞ controllers", *IEEE Transactions on Automatic Control*, vol. 40, no. 5, pp. 853–864, 1995.

[APK 02] APKARIAN P., NOLL D., "Nonsmooth H_∞ synthesis", *Journal of Class Files*, vol. 1, no. 11, pp 1–15, 2002.

[APK 03] APKARIAN P., NOLL D., TUAN H.D., "Fixed-order H_∞ control design via a partially augmented Lagrangian method", *International Journal of Robust and Nonlinear Control*, vol. 13, no. 12, pp. 1137–1148, 2003.

[APK 09] APKARIAN P., "Hinfinity control with stable controllers", available at: http://pierre.apkarian.free.fr, 2009.

[APK 10] APKARIAN P., "Comparison of HINFSTRUCT Matlab Robust Control Toolbox R2010b and HIFOO 3.0 with HANSO 2.0", available at: http://pierre.apkarian.free.fr, 2010.

[APK 11] APKARIAN P., "Non-smooth μ-synthesis", *International Journal of Robust and Nonlinear Control*, vol. 21, no. 13, pp. 1493–1508, 2011.

[APK 12] APKARIAN P., "Tuning controllers for multiple models or operating conditions", available at: http://pierre.apkarian.free.fr, 2012.

[APK 13] APKARIAN P., "Tuning controllers against multiple design requirements", *American Control Conference (ACC)*, pp. 3888–3893, June 2013.

[BIR 03] BIRBIL S.I., FANG S.C., "An electromagnetism-like mechanism for global optimization", *Journal of Global Optimization*, vol. 25, pp. 263–282, 2003.

[BOY 94] BOYD S., EL GHAOUI L., FÉRON E. *et al.*, *Linear Matrix Inequalities in Systems and Control Theory*, SIAM Publications, Philadelphia, 1994.

[BRA 07] BRATON D., KENNEDY J., "Defining a standard for particle swarm optimization", *IEEE Swarm Intelligence Symposium 2007*, pp. 120–127, April 2007.

[BRE 06] BREST J., ZUMER V., SEPESY MAUCEC M., "Control parameters in self-adaptive differential evolution", *Proceedings of the 2nd International Conference on Bioinspired Optimization Method and Their Applications*, Ljubljana, pp 35–44, 9–10 October 2006.

[BUR 06] BURKE V., HENRION D., LEWIS A.S. *et al.*, "Hifoo – a Matlab package for fixed-order controller design and H∞ optimization", *Proceedings of the IFAC Symposium on Robust Control Design*, pp 339–344, 2006.

[CHI 96] CHIPPERFIELD A.J., DAKEV N.V., FLEMING P.J. *et al.*, "Multiobjective robust control using evolutionary algorithm", *IEEE International Conference on Industrial Technology*, pp. 269–273, December 1996.

[CLE 02] CLERC M., KENNEDY J., "The particle swarm-explosion, stability, and convergence in a multidimensional complex space", *IEEE Transactions on Evolutionary Computation*, vol. 6, pp. 58–73, 2002.

[CLE 04] CLERC M., "Une nouvelle métaheuristique pour l'optimisation difficile: la méthode des essaims particulaires", *J3eA*, vol. 3–7, available at www.j3ea.org, 2004.

[CLE 06] CLERC M., *Particle Swarm Optimization*, ISTE Ltd, London, 2006.

[CLE 12] CLERC M., "Standard particle swarm optimization", available at: http://clerc.maurice.free.fr/pso, 2012.

[DAL 08] DALESSIO J., SHUSTER E., HUMPHREYS D.A. *et al.*, "Robust control of resistive wall modes in tokamak plasmas using μ-synthesis", *17th IFAC World Congress*, Seoul, Korea, pp. 6402–6409, 6–11 July, 2008.

[DAS 11] DAS S., SUGANTHAN P.N., "Differential evolution: a survey of the state-of-the-art", *IEEE Transactions on Evolutionary Computation*, vol. 15, no. 1, pp. 4–31, 2011.

[DEV 10] DE VITO D., KRON A., DE LA FONTAINE J. *et al.*, "A Matlab toolbox for LMI-based analysis and synthesis of LPV/LFT self-scheduled H_∞ control systems", *Proceedings of the IEEE Multi-Conference on Systems and Control*, Yokohama, Japan, pp. 1397–1402, 8–10 September, 2010.

[DOY 89] DOYLE J.C., GLOVER K., KHARGONEKAR P.K. *et al.*, "State-space solutions to standard H_2 and H_∞ control problems", *IEEE Transactions on Automatic Control*, vol. 34, no. 8, pp. 831–846, 1989.

[DRÉ 06] DRÉO J., PÉTROWSKI A., SIARRY P. *et al.*, *Metaheuristics for Hard Optimization. Methods and Case Studies*, Springer-Verlag, Berlin, 2006.

[EBE 01] EBERHART R.C., SHI Y., "Particle swarm optimization: development applications and resources", *IEEE Proceedings Congress on Evolutionary Computation*, vol. 1, pp. 81–86, May 2001.

[ELG 94] EL GHAOUI L., BALAKRISHNAN V., "Synthesis of fixed-structure controllers via numerical optimization", *IEEE Proceedings of the 33rd Conference on Decision and Control*, Lake Buena Vista, pp. 2678–2683, December 1994.

[ELG 97] EL GHAOUI L., OUSTRY F., AITRAMI M., "A cone complementarity linearization algorithm for static output-feedback and related problems", *IEEE Transactions on Automatic Control*, vol. 42, no. pp. 1171–1176, August 1997.

[EME 06] EMEDI Z., KARIMI A., "Fixed-order LPV controller design for LPV systems by convex optimization", *Proceedings of the 5th Symposium on System Structure and Control*, Grenoble, France, vol. 5, Part 1, pp. 677–682, February, 2006.

[FAJ 12] FAJFAR I., TUMA T., PUHAN J. *et al.*, "Towards smaller populations in differential evolution", *Journal of Microelectronics, Electronic Components and Materials*, vol. 42, no. 3, pp. 152–163, 2012.

[FEY 13a] FEYEL P., DUC G., SANDOU G., "Fixed-order H_∞ loop-shaping synthesis: a time domain approach", *IEEE Symposium on Computational Intelligence in Control and Automation*, pp. 120–127, 2013.

[FEY 13b] FEYEL P., DUC G., SANDOU G., "Enhancing optimal weight tuning in H∞ loop-shaping control with particle swarm optimization", *5th International Conference on Evolutionary Computation Theory and Applications*, Vilamoura, Portugal, pp. 120–127, September 2013.

[FEY 13c] FEYEL P., *La commande robuste par loop-shaping*, Hermes Science-Lavoisier, 2013.

[FEY 14a] FEYEL P., DUC G., SANDOU G., "Evolutionary fixed-structure H_∞ synthesis for multiple plants", *IEEE MultiConference on Systems and Control*, Antibes, pp. 1208–1213, October 2014.

[FEY 14b] FEYEL P., DUC G., SANDOU G., "μ-synthesis with dynamic *D*-scalings using quantum particle swarm optimization", *IEEE 13th European Control Conference*, Strasbourg, pp. 606–611, June 2014.

[FEY 14c] FEYEL P., DUC G., SANDOU G., "Evolutionary fixed-structure mu-synthesis", *CICA & SDE 2013, IEEE Symposium on Computational Intelligence in Control and Automation*, Orlando, pp. 1–8, December 2014.

[FEY 15] FEYEL P., "Evolutionary fixed-structure LPV/LFT synthesis", *16th IFAC Workshop on Control Applications of Optimization*, Garmich-Partenkirchen, pp. 114–119, Germany, 2015.

[FEY 16] FEYEL P., "Evolutionary fixed-structure controller tuning against multiple requirements", *4th IFAC International Conference on Intelligent Control and Automation Sciences*, pp. 345–352, Reims, France, 1–3 June 2016.

[FIS 13] FISTER I. JR, YANG X.S., FISTER I. *et al.*, "A brief review of nature-inspired algorithms for optimization", *Neural and Evolutionary Computing, Elektrotehniski Vestnik*, vol. 80, pp. 1–7, 2013.

[GAH 94] GAHINET P., APKARIAN P., "A linear matrix inequality approach to H_∞ control", *International Journal of Robust and Nonlinear Control*, vol. 4, pp. 421–448, 1994.

[GAH 11] GAHINET P., APKARIAN P., "Decentralized and fixed-structure H∞ control in Matlab", *50th IEEE Conference on Decision and Control and European Control Conference (CDC-ECC)*, Orlando, 12–15, pp. 8205–8210, December 2011.

[GLO 89] GLOVER K., MCFARLANE D., "Robust stabilization of normalized coprime factor plant description with H_∞-bounded uncertainty", *IEEE Transactions on Automatic Control*, vol. 34, no. 8, pp. 821–830, 1989.

[HIL 08] HILKERT J.M., "Inertially stabilized platforms technology", *IEEE Control Systems Magazine*, vol. 28, no. 1, pp. 26–46, 2008.

[HIR 12] HIRWA S., FEYEL P., DUC G. *et al.*, "Reduced order controllers for inertial line of sight stabilization", *Proceedings of the 7th IFAC Symposium on Robust Control Design*, Aalborg, Denmark, pp. 103–108, 20–22, June 2012.

[HOL 06] HOLST G.C., *Electro-Optical Imaging System Performance*, SPIE Press, JCD Publishing, 2006.

[HOY 91] HOYLE D.J., HYDE R.A., LIMEBEER D.J.N., "An H_∞ approach to two degree of freedom design", *30th IEEE Conference on Decision & Control*, Brighton, pp. 1581–1585, December 1991.

[KAR 07] KARABOGA D., BASTURG B., "A powerful and efficient algorithm for numerical function optimization: artificial bee colony (ABC) algorithm", *Journal of Global Optimization*, vol. 39, no. 3, pp. 459–471, 2007.

[KAR 08] KARABOGA D., BASTURG B., "On the performance of artificial bee colony (ABC) algorithm", *Journal Applied Soft Computing*, vol. 8, no. 1, pp. 687–697, 2008.

[KEN 95] KENNEDY J., EBERHART R.C., "Particle swarm optimization", *IEEE International Conference on Neural Networks*, Perth, pp. 1942–1948, 1995.

[KWA 72] KWAKERNAAK H., SIVAN R., *Linear Optimal Control Systems*, Wiley-Interscience, New York, 1972.

[LAR 12] LARY A., KHORSRAVI A., KARAMNEZHAD R., "A new approach to design a simple structure μ synthesis controller: application to an uncertain DC motor", *7th IFAC Symposium on Robust Control Design*, Aalborg, pp. 672–677, 20–22 June 2012.

[LEE 05] LEE K.S., GEEM Z.W., "A new meta-heuristic algorithm for continuous engineering optimization: harmony search theory and practice", *Computer Methods in Applied Mechanics and Engineering*, vol. 194, pp. 3902–3933, 2005.

[LEI 03] LEIBFRITZ F., COMPleib: COnstrained Matrix-optimization Problem library – a collection of test examples for nonlinear semidefinite programs, control system design and related problems, Technical report, University of Trier, 2003.

[LEW 12] LEWIS A.S., OVERTON M.L., "Nonsmooth optimization via quasi-Newton methods", *Mathematical Programming: A Publication of the Mathematical Optimization Society*, Springer-Verlag, Berlin, 2012.

[LIM 93] LIMEBEER D.J.N., KASANELLY E.M., PERKINS J.D., "On the design of robust two degree of freedom controllers", *Automatica*, vol. 29, no. 1, pp. 157–168, 1993.

[LIN 05] LIN C.C., CHOU Y.S., "A novel algorithm for real mu synthesis", *IEEE ICSS2005 International Conference On Systems & Signals Society*, pp. 734–739, 2005.

[LUS 09] LUSZCZEK P., "Parallel programming in Matlab", available at: www.mathworks.com, 2009.

[MAR 08] MARUTA I., KIM T.-H., SUGIE T., "Synthesis of fixed-structure H∞ controllers via constrained particle swarm optimization", *17th IFAC Control Conference*, Séoul, pp. 7843–7848, 2008.

[MAS 08] MASTEN M.K., "Inertially stabilized platforms for optical imaging systems", *IEEE Control Systems Magazine*, vol. 28, no. 1, pp. 47–64, 2008.

[MAT 10] MATHWORKS, "Matlab Parallel Computing Toolbox", available at: www.mathworks.com, 2010.

[MAT 14] MATHWORKS, "Robust Control Toolbox", available at: www.mathworks.com, 2014.

[MAY 99] MAYZONETTE J.L., *Basc dc radiométrie optique*, Editions Cépaduès, 1999.

[MCF 90] MCFARLANE D., GLOVER K., *Robust Controller Design using Normalized Coprime Factor Plant Descriptions*, Springer Verlag, Berlin, 1990.

[MCF 92] MCFARLANE D., GLOVER K., "A loop shaping design procedure using H_∞ synthesis", *IEEE Transactions on Automatic Control*, vol. 37, no. 6, pp. 759–769, 1992.

[MCK 96] MCKELVEY T., HELMERSSON A., "State-space parametrizations of multivariable linear systems using tridiagonal matrix forms", *35th IEEE Decision and Control Conference*, vol. 4, pp. 3654–3659, December 1996.

[MER 04] MERIKOSKI J.K., KUMAR R., "Inequalities for spreads of matrix sums and products", *Applied Mathematics E-Notes*, vol. 4, pp. 150–159, 2004.

[MEZ 08] MEZURA-MONTES E., REYES-SIERRA M., COELLO COELLO C.A., "Multi-objective optimization using differential evolution: a survey of the state-of-the-art", *Advances in Differential Evolution Studies in Computational Intelligence*, vol. 143, pp. 173–196, 2008.

[MIY 97] MIYAMOTO S., VINNICOMBE G., "Fixed structure H_∞ control design for multiple plant. A coprime factorisation LMI based approach", *European Control Conference*, pp. 1364–1369, Brussels, 1997.

[NEL 65] NELDER J.A., MEAD R., "A simplex method for function minimization", *Computer Journal*, vol. 7, pp. 308–313, 1965.

[OZT 12] OZTURK C., KARABOGA D., GORKEMLI B., "Artificial bee colony algorithm for dynamic deployment of wireless sensor networks", *Turkish Journal of Electrical Engineering and Computer Sciences*, vol. 20, no. 2, pp. 255–262, 2012.

[POP 10] POPOV A., Methods for fixed-structure H∞ controller synthesis – a comparison, Technical report, Institute of Control Systems, Hamburg, 2010.

[PRE 04] PREMPAIN E., "Static H_∞ loop-shaping control of a fly-by-wire helicopter", *Conference on Decision & Control*, vol. 2, pp. 1188–1195, 2004.

[PRE 05] PREMPAIN E., "A comparison of two algorithms for the static H_∞ loop-shaping problem", *16th IFAC World Congress*, vol. 16, part 1, pp. 481–486, Prague, 2005.

[PRI 05] PRICE K., STORM K., LANPINEN J.A., *Differential Evolution – A Practical Approach to Global Optimization*, Springer-Verlag, Berlin, 2005.

[ROC 94] ROCKAFELLAR T., "Nonsmooth optimization", in BIRGE J.R., MURTY K.G. (eds), *Mathematical Programming: State of the Art 1994*, University of Michigan Press, Ann Arbor, 1994.

[SAF 81] SAFONOV M.G., LAUB A.J., HARTMANN G., "Feedback properties of multivariable systems: the role and use of return difference matrix", *IEEE Transactions on Automatic Control*, vol. 26, no. 1, pp. 47–65, 1981.

[SAN 08] SANDOU G., DUC G., BEAUVOIS D., "Optimisation par essaim particulaire du réglage d'un correcteur H∞", *Conférence Internationale Francophone d'Automatique*, CIFA 2008, Bucarest, Romania, 2008.

[SCH 97] SCHERER C., GAHINET P., CHILALI M., "Multiobjective output-feedback control via LMI optimisation", *IEEE Transactions on Automatic Control*, vol. 42, no. 7, pp. 896–911, 1997.

[SED 09] SEDIGHIZADEH D., MASEHIAN E., "Particle swarm optimization methods, taxonomy and applications", *International Journal of Computer Theory and Engineering*, vol. 1, no. 5, pp. 1793–8201, 2009.

[SIA 02] SIARRY P., "Application des métaheuristiques d'optimisation en électronique", *Techniques de l'Ingénieur*, RE 8 – 1, October 2002.

[SKO 01] SKOGESTAD S., POSTLETHWAITE I., *Multivariable Feedback Control. Analysis and Design*, 2nd ed., John Wiley & Sons, New York, 2001.

[STO 95] STORM R.M., PRICE K., Differential evolution – a simple and efficient adaptive scheme for global optimization over continuous spaces, Technical report TR9 -012, International Computer Science Institute, Berkeley, 1995.

[STO 96] STORM R.M., "On the usage of differential evolution for function optimization", *North American Fuzzy Information Processing Society*, pp. 519–523, 1996.

[SUG 05] SUGANTHAN P.N., HANSEN N., LIANG J.J. *et al.*, Problem definitions and evaluation criteria for the CEC 2005 special session on real-parameter optimization, Technical report, Nanyang Technological University, Singapore And KanGAL Report Number 2005005 (Kanpur Genetic Algorithms Laboratory, IIT Kanpur), 2005.

[SUN 12a] SUN J., LAI C.H., WU X.J., *Particle Swarm Optimization: Classical and Quantum Perspectives*, CRC Press, London, 2012.

[SUN 12b] SUN J., WU X.J., PALADE V. *et al.*, "Convergence analysis and improvements of quantum-behaved particle swarm optimization", *Information Sciences*, vol. 193, pp. 81–103, 2012.

[TOF 95] TOFFNER-CLAUSEN S., ANDERSEN P., SOUSTRUP J. *et al.*, "A new approach to μ-synthesis for mixed perturbation sets", *3rd European Control Conference*, Rome, pp. 147–152, 1995.

[TSA 02] TSAI K.Y., HINDI H.A., "DQIT: μ-synthesis without D-scale fitting", *Proceedings of the American Control Conference*, Anchorage, pp. 493–198, 2002.

[VID 85] VIDYASAGAR M., *Control System Synthesis. A Factorisation Approach*, MIT Press, Cambridge, 1985.

[VIN 97] VINNICOMBE G., MIYAMOTO S., "On reduced order H_∞ loop-shaping controllers: a design method and examples of local optima", *European Control Conference*, pp. 1364–1369, Brussels, July 1997.

[VIN 00] VINNICOMBE G., *Uncertainty and Feedback – H_∞ Loop-Shaping and the ν-Gap Metric*, Imperial College Press, London, 2000.

[WAN 12] WANG G., GUO L., DUAN H. *et al.*, "A hybrid metaheuristic DE/CS algorithm for UCAV three-dimension path planning", *The Scientific World Journal*, volume 2012, available at: www.hindawin.com, 2012.

[WIJ 06] WIJNHEIJMER F., NAUS G., POST W. *et al.*, "Modelling and LPV control of an electro-hydraulic servo system", *Proceedings of the International Conference On Control Applications*, Munich, pp. 3116–3121, October 2006.

[YAN 09a] YANG X.S., DEB S., "Cuckoo search via levy flight", *Proceedings of World Congress on Nature & Biologically Inspired Computing*, Coimbatore, IEEE Publications, pp. 210–214, December 2009.

[YAN 09b] YANG X.S., "Firefly algorithms for multimodal optimization", *Stochastic Algorithms: Foundations and Applications, Lecture Notes in Computer Science,* Springer, vol. 5792, pp. 169–178, 2009.

[YAN 10a] YANG X.S., *Nature-Inspired Metaheuristic Algorithms*, Lunivers Press, Cambridge, 2010.

[YAN 10b] YANG X.S., "A new metaheuristic bat-inspired algorithm", *Nature Inspired Cooperative Strategies for Optimization, Studies in Computational Intelligence*, Springer, Berlin, vol. 284, pp. 65–74, 2010.

[YOU 90] YOUNG P., DOYLE J., "Computation of μ with real and complex uncertainties", *29th IEEE Conference on Decision and Control*, Honolulu, pp. 1230–1235, 1990.

[YOU 93] YOUNG P., "Controller design with real parametric uncertainty", *Technical Memorandum*, Control and Dynamical Systems, California Institute of Technology, Pasadena, 1993.

[ZHO 96] ZHOU K., DOYLE J.C., GLOVER K., *Robust and Optimal Control*, Prentice-Hall, Englewood Cliffs, 1996.

Index

Other titles from

ISTE

in

Systems and Industrial Engineering – Robotics

2016

ANDRÉ Michel, SAMARAS Zissis
Energy and Environment
(Research for Innovative Transports Set - Volume 1)

AUBRY Jean-François, BRINZEI Nicolae, MAZOUNI Mohammed-Habib
Systems Dependability Assessment: Benefits of Petri Net Models (Systems Dependability Assessment Set - Volume 1)

BLANQUART Corinne, CLAUSEN Uwe, JACOB Bernard
Towards Innovative Freight and Logistics (Research for Innovative Transports Set - Volume 2)

COHEN Simon, YANNIS George
Traffic Management (Research for Innovative Transports Set - Volume 3)

MARÉ Jean-Charles
Aerospace Actuators 1: Needs, Reliability and Hydraulic Power Solutions

REZG Nidhal, HAJEJ Zied, BOSCHIAN-CAMPANER Valerio
Production and Maintenance Optimization Problems: Logistic Constraints and Leasing Warranty Services

TORRENTI Jean-Michel, LA TORRE Francesca
Materials and Infrastructures 1 (Research for Innovative Transports Set - Volume 5A)
Materials and Infrastructures 2 (Research for Innovative Transports Set - Volume 5B)

WEBER Philippe, SIMON Christophe
Benefits of Bayesian Network Models
(Systems Dependability Assessment Set – Volume 2)

YANNIS George, COHEN Simon
Traffic Safety (Research for Innovative Transports Set - Volume 4)

2015

AUBRY Jean-François, BRINZEI Nicolae
Systems Dependability Assessment: Modeling with Graphs and Finite State Automata

BOULANGER Jean-Louis
CENELEC 50128 and IEC 62279 Standards

BRIFFAUT Jean-Pierre
E-Enabled Operations Management

MISSIKOFF Michele, CANDUCCI Massimo, MAIDEN Neil
Enterprise Innovation

2014

CHETTO Maryline
Real-time Systems Scheduling
Volume 1 – Fundamentals
Volume 2 – Focuses

DAVIM J. Paulo
Machinability of Advanced Materials

ESTAMPE Dominique
Supply Chain Performance and Evaluation Models

FAVRE Bernard
Introduction to Sustainable Transports

GAUTHIER Michaël, ANDREFF Nicolas, DOMBRE Etienne
Intracorporeal Robotics: From Milliscale to Nanoscale

MICOUIN Patrice
Model Based Systems Engineering: Fundamentals and Methods

MILLOT Patrick
Designing Human–Machine Cooperation Systems

NI Zhenjiang, PACORET Céline, BENOSMAN Ryad, REGNIER Stéphane
Haptic Feedback Teleoperation of Optical Tweezers

OUSTALOUP Alain
Diversity and Non-integer Differentiation for System Dynamics

REZG Nidhal, DELLAGI Sofien, KHATAD Abdelhakim
Joint Optimization of Maintenance and Production Policies

STEFANOIU Dan, BORNE Pierre, POPESCU Dumitru, FILIP Florin Gh.,
EL KAMEL Abdelkader
Optimization in Engineering Sciences: Metaheuristics, Stochastic Methods andDecision Support

2013

ALAZARD Daniel
Reverse Engineering in Control Design

ARIOUI Hichem, NEHAOUA Lamri
Driving Simulation

CHADLI Mohammed, COPPIER Hervé
Command-control for Real-time Systems

DAAFOUZ Jamal, TARBOURIECH Sophie, SIGALOTTI Mario
Hybrid Systems with Constraints

FEYEL Philippe
Loop-shaping Robust Control

FLAUS Jean-Marie
Risk Analysis: Socio-technical and Industrial Systems

FRIBOURG Laurent, SOULAT Romain
Control of Switching Systems by Invariance Analysis: Application to Power Electronics

GROSSARD Mathieu, REGNIER Stéphane, CHAILLET Nicolas
Flexible Robotics: Applications to Multiscale Manipulations

GRUNN Emmanuel, PHAM Anh Tuan
Modeling of Complex Systems: Application to Aeronautical Dynamics

HABIB Maki K., DAVIM J. Paulo
Interdisciplinary Mechatronics: Engineering Science and Research Development

HAMMADI Slim, KSOURI Mekki
Multimodal Transport Systems

JARBOUI Bassem, SIARRY Patrick, TEGHEM Jacques
Metaheuristics for Production Scheduling

KIRILLOV Oleg N., PELINOVSKY Dmitry E.
Nonlinear Physical Systems

LE Vu Tuan Hieu, STOICA Cristina, ALAMO Teodoro,
CAMACHO Eduardo F., DUMUR Didier
Zonotopes: From Guaranteed State-estimation to Control

MACHADO Carolina, DAVIM J. Paulo
Management and Engineering Innovation

MORANA Joëlle
Sustainable Supply Chain Management

SANDOU Guillaume
Metaheuristic Optimization for the Design of Automatic Control Laws

STOICAN Florin, OLARU Sorin
Set-theoretic Fault Detection in Multisensor Systems

2012

AÏT-KADI Daoud, CHOUINARD Marc, MARCOTTE Suzanne, RIOPEL Diane
Sustainable Reverse Logistics Network: Engineering and Management

BORNE Pierre, POPESCU Dumitru, FILIP Florin G., STEFANOIU Dan
Optimization in Engineering Sciences: Exact Methods

CHADLI Mohammed, BORNE Pierre
Multiple Models Approach in Automation: Takagi-Sugeno Fuzzy Systems

DAVIM J.Paulo
Lasers in Manufacturing

DECLERCK Philippe
Discrete Event Systems in Dioid Algebra and Conventional Algebra

DOUMIATI Moustapha, CHARARA Ali, VICTORINO Alessandro, LECHNER Daniel
Vehicle Dynamics Estimation using Kalman Filtering: Experimental Validation

GUERRERO José A, LOZANO Rogelio
Flight Formation Control

HAMMADI Slim, KSOURI Mekki
Advanced Mobility and Transport Engineering

MAILLARD Pierre
Competitive Quality Strategies

MATTA Nada, VANDENBOOMGAERDE Yves, ARLAT Jean
Supervision and Safety of Complex Systems

POLER Raul *et al.*
Intelligent Non-hierarchical Manufacturing Networks

TROCCAZ Jocelyne
Medical Robotics

YALAOUI Alice, CHEHADE Hicham, YALAOUI Farouk, AMODEO Lionel
Optimization of Logistics

ZELM Martin *et al.*
Enterprise Interoperability –I-EASA12 Proceedings

2011

CANTOT Pascal, LUZEAUX Dominique
Simulation and Modeling of Systems of Systems

DAVIM J. Paulo
Mechatronics

DAVIM J. Paulo
Wood Machining

GROUS Ammar
Applied Metrology for Manufacturing Engineering

KOLSKI Christophe
Human–Computer Interactions in Transport

LUZEAUX Dominique, RUAULT Jean-René, WIPPLER Jean-Luc
Complex Systems and Systems of Systems Engineering

ZELM Martin, *et al.*
Enterprise Interoperability: IWEI2011 Proceedings

2010

BOTTA-GENOULAZ Valérie, CAMPAGNE Jean-Pierre, LLERENA Daniel, PELLEGRIN Claude
Supply Chain Performance / Collaboration, Alignement and Coordination

BOURLÈS Henri, GODFREY K.C. Kwan
Linear Systems

BOURRIÈRES Jean-Paul
Proceedings of CEISIE'09

CHAILLET Nicolas, REGNIER Stéphane
Microrobotics for Micromanipulation

DAVIM J. Paulo
Sustainable Manufacturing

GIORDANO Max, MATHIEU Luc, VILLENEUVE François
Product Life-Cycle Management / Geometric Variations

LOZANO Rogelio
Unmanned Aerial Vehicles / Embedded Control

LUZEAUX Dominique, RUAULT Jean-René
Systems of Systems

VILLENEUVE François, MATHIEU Luc
Geometric Tolerancing of Products

2009

DIAZ Michel
Petri Nets / Fundamental Models, Verification and Applications

OZEL Tugrul, DAVIM J. Paulo
Intelligent Machining

PITRAT Jacques
Artificial Beings

2008

ARTIGUES Christian, DEMASSEY Sophie, NERON Emmanuel
Resources–Constrained Project Scheduling

BILLAUT Jean-Charles, MOUKRIM Aziz, SANLAVILLE Eric
Flexibility and Robustness in Scheduling

DOCHAIN Denis
Bioprocess Control

LOPEZ Pierre, ROUBELLAT François
Production Scheduling

THIERRY Caroline, THOMAS André, BEL Gérard
Supply Chain Simulation and Management

2007

DE LARMINAT Philippe
Analysis and Control of Linear Systems

DOMBRE Etienne, KHALIL Wisama
Robot Manipulators

LAMNABHI Françoise *et al.*
Taming Heterogeneity and Complexity of Embedded Control

LIMNIOS Nikolaos
Fault Trees

2006

FRENCH COLLEGE OF METROLOGY
Metrology in Industry

NAJIM Kaddour
Control of Continuous Linear Systems

Printed and bound by CPI Group (UK) Ltd, Croydon, CR0 4YY
15/09/2026
14962356-0006